AF676342

Topical Collection "Challenges and Opportunities in Applied System Innovation"

Topical Collection "Challenges and Opportunities in Applied System Innovation"

Editor

Christos Douligeris

MDPI • Basel • Beijing • Wuhan • Barcelona • Belgrade • Manchester • Tokyo • Cluj • Tianjin

Editor
Christos Douligeris
University of Piraeus
Greece

Editorial Office
MDPI
St. Alban-Anlage 66
4052 Basel, Switzerland

This is a reprint of articles from the Topical Collection published online in the open access journal *Applied System Innovation* (ISSN 2571-5577) (available at: https://www.mdpi.com/journal/asi/topical_collections/Chall_Oppor).

For citation purposes, cite each article independently as indicated on the article page online and as indicated below:

LastName, A.A.; LastName, B.B.; LastName, C.C. Article Title. *Journal Name* **Year**, *Volume Number*, Page Range.

ISBN 978-3-0365-4401-4 (Hbk)
ISBN 978-3-0365-4402-1 (PDF)

© 2022 by the authors. Articles in this book are Open Access and distributed under the Creative Commons Attribution (CC BY) license, which allows users to download, copy and build upon published articles, as long as the author and publisher are properly credited, which ensures maximum dissemination and a wider impact of our publications.
The book as a whole is distributed by MDPI under the terms and conditions of the Creative Commons license CC BY-NC-ND.

Contents

Preface to "Topical Collection "Challenges and Opportunities in Applied System Innovation" "

This book contains eleven articles published in Applied System Innovation which stand out due to the way they approach the issues they address. The main idea that connects these articles is that innovation is occurring in a variety of industries and that there is a strong interdependence on the introduction of new technologies and their acceptance in various societal functions. Thus, this book introduces and provides solutions to a variety of problems faced by society, companies and individuals in a quickly changing and technology-dependent world.

The wide acceptance of artificial intelligence (AI), the upcoming fourth industrial revolution and the newly designed 6G technologies are seen as the main enablers and game changers in this environment. The book considers these issues not only from a technological viewpoint but also on how society, labor and the economy are affected, leading to a circular economy that affects the way people design, function and deploy complex systems.

The main focus of the first three articles is artificial intelligence. The first article considers how AI will influence modern societies, while the other two articles discuss specific applications in the provision of sustainable development solutions and fast, accurate and ready-to-deploy applications in problems faced in traditional bridge problems.

Even though 5G has only recently become part of our everyday lives, new 6G ideas are explored in the next two articles to further expand the Internet of Things (IoT) concept into the Internet of Everything (IoE) paradigm. The need for a theoretical framework that would allow researchers and practitioners to design effective and efficient systems in the healthcare sector is emphasized.

The advancements in Information and Communication Technologies (ICT) have enabled a variety of innovations and have revolutionized many industrial settings. Three-dimensional (3D) printing technologies and cognitive manufacturing are two ideas that are explored in the next two articles. However, this revolution has raised new questions in terms of the problems of labor productivity and employee retention. Data science solutions as well as innovative approaches based on human factor research are explored in the following articles.

All of the ideas discussed above would definitely benefit from new modeling structures that ICT technologies allow. The digital twins concept is one such structure—the article entitled "Digital Twin: Origin to Future" provides a pathway to this concept with a strong emphasis on solving crucial and important problems in the manufacturing research and development area.

Even though these developments are rapid and innovative, their environmental effect should not be ignored. An article within this book, which discusses the circular economy, can be seen as a guideline for all industries and organizations, even though it only discusses buildings.

I hope that this collection of articles will provide a background for fruitful discussions, a plethora of idea exchanges and the basis for further research and development. Innovation is a never-ending process, full of surprises and exciting new ideas.

Christos Douligeris
Editor

applied system innovation

MDPI

Article

Chances and Risks of Artificial Intelligence—A Concept of Developing and Exploiting Machine Intelligence for Future Societies

Reinhard Hutter [1,*] and Marcus Hutter [2]

1 Centre for European Security Strategies, 81545 Munich, Germany
2 Research School of Computer Science (RSCS), Australian National University, Canberra, ACT 2600, Australia; marcus.hutter@gmx.net
* Correspondence: Reinhard.hutter@gmx.de or hutter@cess-net.eu; Tel.: +49-89-69377147

Abstract: Artificial Intelligence (AI): Boon or Bane for societies? AI technologies and solutions—as most revolutionary technologies have done in the past—offer negative implications on the one hand and considerable positive potential on the other. Avoiding the former and fostering the latter will require substantial investments in future societal concepts, research and development, and control of AI-based solutions in AI security while avoiding abuse. Preparation for the future role of AI in societies should strive towards the implementation of related methods and tools for risk management, models of complementary human–machine cooperation, strategies for the optimization of production and administration, and innovative concepts for the distribution of the economic value created. Two extreme possible "end states" of AI impact (if there is ever an end state) that are being discussed at present may manifest as (a) uncontrolled substitution by AI of major aspects of production, services, and administrative and decision-making processes, leading to unprecedented risks such as high unemployment, and devaluation and the underpayment of people in paid work, resulting in inequality in the distribution of wealth and employment, diminishing social peace, social cohesion, solidarity, security, etc., or, on the contrary, (b) the freeing of people from routine labor through increased automation in production, administration and services, and changing the constitution of politics and societies into constituencies with high ethical standards, personal self-determination, and the general dominance of humane principles, as opposed to pure materialism. Any mix of these two extremes could develop, and these combinations may vary among different societies and political systems.

Keywords: artificial and human intelligence; security; risks and risk management; quality of life; common welfare; socio-political assessment

Citation: Hutter, R.; Hutter, M. Chances and Risks of Artificial Intelligence—A Concept of Developing and Exploiting Machine Intelligence for Future Societies. *Appl. Syst. Innov.* **2021**, *4*, 37. https://doi.org/10.3390/asi4020037

Academic Editor: Andrzej Białas

Received: 3 April 2021
Accepted: 12 May 2021
Published: 2 June 2021

Publisher's Note: MDPI stays neutral with regard to jurisdictional claims in published maps and institutional affiliations.

Copyright: © 2021 by the authors. Licensee MDPI, Basel, Switzerland. This article is an open access article distributed under the terms and conditions of the Creative Commons Attribution (CC BY) license (https://creativecommons.org/licenses/by/4.0/).

1. Introduction

Artificial intelligence (AI) and all its theoretical, technical and operational ramifications offer a vast and rapidly developing potential to serve humankind. The authors were motivated by realizing that most discussions of AI stress either technical matters or the possible negative impacts of AI's expansion into society. The ideas presented here deviate to from today's mainstream public discussions regarding the future role of AI. Although they are a bit revolutionary, or at least unusual, the concepts presented here promise a high payoff for societies derived from AI, while, at the same time, avoiding quite a high number of risks of and controversies. Figure 1 illustrates that, sooner or later, all relevant societal spheres will be impacted by AI-based solutions, and that, in turn, the path of AI development will require guidance and control. This will require an unprecedentedly concerted effort among all societal forces with respect to the preparation, introduction, operation and maintenance of AI solutions, and with respect to the principles that should guide their development and use.

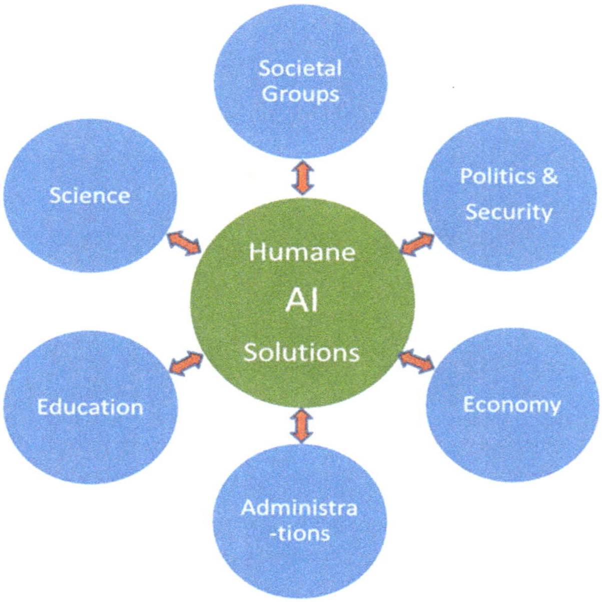

Figure 1. Cooperation model.

Developments may imply significant improvements to the human condition, but may also constitute risks of the contrary, including abuse or bias from groups with competing interests. Furthermore, the ultimate variants in future scenarios could include the risk of full takeover of societal governance by AI and robots, relegating mankind to a remnant of the past, enslaved or tolerated at best, or, at the other extreme, the creation of a full biblical Land of Plenty, with AI services and robots providing a paradise for humans to live in.

The novelty presented in this paper is that it takes a realistic view, while trying to develop an optimistic vision: a feasible strategy that would improve security, quality of life and human cohabitation, while maintaining dignity and self-determination and diminishing the fear and uncertainty of people, societies and politics when discussing how to manage the risks that may emerge from these concepts. The accompanying benefits could and should include climate control, optimal use of natural resources and the sustainability and recreation of biological diversity, along with improved overall safety and security and adequate risk and crisis management.

In a nutshell, AI is developing abilities that will allow for progress in many directions. The path towards which mankind strives varies substantially depending on region, cultural background, political system and principles, available resources, and environmental conditions. The manner in which politics, economy, science and societies prepare themselves for, agree upon and manage the elementary prerequisites for the transformation and shaping of future "cyber societies" will be essential. The basics need to be determined sooner rather than later, and this should be started now, at both national and global scales.

Section 2 provides a brief overview of how AI has developed over the decades, and shows that it is now at a stage where collaborative (CI) models between artificial and human intelligence are key to future societal applications and their expected effects, as discussed in Section 3. Section 4 discusses the impacts of these CI solutions on the economy, on politics, and on data requirements. Sections 5 and 6 discuss promising potential future applications of CI methods, while in Sections 7 and 8, the benefits and possible risks of such solutions are evaluated. As these ideas are rather challenging, the prerequisites are discussed and their feasibility is assessed in Sections 9 and 10, respectively.

2. Starting Point and Background

Having followed and being involved in public and scientific discussions and the professional development of AI for some 25 years, we believe that its potential has reached a level that will make a big difference in future applications. We, therefore, derive some ideas and draw some conclusions that could be useful for future debate and progress in the AI domain, and that hopefully will be substantial for future human society at large.

AI started some 60 years ago with the main intention to somehow understand human rationality and cognition technically, and to build human-level AI. In the 1980s, it developed into replicating human rationality and cognitive power in rather limited application areas. More recently, AI R&D has strived towards a solution to complex problems and generalizing theories and practices beyond the limits of human capability [1]. We thus put forward the question of whether AI and its extension to Artificial General Intelligence (AGI) is the intelligence of a machine that could successfully perform any intellectual task that a human being is capable of; see, e.g., https://en.wikipedia.org/wiki/Artificial_general_intelligence, accessed on 20 December 2020) should be pushed more, in the sense that it will design ways to solve problems in areas where human organizations are limited in speed, capacity, and ability, as well as manage the involved risks and create optimum solutions; in other words, problem areas which, in present political systems, are constrained through societal and political restrictions by the need for compromise and by the limitations in designing and executing sustainable long-term concepts. In this sense, after phases of both enthusiasm and nightmares (such as George Orwell: "1984"), we feel it is time to think of a positive far-end of AI.

Past public and political recognition and discussions of AI have occurred in waves. In the 1980s, hope and enthusiasm was concentrated on the prospects of extending computer ability to problems that are logically structured and rather deterministic (e.g., Expert Systems). The disappointments caused by the huge cost of the required computer power and limited scientific experience with complex logical networks and the self-learning ability of systems, etc., resulted in disillusionment. Only a few, rather modest, practicable solutions evolved and survived.

The present public discussion focuses more on what AI really means for our future society. Some effort has been spent defining what AI is really meant to be. Comparing AI with, and trying to simulate, Human Intelligence (HI) alone is, to some extent, frustrating, and may lead us in the wrong direction, at least as long as there are still so many loose ends regarding how human cognition and rationality really work, let alone their dependence on basic brain structure, genetically determined or not, for the interaction with sensory input, influence of mental/psychological states and the inter-operation of all these [2]. AI research and brain research are delivering a number of insights that will help to improve AI approaches and efficiency (e.g., neural networks, interfacing machines with biological neural systems, complex pattern recognition, language interpretation and many more). Scientific progress in artificial neural networks and machine learning have also contributed. We have a deep respect for the achievements reached with systems such as Deep Blue, AlphaGo/AlphaZero or Watson, and robotics that increasingly adapt or exceed human capabilities. However, our core question has arisen as to whether these should be the priority objectives of AI R&D and if they are the right approaches and exploitations for the more general benefit of mankind. Except for the sake of scientific progress, the core driver of and the related investments into RTD on AI should not primarily deal with, e.g., fancy dancing robots or playing games, to name only two of the most spectacular examples at present, which we, a bit disrespectfully, call "toys".

We do not want to arrogantly disregard any achievements gained in AI to date. However, AI, in the future, from a political and socio-economic perspective, should focus more on solving the problems of mankind, societies, groups and individuals; in short, on improving the quality of life and common welfare. Compared to the "toys" referred to above, there are already fascinating and very helpful AI-based applications, such as pattern recognition (speech, face), language translation or improvements in self-driving cars. However, what we are advocating here is to further focus investments, public attention and success stories on solutions that more strictly follow a humankind-oriented approach. AI should not concentrate too much on things that the human brain can do already, but on things that humans cannot perform, or perform too slowly or imperfectly, such as solutions to over-population, fairness in the distribution of resources and wealth, sustainable traffic and transportation concepts or genome sequencing for healthcare purposes. This, however, is a moving target, as ultimately AI may perform better than humans at almost everything. Nevertheless, there seem to be qualities in HI, such as "Emotional Intelligence" [3] or intuition, that may never, or only in an undefined future, be achieved by machine intelligence. Of course, this is not a black-and-white issue. Recognizing faces, automating production lines, improving security through tamper-proof identification, or optimizing the search for a parking lot have personal and economic benefits. The bottom line, however, of our admittedly rather fundamental considerations is that machine intelligence should no longer focus on imitating HI, but better capitalize on tasks that cannot, or can only rather sub-optimally, be handled by humans, in particular tasks for whole societies, politics, and large constituencies, such as the global financial system, commerce, trade, and alleviating boring work, risks, duties and suffering for mankind on a larger scale. "Too many problems (in most urgent real-world challenges) suffer from painfully slow progress" [4].

The potential of big payoffs for the advocated future AI applications should not be created via blunt automation alone, but involve cooperation between human and machine intelligence. It should not be a substitution for human intelligence and labour, but complement and improve humans' mental and physical skills. In that sense, we will refer to CI. CI here stands for Complementary, Collaborative Intelligence, not Computer Intelligence as it is also sometimes used.

3. CI Applications Effects

Research and industry should concentrate CI on the exploitation of machine properties, such as basic computing power, fast networking and algorithmic concepts and solutions to complex problems. These CI application areas should be prioritized over those which obviously have not, or have very unsatisfactorily, been solved by humans to date, but for which promising solutions and payoffs are developing in AI (see also Table 1: Some Examples of Promising CI). From there, application domains should be identified where CI can produce help and solutions that will be significantly superior to human-made approaches tried in the past. The payoff will not just be high economically and commercially, but also in terms of enhanced societal qualities, including risk management.

Table 1. Some Examples of Promising CI (for further examples, see also [3]).

Subject Domain	Contribution by Human Intelligence (HI)	Contribution by Complementary Intelligence	Guardrails and Precautions Needed for Risk Management and Optimum Use
Medical diagnostics	Experience from dedicated cases; personal consultation	Systematic diagnosis based on large data bases; Improve focus and reliability of diagnoses	Privacy and data protection
Medical therapy	Control of therapy and results; Lessons learned feedback	Therapy suggestions from large data bases; robot support in, e.g., physical therapy; tele-therapy; reduce human effort in therapy (see, e.g., developing medical APPs)	Priorities and risk assessment; AI liability
Surgery and transplantation	Fallback and advice in critical situations; Cross-fertilization from other disciplines, e.g., materials research	Robot and tele-surgery support and even substituting human intervention; design and generation of synthetic body parts and organs; neuro-computer- interfacing	Insurance policy and accountability; help for the handicapped

Table 1. *Cont.*

Subject Domain	Contribution by Human Intelligence (HI)	Contribution by Complementary Intelligence	Guardrails and Precautions Needed for Risk Management and Optimum Use
Biotechnology	Conventional breeding; Concepts for accelerated evolution; survival strategies of biological life, diversity, etc.	Sequencing and optimization of genomes of food plants, animals, humans, and implementation control	Ethical criteria; rules on equality vs. preferences
Nanotechnology	Prioritizing and selection of candidate applications, e.g., in medicine; novel and "intelligent" materials design	Development of intelligent technologies, e.g., micro-robots and sensors, for continuous monitoring of physiological parameters and disease indicators; novel and "intelligent" bio-materials	Close cooperation between Bio/Nano and IT diciplines Methods for evaluating and supervising ethical principles
Social security and care systems, e.g., health insurance and pension systems	Transformation of political principles, social preferences, culture and constituents. Implementation of improved concepts and provisions for political acceptance	Development and control of sustainable mid- to long-term social security models. Incremental, periodic validation against reality	Confidentiality and data privacy; rules on fairness, equality
Climate Change	International agreement on facts, on objectives and action plans and their implementation	Suggested ways of optimum strategies and of compromising between diverging interests	Priorities; international frameworks and treaties
Ground and air traffic, Transportation, logistics	Implementation of improved concepts, related legislation and exception handling. New models of car (or other type of vehicle) sharing.	Concepts of information netting and automation (self-driving, alerting, parking management, airborne area logistics). Optimized mixes of public and private transportation and delivery systems. Automated modular area-wide public transportation systems. Dispatching, logistical and financial optimization of resources	Rules for optimum and fair use
Education	Setting of new foci, e.g., on rationality, but also on societal, cultural and humane skills such as philosophy, arts, creativity and critical thinking and inquiring; supervision of data abuse, data quality control principles and organization. Avoidance of abuse of personal data and profiles and if fake information.	Provision, searching and distribution of contextual knowledge; Matching of personal profiles and academic and/or vocational education. Support of quality control of education results and of intelligent search, interpretation and validation of data and of using media. Software support of teaching and learning processes. Creation of symbiotic ideas for arts, e.g., fine arts, music.	Creating and supervising of new/changing education objectives and methods. Monitoring of societal impact.
Law and law enforcement	Regarding of individual circumstances in lawsuits, e.g., psychological factors, special societal situations. Basics on responsibility and liability of AI actors	Identification of and conclusions from applicable law. Identification of historical role cases. Care for equality of law enforcement/jurisdiction rules	Sharing of responsibilities between human and automata. Legislation regarding autonomous AI agents
Crime investigation	Handling of complexity issues and contradictions; credibility of witnesses	Tools for exploiting case based knowledge and forensics analysis; Improvement of credibility of facts and witnesses	Balancing between AI findings and human jurisdiction
Financial systems and taxation	Handling of special cases and exceptions. Enforcing of measures to early mitigate financial crises already at indication and warning status	Automation of taxation processes; Local and global crises to be indicated early and mitigation measures to be recommended. Models of financial behaviour in support of improved tax fairness and fraud avoidance	Fairness rules and governance legislation
Security and defence	Evaluation of basic political strategies and options. Interpretation of gaming results, setting of priorities and rules in decision making, procurement, etc.; principles of data security and privacy	Simulation of complex conflicts, wars, terroristic acts and other catastrophic scenarios (e.g., by gaming and exercising); analysis of results; robots in dangerous environments; Command and Control and decision support,	Trustworthiness, controllability, responsibility, accountability [5]; privacy and data protection, confidentiality rules

Table 1. *Cont.*

Subject Domain	Contribution by Human Intelligence (HI)	Contribution by Complementary Intelligence	Guardrails and Precautions Needed for Risk Management and Optimum Use
Socializing and partnering	Organizing and performing human exchange. Discussion of personnel selection parameters; psychological advises	Finding better matching partners; support groups of common interest. Intelligent partner matching of not only the "equal" but also the "different"	Data protection and privacy rules; psychological support
Job creation and assignment	Setting the basic personal preferences and assessing personal restrictions	Find the optimum job around the globe. Provide tradeoffs, e.g., between location, income, perspective, family situation.	Amelioration/harmonization of national, gender, racial, etc., restrictions and differences
Communications	Sharing of private and public authority over networks, and network services	Undecryptable coding, Real-time language translation, intelligent use of social networks	Control of information contents, e.g., of fake news, illegal contents, of profiling and abuse of algorithms
Global challenges	International agreements on objectives and control of efforts	Optimum strategies for climate control, exploitation of natural resources, bio diversity and sustainability; Suggestions for better governance of treating global challenges	Rules of fairness of distribution of benefits and of impacts
Managing the future	Identification of subject priorities, e.g., climate, bio-sphere, population, migration and involved problems; setting of priorities	Intelligent analysis and interpretation of mass data. Improved forecasting confidence and reliability; support in choosing priority subjects	Legal frameworks; international treaties and control mechanisms

Of course, related concepts (and, later, the solutions) need to be assessed against a variety of impacts they will undoubtedly have, including:

- E1 Compliance with Ethical standards;
- E2: Gross impact on Economy;
- L: Conformity with national and international legal settings and treaties, and possibly the need to modify them;
- P: Compliance with basic political agendas of "leading" nations and international regimes such as the UN, EU, OECD, IMF;
- S: The impact on and perception of such impacts by societies.

The basics of such an "EELPS" assessment have been developed in several international research projects and are further discussed in [6].

The superiority of CI then would mean the superiority of solutions, possibly superiority over human abilities, but also superiority over legacy organizations and political systems. Debating threat scenarios where AI substitutes the human (labor, cognition, rationality) "enjoys" huge attention from, and has a huge attraction to, the public and media at present (see, e.g., [7], strategy # 1; 7; 10). Instead of this, a positive and constructive concept of AI and CI should aim for solutions that are developed for, and implemented and operated in, cooperative models: here called Intelligent Complementary Human-Machine Collaboration (ICHMaC or CI). For the time being, we leave open the question of whether this will materialize in direct bio–machine interaction, organization of close cooperation, or the complementary use of brain and machine intelligence. The degree of integration of HI and AI will depend on the selected problem area and the related technical solutions being developed, as well on the risks and political will.

4. Validation and Consequences

4.1. Human vs. Machine Capabilities

When following this basic approach, there would be no need for further (mostly cramped) discussion on what is human and what is artificial. In CI concepts, they simply coexist and cooperate. This will, however, require a kind of general "societal contract" or agenda on the weighting of societal values and human skills relative to AI. As AI will become increasingly smarter, this is not a one-time fixed agreement; rather, it needs to become a continuous process, including "sandboxes" for experimenting with all related societal groups, and clear criteria for the validation of benefits and risks.

There may be necessary limitations, beyond which AI has no say. Then, the somehow still fictitious discussion of whether AI is, or will ever become, superior to HI, and even endanger human existence, should become irrelevant. Inherent to the concept should be regulations to avoid false or counterproductive competition between AI and humans. False and counterproductive here means with new risks or insufficient compensation for those people or organizations affected (see also Sections 7 and 8 on benefits and risks). In the longer term, and from a more philosophical perspective, as in our concept, issues which are not so relevant today, but may possibly in the future, should be discussed, such as whether machines may ultimately acquire properties of consciousness, free will, a "spirit", or whatever "mental" characteristics may be attributed to AI above or beside rationality. Attributing consciousness to AI systems, however, would require a clear definition and understanding of what consciousness really means, physically and logically [8]; this is not available at present, and it could take a long time to fully understand (see e.g., [9]). The consciousness of AI, then, would lead to extended elaborations on ethical impact of AI, and the evolution of superintelligence, transhumanism and digital "Singularity" [10], terms that form a different level of philosophy, which we will not further detail here.

However, these speculations will lose their menacing nature when societies realize the boons of AI and CI. "These challenges may seem visionary, but it seems predictable that we will encounter them; and they are not devoid of suggestions for present-day research directions" [11]. More on this topic is also available in [12,13].

Further practical pros and cons should be considered in socio-political discussions on future CI concepts, including:

- The goals of AI should not primarily strive to substitute human labour due to rationalization, but should rather liberate mankind from "stupid", routine jobs, enabling humans to concentrate more on care and socializing;
- AI "machines" may act as autonomous, responsible and liable agents;
- AI actors may become morally superior to humans, e.g., in critical incident risk assessment;
- AI solutions may, and hopefully will, in many areas, interact with much lower failure rates than humans;
- AI should be free from negative emotional motivators which prevail in humans.

In a cooperative AI concept, as sketched here, there will be no need to fear risks of AI (CI) becoming adverse or hostile to humanity. The precautions of creating only "benign" systems and for preventing adverse, uncontrollable or dangerous AI solutions will be, and must be, an integral part of this CI, and of any AI concept [13].

CI will leave humans to complete those tasks that he/she can do better than a machine, which require abilities and skills which are hard to implement in computers (e.g., intuition, life experience, highly sophisticated skills and knowledge such as medicine, creativity, arts, and decision-making under contradicting arguments). This will be a dynamic process, as AI/AGI skills will develop and further capturing these human properties, and AI skill profiles and human societies' trust in AI will evolve over time and integrate with human abilities to form advanced cooperative models, e.g., virtual reality arts.

4.2. CI and Economy

Economic and socio-political models of human–machine cooperation and partnerships, other than models of competition between human labour and machines, need to be created and established (sometimes called "Centaur Teams" with "Hybrid Intelligence", e.g., in [3]), controlled and maintained. Interdisciplinary collaboration of the AI community with economics, and political, social, psychological and philosophical science will become indispensable in the design, development, implementation and operation of those models. Their benefits and shortcomings continuously need to be monitored and evaluated from these very different disciplinary perspectives (see also Sections 3 and 7).

A cost–benefit assessment (short-, medium- and long-term CBA) needs to be established and regularly applied. As these CI concepts will create, or at least facilitate, rather innovative business models and political scopes of action, classical CBA will no longer be adequate. New assessment methods need to be created that have to include in their equation—besides the monetary/materialistic payoffs—the socio-economic, socio-political and even ethical implications of CI concepts in society [6].

4.3. CI, Politics, Taxation

As we have seen in the industrial revolutions of past centuries, automation is key to progress. First, industrial machines replaced raw labour, then refined machines produced goods, then general-purpose industry robots could diversify and adapt quicker to market needs, then the computer helped further automation, then the advent of the internet (not an automation tool in itself, but a productivity booster) and, finally, AI, have allowed for the creation of an increasing number of goods and services, much more cheaply than was possible with human labour.

Any country not investing in AI technology or trying to slow it down will seriously fall behind, as did any country that missed the past industrial revolution, probably even more so. In the future, computers in general, and AI in particular, will dominate the economy of every developed country.

Some believe that AI will take our jobs and cause mass unemployment and poverty. Others believe that, for every job taken over by AI, new jobs will be created. We believe that both sides of the debate are misguided. The primary goal is not to have a job and for businesses to maximize profit, but to gain a decent standard of living for as many people as possible. Un(der)employment should not be the focal problem; one may even argue that it is desirable and affordable (for more rationale see e.g., [14]). Crucial to maintaining or increasing global wealth is that the total amount of quality goods and services that are created stays the same or increases, respectively. If a robot or an AI replaces a human worker, the same amount of goods and services are produced, and likely even more, or fewer, but with higher quality. It is then "just a political matter" to redistribute these goods and services fairly, ensuring that nobody is worse off—most likely, everyone will be better off and, in the long run, much better off. The importance of immaterial benefits will increase. Few people seem to appreciate this simple economic fact today.

This redistribution can be done in a number of different ways:

- One could continue to decrease working hours for all (e.g., rather than one person working 40 h and one person having no paid job, both could work 20 h with the same salary as before, then 10 h, 5 h, etc.);
- Retirement age could be reduced, instead of the current trend of increasing it;
- We could have universal basic income schemes, funded by higher tax rates, e.g., for those still employed, on productivity and on unproductive financial transactions;
- The government could guarantee every shares in companies for every citizen (maybe initially in companies they have been laid off from), so that they profit rather than suffer from the automation;
- Robots/AI could be taxed, though one needs to be careful with taxing productivity gains to make sure innovation is not stifled. Preferably, one would only tax the negative side effects of a particular automation, e.g., the need for re-education and temporary unemployment costs of laid-off workers.

Probably, a well-thought-through combination of these and other measures, suitably adjusted to the increasing automation over time, could avoid any larger outfall, as we have also largely managed technological advances in the past. Of course, the right government and governance is necessary for this transition to be possible and peaceful, but if this occurred, nearly everyone would profit from it (see more in Sections 9 and 10). Generally, policies should be set and managed in a way that the poor could become richer and others, such as investors and entrepreneurs, would profit, but not unduly, from this new societal paradigm.

The replacement or improvement of human work with IT has been ongoing for more than half a century, with benefits for all, causing temporary layoffs but generally producing new jobs and societal and economic benefits [15]. Tendencies, intentions and plans, however, to replace human jobs with AI, without adequate compensation and tradeoffs between benefits and losses, need to be avoided through political and economic precautions, and substituted with ICHMaC concepts. Supporting social programs and lifelong academic and/or vocational training will need to accompany most career profiles. The primary goal of human existence will no longer be to have a job, but to have a decent standard of living. The required human working skills will shift from routine to creativity, from standard work, replaceable by robots, to higher quality work that exceeds the abilities of AI, that will not be bound to a fixed number of working hours, but will require high flexibility from employees, innovation and the abandoning of standard procedures. CI will alleviate the compulsion for gainful occupation. Unemployment must no longer be considered a problem; it may even become desirable and affordable! Politicians, economists and entrepreneurs, however, will need to adapt to these changes in societal paradigms and agree on different taxation and social security models that will avoid "arms races" and other risks of conflict between human and complementary machine intelligence.

4.4. CI and Data

AI/CI concepts will rely on the availability of data to learn from and use in their applications, usually vast amounts of data. This requires the preparation and assessment of available or emerging databases and the creation of the necessary:

- Organizational and technical structures to maintain and communicate the relevant data;
- Political/legal prerequisites, including those on data propriety and privacy regulations, access rights and responsibilities;
- Technical solutions to harmonize data from various sources;
- Verification and validation of data;
- Intelligent compression of large amounts of data.

Data compression and information theory are not only important for saving storage space and channel capacity for transmissions, they also play a key role in modern approaches to artificial intelligence [16].

We will not be able to discuss AI using complex data sources within this article. Nevertheless, it is common understanding that all practical and useful AI applications begin with data. As an example of AI in manufacturing, see, e.g., https://www.forbes.com/sites/willemsundbladeurope/2018/10/18/data-is-the-foundation-for-artificial-intelligence-and-machine-learning (accessed on 20 May 2021).

A typical classification scheme of "big data", as seen today in the era of internet with social media, with cloud services and many others, is depicted in Figure 2.

The top categories in this picture, Velocity, Volume and Variety of data, are only three of the most important technical classification criteria for mass data characterization.

Further data characteristics need to include:

- Data sources and capturing, for instance, from sensors (cameras, RFID, etc.), organizations, individuals, software systems;
- Processes of capturing, transferring, structuring and storing data;
- Processing, analyzing and validating contents;
- Maintenance and quality management, including authenticity, correctness, reliability and scalability;
- Security and data protection, which are one of the most sensitive precautions needed (see also Section 6.2 on security);
- Socio-political aspects of data use; see, e.g., Elizabeth Denham (2017), big data, artificial intelligence, machine learning and data protection, Version: 2.2 [18];
- Scalability and durability of databases.

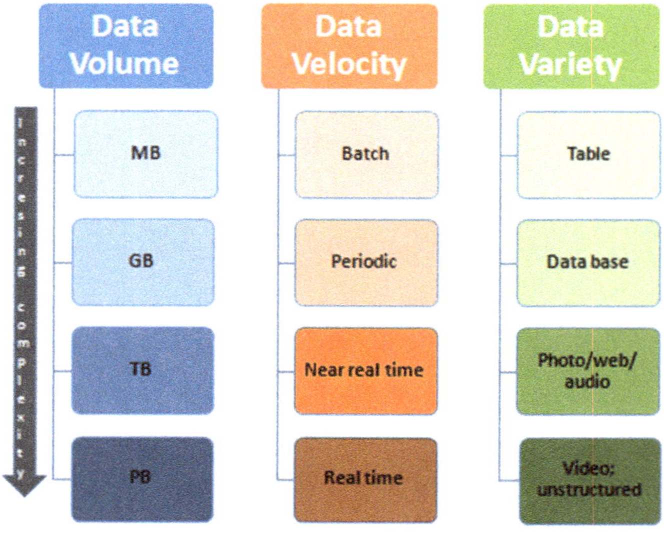

Figure 2. Big Data Classification Parameters (adapted from: [17] where, in addition to the Figure, a full suite of big data characterstics are defined).

Among this rich spectrum are the key requirements concerning data that will be used for AI. When we discuss the applications of AI/CI in the social context, as we do in this article, we believe that, for preparation at the socio-political level, certain elements are indispensable:

1. Public authorities: Public institutions and actors such as healthcare, security and transportation must agree and prepare to apply their data to novel solutions in AI;
2. Public–private cooperation: A consensus should be reached among public authorities with private big data resources. This, hopefully, will be reached in a consensus-creating environment. In the best case scenario, we hope for public–private win–win strategies, but reinforcement by regulations may become necessary in order to create and maintain a balance of power between governments and private industries;
3. International agreements, in particular when preparing for global AI applications, e.g., for security, climate control, population growth control, and the distribution of global resources or global rules for market and finance.

These efforts require political initiatives and need to be started now and sustained and supported by impartial research capacities.

5. Beyond CI

Futuristic visions are challenging and their risks need to be perceived and managed. This paper assumes the continued existence and (hopefully) the continued exploitation and improvement of political systems such as "western" democracies. Of course, changes may occur, even major ones, of constitutions, of priorities and rankings in economy, and of society and the political system, some of which will be recommended here. More revolutionary scenarios, however, such as singularities in the sense of [10,19] machine takeover or conscious machines [12,20], and their possibilities and probabilities, lie beyond the scope of this article.

Computers have always partially outperformed human performance by factors of millions and more, e.g., in calculation and communication speed, automation of office and production labour, and nobody has cared too much or would describe this as revolutionary rather than an evolution, mainly to the benefit of society (for many examples, see, e.g., [15]).

"AI Takeover" scenarios by "hyper-intelligence" and "superhumans" are likely to remain, for decades and maybe longer, in the category of Science Fiction. The basic question, at present but particularly in the future, is where to draw the line between technology's blessings and curse, how to detect or forecast them and how to reflect on them. If the line drawn too narrow, we may forfeit chances for higher living standards—not just in terms of monetary gain, but also possibly a cure for cancer, the elimination of hereditary diseases, improvements in longevity, the prevention of war, and/or saving the environment and the climate. If drawn too wide, a "singularity" or other catastrophic developments may happen, the consequences of which we, and even their proponents, cannot assess and predict with the present way of thinking about the future.

6. Some Promising Examples of Future CI

6.1. Survey

The spectrum of applying AI in the CI sense, as discussed here, appears to be almost unlimited. The few selections of promising applications given in Table 1 are rather arbitrary candidates, however likely, and not exhaustive. Columns 2 and 3 approximately indicate how work could be shared between human and AI partners in different domains. This should demonstrate that the basic principle of CI is for HI and AI to complement rather than compete with each other. The chances, challenges and benefits are left to the judgement of the readers or experts in the subject domain. Requirements, risks and benefits in the implementation of CI will need detailed tradeoffs. However, some basic guiding rules for the implementation are mentioned in the right-hand column. Solutions and their realization will require extensive research and analysis, planning and consensus-building among all involved and/or affected stakeholder groups.

6.2. CI for Risk Management and Security

The present discussion about security comprises, in principle, three different aspects:

(1) Security of AI/CI solutions themselves;
(2) Risks from AI abuse or malfunction;
(3) Security as a socio-political application domain for CI.

Here, we elaborate on the latter: security as a promising application area of CI, one of the many candidates shown in Table 1.

Security is the most fundamental need of humankind, following the basic physiological needs for food and water (https://www.simplypsychology.org/maslow.html#gsc.tab=0) (accessed on 20 May 2021); this is true for societies, groups and individuals. The protection and integrity of lives and possessions are deeply anchored in the psyche of mankind, due to evolution's survival and selection principle, and are present in mankind's organizational constructions. Treats are usually classified as natural disasters, technological risks and man-made disasters. The basic prerequisites for establishing and maintaining security are subsumed in the term Risk Management. A typical organization scheme of risk management is presented in Figure 3.

Almost all tasks in this activity cycle require considerable skills and resources. Despite various approaches to security and an enormous, historical trove of experience from wars and other disasters, precautions for and measures to maintain security have always been all but optimal. This is due to the huge complexity of the various problem areas, the fuzziness of data and the short reaction times required, as well as the diverging or conflicting interests of the individuals and groups involved, a lack of incentives for long-term planning in government and industry policies, and the tendency to ignore very rare but global, truly catastrophic events that could endanger humanity in its entirety [21].

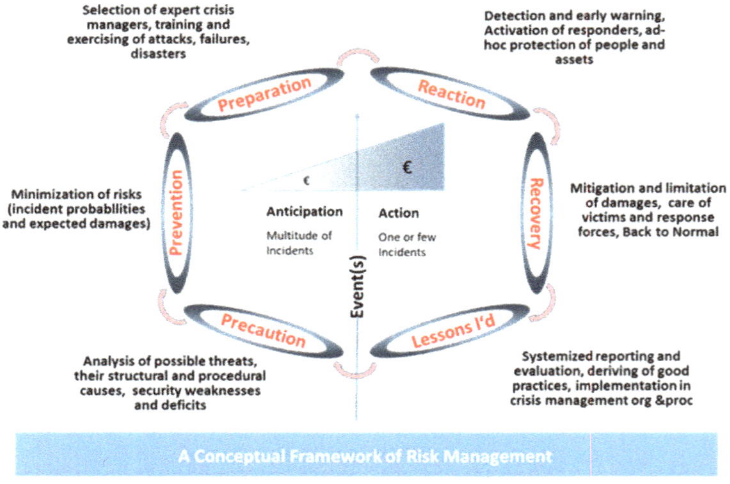

Figure 3. Security and Risk Management.

Even during the last 20 years, in which security increased in priority due to the 9/11 catastrophe and follow-up terrorist activities, measures taken to establish or modify organizations, in terms legislation and responsive actions, often appear ill-designed, or lacking in rationality. The authors appreciate all efforts taken and do not claim to know better, but with AI/CI, there is hope that we can do much better.

Nevertheless, the skills required in risk management, including data acquisition and analysis, organization, plans and actions, situation monitoring, and inferring and applying best practices, distinguish security as a preferred candidate for the extensive application of AI and CI. Arguments for this focus on security include:

- Security has to cover all areas of basic societal relevance, including ethical, social, legal, political and economic ones. Intelligent algorithms could improve data and privacy protection;
- Security requires a fast reaction in response to changing threats and disasters. AI will provide solutions to provide early warning, situation awareness and the fast creation of reaction options for responding units;
- These include scenarios of high complexity, with many uncertainties;
- Security will directly and visibly save lives and high-value assets, as well as improve well-being, both in reality and via fast-running, synthetic games and exercises;
- This will have a high positive impact on many other areas of society. CI security solutions will be appreciated and accepted in society, more than many other AI application areas;
- This has a high potential to increase trust and confidence in political systems and security measures;
- This will require intensive cooperation, trust and responsibility, from and between human and AI operations and resources (as has always been the case in security with technical resources such as rescue equipment or weapons);
- CI will substantially contribute to the reduction in risks for human reaction forces;
- The solutions will have to cover a huge variety of missions.

Overall, the delineated spectrum of tasks and potential make security a prime candidate for the exploitation of convincing AI/CI solutions. They could range from intelligent monitoring and detection and the assessment of warning indicators to the use of robotics instead of humans in dangerous areas; from optimal strategies to deal with pandemics to proven good practices in allocating patients to medical resources in disaster situations;

from intelligent analysis of operational situation data to optimal strategies for damage reduction and mitigation. There is an urgent need to proceed in this direction.

Elaborating on the examples of Table 1 would exceed the limits of this article and go beyond the focus of this special volume on Risk and Crisis Management. The scheme of CM discussed here, however, may serve as a role model for other subject domains listed in Table 1, many of which also address the security of societies in a wider sense.

As discussed at the very beginning of this chapter, security is a fundamental need. The other side of this coin, however, offers a huge potential for abuse: of all domains of political responsibility, and broadly independent of the political system in place, security implies the most traps and loopholes, to circumvent warp legal and societal control mechanisms. This is because the need for security often overrides the need for assessment by individuals as well as administrations, and limits rational discussions and the balancing of diverging interests and alternative options. Numerous such cases, particularly after 9/11, have occurred in security, leading to:

- Legislation conflicting with basic laws like constitutions;
- Unauthorized surveillance;
- Violation of data privacy;
- Breach of human rights;
- Abuse of data for unauthorized applications;
- Violation of basic ethical principles;
- Lacking proof of cost-efficiency and wasting money (for more examples, see, e.g., [22]).

Societies and politics need to support the preparation abilities of AI/CI in security to prevent rather than support the phenomena which are counterproductive to a prosperous and peaceful society, which is the aim of this contribution.

7. Evaluation of Benefits

Human brains and HI are wonders of evolution. Nevertheless, they also embody huge shortcomings that manifest in conflicting interests and controversies among institutions and decision makers, emotion-based preferences, and personal struggles with mental and/or physical force and injuries, which can lead to wars and global conflicts with apocalyptic outcomes, such as weapons of mass destruction, serious environmental impacts and social disruptions. The technical deficits in HI lie in the operating speed, replication accuracy of knowledge, diverging preferences of individuals and groups, limited objectivity, limitations in solving complex problems, and finding, introducing and sustaining optimal solutions to complex societal issues. Additionally, the limited sustainability and replication accuracy of human and societal memory and of the longer-term planning, control and maintenance of decisions and their impacts can cause problems Religions (this in no way intends to diminish the beneficial effect of religion) and un-scientific or biased mindsets, conspiracy theories, etc., also contribute to the deficits in rationality [15]. Other shortcomings produce emotional drama or allow for the creation of ideologies that may be catastrophic for mankind or for groups and individuals.

Generally, the mankind's ability to anticipate, manage and mitigate risk seems to be inadequate, or at least suboptimal, in today's sophisticated, technological and organizational, highly developing world [21].

The CI approach presented here should concentrate on compensating for, or at least ameliorating or attenuating, these kinds of deficit. It is the basic postulation of this concept that this should produce the highest benefits for the well-being, standard of living, quality of life and the further development of societies into civilizations of the highest standards (some visions of what a "highly developed society/civilization" could actually mean are given in [23] (some 400 years ago) and [24]). Monetary profits are included, but should not dominate the proposed benefits. These should also entail the adequate control of climate, environmental exploitation and biodiversity. The possible benefits are obvious; however, they need to be verified and evaluated in a continuous process. Some promising research ends are given in the Stanford Human-Centred AI Initiative, stating that "the ultimate purpose of AI should be to enhance our humanity, not diminishing or replacing it" [25]. Other initiatives follow similar principles, e.g., [14,26].

The benefits will be a mixture of material/monetary and immaterial/qualitative gains. Advanced methodologies need to be developed to assess the benefits, particularly life-cycle models and mixed qualitative/quantitative cost—benefit tools to balance the investments and benefits. This will also create new paradigms in the definition of "cost" and "benefits". For example, if a "robot" can do what a human can for half the cost, it is "just" a political matter to ensure that everyone benefits, including those people that would otherwise lose their jobs and income.

In an earlier European Union project [27], the principles of a balanced assessment of measurable benefits, cost and societal impacts were developed. Different kinds of computer-based model focused on:

- Security risk assessment (probability of threats and damages);
- Life-cycle cost analysis (investment, operational, maintenance, update and replacement costs);
- Socio-political impact assessment, using mostly qualitative criteria (see Section 3).

These were defined, implemented, tested and successfully demonstrated in a number of use cases. The final challenge was to integrate the results on risk assessment, cost, and societal impact evaluation into an integrated decision support tool. Although, in that project, the applications were limited to investments in security, the basic methodology and plurality of the evaluation criteria can be viewed as a baseline, which can modified to evaluate future CI concepts. Unfortunately, the willingness of public institutions to submit their decisions to such a systematic assessment was rather limited.

Nevertheless, using a comparable methodology, such wide balances between autonomy and controllability, as well as between investment, risks and benefits, need to be found for all CI solutions. This leads to the discussion of the risks of CI solutions.

8. Risks of CI

There is no technology without risk. The risks from fast-developing and internationally/globally spreading technologies include those to and from AI, as AI will increasingly penetrate all domains of society worldwide. "Global problems need global answers" [28]. Apocalyptic nightmares about the future risks of AI are frequently publicized; however, they are usually biased and based on speculation rather than sound analysis and, therefore, are rarely helpful or likely to happen. Nevertheless, precautions need to be considered and implemented to ensure AI development and employment remains "benign" and avoid hostile or dangerous applications. An example of a potential risk is that deep learning AI system algorithms may one day have the potential to go beyond human control, or at least human comprehension of what they are actually doing, with the potential to abuse economic and political power, or even sabotage and undermine democratic constituents. Even worse could occur if such systems are hacked and used to cause military controversies or abused for large-scale criminal or terrorist purposes. This means that CI regimes will require internationally agreed-on standards, validation and certification rules, and a reliable regime of governance, before they are established. Some examples of CI applications and precautions are given in Table 1. Promising developing role models for governance can be seen in the European Union's CIP, European Programme for Critical Infrastructure Protection policy (ECIP) (https://ec.europa.eu/home-affairs/e-library/glossary/european-programme-critical_en, accessed 20 May 2021) or in the General Data Protection Regulation (GDPR) of the European Union, which is compulsory for EU member states but appears to be becoming widely accepted as an international standard [29]. Another positive example comprises the agreements on cyber-security, which was first developed in the 1970s and has resulted in a number of internationally accepted standards in IT security (for the most prominent examples, see [30]). It should, however, be noted that the processes behind these initiatives took decades, from the initial idea to putting the regulations into force; we may not have these timeframes with AI.

Political prerequisites, chances and obstacles regarding the further development of AI technology, and governance of CI solutions, are discussed in the next chapter.

9. Prerequisites for Success of CI Concepts

This CI concept could develop into a peaceful revolution of the human condition into the next generation of the socio-economy, related politics and modified constituents, and should be incorporated in state constitutions and national and international agendas (e.g., the UN). The basic vision of CI that should be created and communicated to all societal groups, organizations and their people is the creation of win–win situations between society and technology; between economic interests and those of the "working people"; between the maximization of consumption and the "higher values of life"; of a controllable financial sector with the aim to fairly balance the distribution of monetary resources among societies.

The classical neo-liberal concept (driven by capital, market and profit), to a large extent, needs to be replaced by, or transformed into, a concept of optimum benefit for societies (driven by social and humane values). The intent of this concept and its risks and problems is not, in any respect, to limit or hamper work on AI. Instead, it aims to employ and exploit AI to the mankind's benefit. Defining what is "best" for humanity, however, is an even more complex, manifold, ambitious endeavor. Different cultures and political systems have different values and behavioral rules and processes; the interests of states or regions vary between nations, and often within nations. The same is true for different groups, be they the financial sector, associations of interest such as unions, religions, environmentalist movements and many more.

Nevertheless, we believe that the basic statements of the Universal Declaration of Human Rights of the United Nations (1948) and the European Convention on Human Rights (1953) could set basic guidelines, to be complemented by AI-specific aspects, e.g., those based on Isaac Asimov's "Laws of Robotics" and adjacent principles, discussed by other great thinkers [31]. Without placing AI at the same level of risk as nuclear technology (civil and military), the international negotiations, treaties and organizations on nuclear control could provide useful lessons for a future international AI control regime. For a basic outline, see [21], chapters. 15 and 18. Beyond, and supporting, principles of humanity, cooperative CI models also need to define governance principles and roles for human–machine interfacing and partnerships, authority and hierarchy, management, and command and control loops.

Collaborative CI concepts will only prove successful if designed, developed and implemented in a joint effort of political, economic, commercial and societal forces, with massive support and participation from the scientific disciplines, including philosophy, sociology, psychology, economy, commerce and the Cyber and AI research community. This integrated concept needs to be established by an "elite movement", furnished with resources and authority. Basic rules need to be developed and agreed on, starting now. It is, however, hard to imagine that the adequate organizations and processes could be implemented by present-day political, administrative and economic systems, and the people in power (a typical, negative example of such organization is the final disposal of high-intensity radioactive waste, and the GE example: political discussions and temporary experiments for 70 years, a promised time horizon of site selection in the 2030s, site establishment "starting in the 2050-ies", i.e., roughly 100 years after the technology was introduced). A common understanding among all relevant groups will be necessary in all the discussed areas. Only a common agreement on the objectives of an ICHMaC strategy will facilitate solutions and better decisions than those of today. Political decision makers, legislation and responsible administrations need to accept the fact that CI can produce systems in support of many societal domains, superior to the possible solutions in present economic–political systems.

Consequently, the basic objective functions in the economy and politics need to be changed, to emphasize care for people and society, instead of saving traditional jobs, maximizing material wealth and becoming re-elected. Economic systems will need to focus on human–machine collaborative models, instead of maximizing profit at the expense of the "working", as is common today in terms of a too-high divergence in income levels, opportunities and social states. The financial system needs to be regulated and controlled.

This all needs to begin with a modified vision of quality of life that coincides with the AI vision of relieving humans from the current primary need for jobs rather than to live much self-determined lives. This is not a "neo-communist" utopia in a new tapestry. Such basic ideas on quality of life were discussed by ancient Greek philosophers and in the vision of Thomas More [23] in the early 16th century. Present-day visions are also available, e.g., in [14]. This will, however, require new ways of thinking about societal constituents, modified or new models of creating and maintaining jobs and incomes, different concepts of education and lifelong learning that keeps pace with technological and societal developments, and modified or new concepts of general social security and care.

The income profiles of governments need to be adapted. Classical income taxes on wages and other earnings, consumption, etc., may shrink, and need to be ramped up in other ways, as discussed in Section 4.3. Alternative sources may be needed for unproductive speculation, real estate, the exploitation of rare natural resources, etc. Otherwise, the diverging distribution of income between the current, so-called "elite", i.e., those who would "own" power over production, data, intelligence and capital goods, and the poorer and less privileged would accelerate, possibly to a ratio that may endanger social peace.

International organizations (such as the EU, UN, IMF, World Bank, OECD, G8 and global finance and commerce) need to support the ICHMaC concept. Political and scientific support will require increased funding of these organizations. In the same order of relevance, for the protection of the earth's environment, biosphere, resources and climate, ICHMaC needs to establish an international planning regime and binding agreements, which will be sustainable over periods far exceeding the current electoral/legislative intervals and planning horizons.

10. Conclusions and Feasibility

The basic overall requirement must be insight into, and propagation of, the idea that CI will be able to overcome the flaws in past and present socio-political systems, and the consensus and will of "all" that better and collaborative solutions can be achieved and create unprecedented win–win-models.

Future solutions for society in the cyber world will require international, improved global rules and arrangements. This is true for the cyber domain including AI, as well as for climate protection and compliance with human rights. Unfortunately, there are always reverse tendencies, with populist movements prospering and propagating absolutistic, traditionalist illusions of "the past was better" agendas. With their limited appreciation of science and their backwards orientation, they will support the CI ideas. Instead, they will obstruct them.

Concerning the diligent treatment and exploitation of AI, we have only briefly outlined an option for the future. Its realization would need to start now, with an assessment of the prerequisites discussed above. Implementation of the concept or subsets of AI should be smooth and peaceful. However, realizing the pace of development in all cyber areas, including AI, we strongly suggest that politics and the economy prepare for such solutions sooner rather than later. Regarding the coming challenges and enormous potential of AI, present "western-type" governments are more or less dormant; real strategic thinking and political concepts are missing, and transferring progressive ideas into practice takes too long and is hampered by bureaucracies. "We are not a pre-emptive democracy, we are a reactive one" should be seen as a wakeup call [32]. The concepts discussed here are rather revolutionary. Therefore, policies need to follow precautionary and pre-emptive strategies. Unfortunately, more autocratic regimes, in principle, might be better-suited to enforce, implement and sustain them, provided their basic objective and agenda is the creation of better conditions for all. However, historically, this has rarely been the case. Less desirable "motivators" for change could be global catastrophic risk (two examples: Fukushima; Coronavirus pandemic), forcing politicians to act. In any political framework, CI will require new and more efficient ways of consultation and consensus-building between peoples, governments, industry and research.

There are two extreme options, at the opposing ends of the spectrum, and many in-between. Things driven by AI, automation and robotics could cause deep tensions and further inequality in societies, and the government and economy will face increased pressure in a "3rd economic revolution". Alternatively, politics and societies may become better-prepared for the substantial changes, and be ready to envision and plan a better world.

The first option may result in the further fragmentation of society, with people losing trust in existing political systems and technologies, feeling lost and no longer represented, needed, or taken care of. Increasing sympathies with the regional, autocratic systems and occurring at present are indications of this direction. At worst, this may lead to societal catastrophe, revolutionary mass movements, damage to democracy, and more (we have seen this in the 20th century and before).

The more optimistic second option, which is hopefully more likely and, of course, possible, would require unprecedented effort from political and economic sectors to create solutions guided by the prerequisites discussed above (and more). This would need to start now [33]. Unfortunately, there are almost no signs in politics, particularly not in education, that this will be prepared in time. As long as the high-level political discussion—as in Germany—is concentrated on internet access bandwidth, some local and rather modest investment in AI R&D, or the question of whether and how to allow or forbid smartphones in schools, we are tickling the little toe of the giant.

Speaking of giants, the largest ones at present, such as Google, Apple, Facebook, Amazon, Microsoft (GAFAM) and more, e.g., in China, are moving ahead. They have visions of future "cyber societies" and invest enormous amounts, including in AI (see e.g., [34]). Time will tell whether or not their results benefit mankind and individual societies. The truth is that their (GAFAM's) power is increasing, while that of democratic governments is shrinking, although we are all aware that IT and AI are THE drivers of the present and future economy and everyday life. In any case, the future role of AI/CI in societies cannot be left to the forces of the market.

This may sound like pessimism, with two apocalyptic options. Nevertheless, through the influential progress of scientific and technological development, political systems will either be forced to adapt to new governance models and rules, or industry power will increasingly dominate, or even dwarf, the decision space of government plans and action. Seneca had a fitting recommendation for such a phenomenon 2000 years ago: *"Fata ducunt volentem, nolentem trahunt"* (Fate leads the willing and drags the unwilling). Only strong coalitions between the IT giants, governments and international organizations will be able to create future CI-based societies, as stressed in this article. As HI and AI should be viewed as two complementary abilities, the political systems need to view societal regimes such as quality of life and the economy as integrated parts of the same medal, rather than competing ones, as typically occurs at present.

An easier transformation into CI—and transformation is what we are discussing—should follow the "Think big, start small" principle: beginning an "Evolution" with rather small, non-spectacular, and politically and societal non-controversial sample projects (e.g., legalizing self-driving cars, improving healthcare systems, further automating routine bureaucracy, optimizing traffic flows and public transportation, fully automating taxation), learning from them and incrementally evolving politics and societies into CI domains of a larger size and greater impact (e.g., in law enforcement, legislation, rationalized political decisions guided by science, optimized and customized schooling, autonomous systems for safety and security, control of the finance system). In fact, cooperative IT-based models, on a smaller scale, are already on their way, e.g., in software, for better administration, or robotics in production lines and supply chains (The Australian, 25 Jan 2019, pg. 18 on the automation of warehouse supply chains: " ... more sites are combining robots and humans."). The current and oncoming AI evolution, however, will not work at a biological pace, taking hundreds of years or more for small changes to evolve. We will have to cope with big changes within a few decades.

As realist–optimists, however, we are sure that the human race and society will survive for a long time, even without CI. Nevertheless, the question is HOW. That is what we are trying to advocate for, with some ideas for the betterment of society.

Author Contributions: R.H.: conceptualization of article, evaluation of mid- to long-term effects of AI on, and its potential for, future society; M.H.: AI methodologies and models, overall validation. Both authors have read and agreed to the published version of the manuscript.

Funding: This research received no external funding.

Institutional Review Board Statement: No institutional/ethical review required; no studies involving humans or animals.

Informed Consent Statement: No studies involving humans or animals.

Data Availability Statement: No supporting data beyond those of the references were needed.

Conflicts of Interest: The authors declare no conflict of interest existing.

Abbreviations

AGI	Artificial General Intelligence	
AI	Artificial Intelligence	
Capability	Quality subsuming intelligence and required time and resources	
CBA	Cost-Benefit-Analysis or -Assessment	
CI	Collaborative or Complementary Intelligence	
CIP	Critical Infrastructure Protection	
CM	Crisis Management	
EELPS	Ethical, Economic, Legal, Political, Societal	Criteria and methodology for socio-political evaluation
EPCIP	Europeam Programme for Critical Infrastructure Protection	
EU	European Union	
GAFAM	Google, Apple, Facebook, Amazon, Microsoft	Here used to represent powerful IT industries
GDPR	General Data Protection Regulation	
HI	Human Intelligence	
ICHMaC	Intelligent Complementary Human-Machine Collaboration	Processes implemented in CI
IMF	International Monetary Fund	
IT	Information Technology	
MI	Machine Intelligence	Here used synonymously with AI
OECD	Organization of Economic Cooperation and Development	
RM	Risk Management	
R&D	Research and Development	
RTD	Research and Technology Development	
UN	United Nations	

References

1. Chomsky, N. Where Artificial Intelligence Went Wrong. 2012. Available online: https://www.theatlantic.com/technology/archive/2012/11/noam-chomsky-on-where-artificial-intelligence-went-wrong/261637 (accessed on 5 May 2021).
2. Pinker, S. *The Blank Slate: The Modern Denial of Human Nature*; Penguin: Random House, UK, 2016.
3. Bush, B. How Combined Human and Computer Intelligence Will Redefine Jobs. 2016. Available online: https://techcrunch.com/2016/11/01/how-combined-human-and-computer-intelligence-will-redefine-jobs (accessed on 5 May 2021).
4. Deep Mind Mission. Available online: https://deepmind.com/about (accessed on 25 December 2020).
5. Fraunhofer Group for Defense and Security. Rise of Artificial Intelligence in Military Weapon Systems. 2020. Available online: http://publica.fraunhofer.de/starweb/servlet.starweb?path=epub.web&search=N-596967 (accessed on 15 May 2021).
6. Hutter, R.; Blobner, C. Assessing the Socio-Political Impacts of Safety and Security Measures. In *Safety Risk Management: Integrating Economic and Safety Perspectives*; DeGruyter: Berlin, Germany; Boston, MA, USA, 2020; Chapter 13; ISBN 978-3-11-063378-8.
7. Chomsky, N. 10 Strategies of Manipulation. 2013. Available online: http://www.99posto.org/english/10-strategies-of-manipulation-by-the-media (accessed on 15 September 2020).

8. Chalmers, D. Philosophy of Mind. In *The Conscious Mind, in Search for a Fundamental Theory*; Oxford Paperbacks: Oxford, UK, 1996; ISBN 8601419222391.
9. Aaronson, S. *Quantum Computing since Democritus*; Cambridge University Press: Cambridge, UK, 2019; pp. 34–42.
10. Kurzweil, R. *The Singularity Is Near*; Penguin Books: New York, NY, USA, 2014.
11. Bostrom, N.; Yudkowsky, E. *The Ethics of Artificial Intelligence*; Cambridge University Press: Cambridge, UK, 2014.
12. Bostrom, N. *Superintelligence: Paths, Dangers, Strategies*; Oxford University Press: Oxford, UK, 2014; ISBN 978-0-19-967811-2.
13. Yudkowsky, E. Artificial intelligence as a positive and negative factor in global risk. *Glob. Catastrophic Risks* **2008**, *1*, 184.
14. Welzer, H. *Alles Könnte Anders Sein: Eine Gesellschaftsutopie für Freie Menschen*; Futurzwei, Stiftung Zukunftsfähigkeit: Berlin, Germany, 2020; ISBN 9783103974010. Available online: https://futurzwei.org (accessed on 15 August 2020).
15. Pinker, S. *Enlightenment Now*; Penguin Random House: New York, NY, USA, 2018.
16. Hutter, M. Human Knowledge Compression Prize. Open Ended. 2006. Available online: http://prize.hutter1.net (accessed on 10 September 2020).
17. Available online: https://en.wikipedia.org/wiki/Big_data# (accessed on 5 March 2021).
18. Denham, E. Big Data, Artificial Intelligence, Machine Learning and Data Protection. 2017. Available online: https://rcn.fpf.org/2017/08/28/big-data-artificial-intelligence-machine-learning-and-data-protection/ (accessed on 15 January 2021).
19. Hutter, M. Philosophical Issues. In *Universal Artificial Intelligence*; Springer: Berlin/Heidelberg, Germany, 2005; Chapter 8.6; ISBN 978-3-540-26877-2.
20. Hutter, M. Conclusions. In *Universal Artificial Intelligence*; Springer: Berlin/Heidelberg, Germany, 2005; Chapter 8.7.
21. Bostrom, N.; Cirkovic, M.M. *Global Catastrophic Risks*; Oxford University Press: New York, NY, USA, 2008; Chapter 15 and 18.
22. Snowden, E. *Permanent Record*; Macmillan: London, UK, 2019; ISBN 978-1-5290-3565-0.
23. Thomas, M. In *Utopia*; Reclam: Stuttgart, Germany, 2012; ISBN 978-3-15-018875-0. Latin/German Version, Original Source 16th Century. Available online: https://www.buchkatalog.de/product/4099276460840799620/Buecher_Romane/Thomas-Morus/Utopia (accessed on 31 May 2021).
24. Lem, S. Star Diaries, the eighth voyage. In *Which Ijon Tichy Represents Earth to Petition for Its Admission to the Congress of United Planets*; Harcourt Inc.: Orlando, FA, USA, 1976.
25. Stanford University. The Stanford Human Centered AI Initiative. 2018. Available online: https://hai.stanford.edu/welcome (accessed on 10 December 2020).
26. HIIG: Humbold Institut für Internet und Gesellschaft. Available online: https://www.hiig.de (accessed on 20 May 2021).
27. ValueSec. EU FP7 Project on "The Value of Security", Deliverables D5.3 and D6.3. 2011. Available online: https://cordis.europa.eu/project/id/261742 (accessed on 15 May 2021).
28. Harari, Y.N. *21 Lessons for the 21 Century*; Penguin Random House: UK, 2018.
29. Hearings of Mark Zuckerberg, CEO of Facebook, before the US Senate and Congress (10.04.2018) and the EU Parliament, 22 May 2018. Youtube. 2018. Available online: https://fortune.com/2018/05/22/facebook-mark-zuckerberg-eu-parliament/ (accessed on 3 April 2021).
30. Security Cyber Security Standards. Available online: https://en.wikipedia.org/wiki/Cyber_security_standards (accessed on 20 May 2021).
31. Available online: https://en.wikipedia.org/wiki/Laws_of_robotics#EPSRC_/_AHEC_principles_of_robotics (accessed on 20 May 2021).
32. Koppel, T. *Lights Out*; Penguin Random House: New York, NY, USA, 2015; Chapter 10; ISBN 978-0-553-41998-6.
33. Burchardt, A. Die Zeit #52/2020. Aljoscha Burchardt: "Steckt dSie KI in Den Sandkasten" (emgl.: "Put Them AI into the Sandbox"). Available online: https://www.genios.de/dosearch?dbShortcut=ZEIT&ZG_PORTAL=ADWPresse&gclid=EAIaIQobChMI3M6ru6b28AIVWpnVCh3jPgIVEAAYASAAEgLOS_D_BwE (accessed on 20 May 2020).
34. Ethics of Artificial Intelligence. Available online: https://en.wikipedia.org/wiki/Ethics_of_artificial_intelligence (accessed on 20 May 2021).

applied system innovation

MDPI

Review

Influence of Artificial Intelligence in Civil Engineering toward Sustainable Development—A Systematic Literature Review

Bilal Manzoor [1], Idris Othman [1], Serdar Durdyev [2,*], Syuhaida Ismail [3] and Mohammad Hussaini Wahab [3]

1 Department of Civil & Environmental Engineering, Universiti Technology PETRONAS, Seri Iskandar 32610, Perak, Malaysia; bilal_18003504@utp.edu.my (B.M.); idris_othman@utp.edu.my (I.O.)
2 Department of Engineering and Architectural Studies, Ara Institute of Canterbury, Christchurch 8011, New Zealand
3 Razak Faculty of Technology and Informatics, Universiti Teknologi Malaysia, Jalan Sultan Yahya Petra, Kuala Lumpur 54100, Malaysia; syuhaida.kl@utm.my (S.I.); hussaini.kl@utm.my (M.H.W.)
* Correspondence: Serdar.Durdyev@ara.ac.nz

Abstract: The widespread use of artificial intelligence (AI) in civil engineering has provided civil engineers with various benefits and opportunities, including a rich data collection, sustainable assessment, and productivity. The trend of construction is diverted toward sustainability with the aid of digital technologies. In this regard, this paper presents a systematic literature review (SLR) in order to explore the influence of AI in civil engineering toward sustainable development. In addition, SLR was carried out by using academic publications from Scopus (i.e., 3478 publications). Furthermore, screening is carried out, and eventually, 105 research publications in the field of AI were selected. Keywords were searched through Boolean operation "Artificial Intelligence" OR "Machine intelligence" OR "Machine Learning" OR "Computational intelligence" OR "Computer vision" OR "Expert systems" OR "Neural networks" AND "Civil Engineering" OR "Construction Engineering" OR "Sustainable Development" OR "Sustainability". According to the findings, it was revealed that the trend of publications received its high intention of researchers in 2020, the most important contribution of publications on AI toward sustainability by the *Automation in Construction*, the United States has the major influence among all the other countries, the main features of civil engineering toward sustainability are interconnectivity, functionality, unpredictability, and individuality. This research adds to the body of knowledge in civil engineering by visualizing and comprehending trends and patterns, as well as defining major research goals, journals, and countries. In addition, a theoretical framework has been proposed in light of the results for prospective researchers and cholars.

Keywords: artificial intelligence; sustainable development; construction; civil engineering; machine learning; construction engineering

Citation: Manzoor, B.; Othman, I.; Durdyev, S.; Ismail, S.; Wahab, M.H. Influence of Artificial Intelligence in Civil Engineering toward Sustainable Development—A Systematic Literature Review. *Appl. Syst. Innov.* **2021**, *4*, 52. https://doi.org/10.3390/asi4030052

Academic Editor: Teen-Hang Meen

Received: 29 May 2021
Accepted: 30 July 2021
Published: 6 August 2021

Publisher's Note: MDPI stays neutral with regard to jurisdictional claims in published maps and institutional affiliations.

Copyright: © 2021 by the authors. Licensee MDPI, Basel, Switzerland. This article is an open access article distributed under the terms and conditions of the Creative Commons Attribution (CC BY) license (https://creativecommons.org/licenses/by/4.0/).

1. Introduction

Artificial intelligence (AI) is playing a critical role in civil engineering in the direction of digitalization and intelligence, allowing for substantial increases in automation, performance, and reliability, as well as establishing an active connection between physical and digital construction [1,2]. As a large economic sector, construction can influence national growth and development as well as help long-term growth [3]. According to a McKinsey Global Institute survey conducted in 2017, the global construction industry accounts for roughly 13% of global GDP, with that amount projected to rise to 15% by 2020 [4]. Furthermore, governments around the world, as well as the world's technology leaders, are putting more effort into the implementation of AI in order to gain a competitive advantage [5]. For example, the U.K. government recently signed an agreement to put the country at the forefront of the AI industry by investing an additional €1.8 billion to make the country more creative and innovative [6,7]. Similarly, France intends to spend €1.5 billion on AI science. First, such investments provide advanced economies with a competitive advantage at the national level; however, they can have a negative effect on the globalization of production and services [8,9]. In the current scenario, industrialized economies outsource resources, such as call centers and manufacturing, to developing economies due to cost advantages [10]. Companies that fully use the capabilities of AI, on the other hand, will no longer have to consider outsourcing costs or living wages [11,12]. Furthermore, rising overhead costs would limit or eliminate the outsourcing of services and output in the developing market, bringing services in-house. This change could result in a reduction in product prices and other company expenditures [13–16].

AI can be described as a distributed, statistical, and symbolic technique that focuses on simulating human functions before moving on to empathy and is making an effort to mimic acceptable communication and focuses on cognitive tasks before moving into the domain of previous approaches, such as deciding analytical ones, and devotes prior effort to modeling [17,18]. In addition, there are three levels of AI, as follows: (a) artificial narrow intelligence (ANI), in which machines only perform one mission, and thus are capable of making decisions on that one aspect of their environment, (b) artificial general intelligence (AGI), the categories of "AGI", which are also known as "powerful AI", "human-level AI", and "real synthetic intelligence" are computers that can find and comprehend the way, think abstractly, plan, figure out problems, learn quickly, and acquire experience quickly. Technologically, economically, or environmentally efficient vehicles and (c) artificial super intelligence (ASI), an intelligence that is far superior to the best human brain in all areas, including scientific ingenuity, general knowledge, and social skills [19–22].

Despite the fact that emerging innovations are opening up new avenues for community participation and governance, cities are still a long way from being genuinely smart and sustainable due to issues of disengagement and exclusivity [23,24]. Despite technological advancements (particularly in computer vision, robotics, and speech recognition), scientists, executives, and government officials have recently expressed concern that AI might endanger the livelihoods and capabilities of businesspeople, automate military operations, and potentially undermine the latter's pride in human superiority [25,26]. The world is entering a new period of sustainable growth, in which nations must work together to solve the most intractable issues such as persistent global poverty, social exclusion, economic inequality, poor governance, and environmental degradation [27,28]. AI is rapidly opening up a new frontier in the disciplines of sustainable development as a result of the employment of machines and robotics with deep learning capabilities. Deep learning skills have both disrupted and enabled governments and society [17]. They also have an impact on broader global sustainability trends. As AI affects our society, it has the potential to usher in a constructive future in which people cohabit peacefully with robots. More immediately, AI can help us make better progress toward the United Nations (UN) Sustainable Development Goals (SDGs) and push us even further toward sustainability goals [29].

A collection of ideas for understanding and caring for long-term prosperity, in addition to economic growth, with a focus on social security and environmental sustainability [30]. Much research has been conducted on the rise of AI from the beginning of the modern era to the present [31–34]. Throughout the 2000s, with their increasing contributions to AI exploration in the world of concepts, the media and academic journals added AI to a vast body of data, information, and solutions to problems in fields all over the world [35–37]. In addition, AI can use sophisticated algorithms to learn from big data and then apply what it has learned to help the construction industry [38]. Furthermore, AI offers numerous opportunities for substantial efficiency gains by analyzing large amounts of data rapidly and accurately [39]. Furthermore, AI systems and technologies can deal with complex, nonlinear functional problems and, once educated, can make predictions and generalizations at a high rate in digital construction [40,41]. AI has sparked significant interest in a wide variety of areas, including computer science, mechanical engineering, and civil engineering, attracting the attention of researchers [42,43].

Some reviews on this subject have been published as a result of the rapid growth of AI applications in civil engineering [44–47]. However, the majority of them only emphasize the importance of AI in a particular subfield, such as structural engineering [48], building information modeling (BIM) [49], automated construction manufacturing [50], computer vision [51], and others. Furthermore, Darko et al. [52] used scientometric analysis to perform a study on AI in the architecture, engineering, and construction (AEC) industry in order to increase awareness of AI in AEC. However, it does not include a thorough introduction to AI techniques and realistic AI applications in civil engineering. Yan et al. [53] conducted a literature review focusing on data mining in the construction industry. Data mining, on the other hand, is a subset of AI that is used to automatically process data and extract valuable insights. A bibliometric analysis of AI engineering applications was recently carried out [54]. However, their research is limited to only one journal's publications and provides a summary of previous work without offering directions for future work.

Therefore, the objectives of this study are:

1. What is the annual publications trend of AI in civil engineering toward sustainable development from 1995 to 2021 (April)?
2. What are the leading journals' contributions in the direction of AI in civil engineering toward sustainable development?
3. Investigating the countries where the AI-related studies were performed in the domain of civil engineering and establishing a comparison between developed and developing countries in terms of sustainable development;
4. What are the civil engineering activities, features of civil engineering, and sustainable assessment toward development?
5. What are the future directions recommended on the basis of this study analysis?

In order to overcome this research gap, a systematic literature review (SLR) was used to seek responses to these questions for the benefit of academia and industry. It is anticipated that the answers to these questions would give rise to contributing knowledge in literature.

2. Research Methodology

A SLR was carried out through a systemic analysis to achieve the objectives suggested in the introduction of this paper. This approach has been chosen to reduce the risk of errors and make reliability possible [55]. The evaluation was conducted in five stages in accordance with the guidelines of Tranfield et al. [56] and Moher et al. [57], also known as PRISMA statement (i) selection of objectives; (ii) selection of databases; (iii) identification of the keywords; (iv) compatible papers selection; and (v) extraction of data.

Regarding the (i) objective, this study is to explore the influence of AI in civil engineering toward sustainable development as well as civil engineering awareness, civil engineering activities, features of civil engineering, and sustainable assessment. As to (ii) database selection, the authors decided to search within the "*Scopus*" database. The reason for using the Scopus database is that scientific journal papers are collected more widely than other databases. (iii) keywords were searched through Boolean operation "Artificial Intelligence" OR "Machine intelligence" OR "Machine Learning" OR "Computational intelligence" OR "Computer vision" OR "Expert systems" OR "Neural networks" AND "Civil Engineering" OR "Construction Engineering" OR "Sustainable Development" OR "Sustainability". The first search provided the result of 3478 Scopus papers. (iv) the selection process began, and documents were omitted other than engineering, for example, in connection with medicine, agriculture Moreover, the papers related to publishing in conferences were also omitted because the conferences papers were not gone through the peer-review process. The number had now been reduced to 320 Scopus documents. Now, another screening process was conducted by evaluating the suitable title and abstract. The final outcome was 105. Afterward, (v) data extraction was subsequently performed. The detailed theme of research methodology is shown in Figure 1.

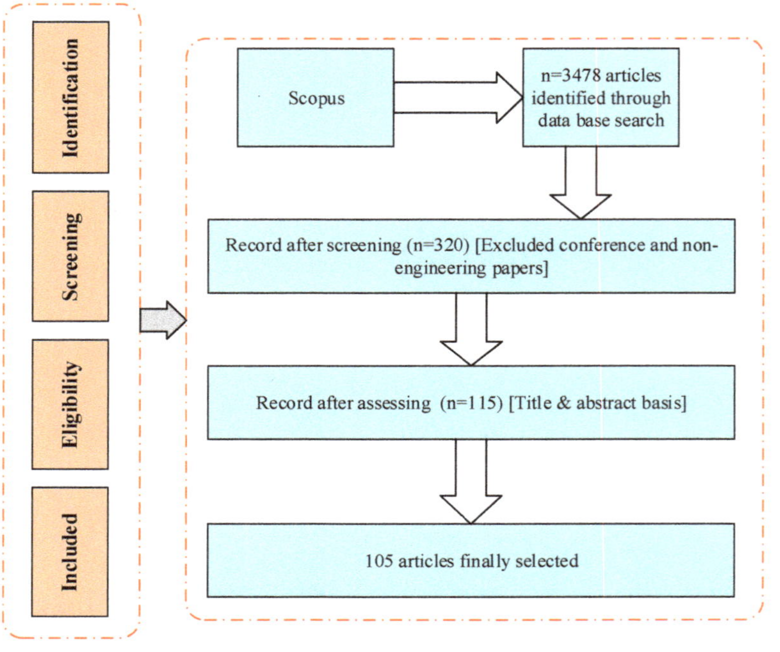

Figure 1. Research flowchart design.

3. Results and Discussion

The follow-up section explores in detail the annual publications trend of AI in civil engineering toward sustainable development, contributions of leading journals in terms of AI in civil engineering toward sustainable development, geospatial distribution, and comparison between developed and developing countries, civil engineering activities toward sustainability, features of civil engineering toward sustainability, AI, and sustainable assessment in order to fulfill the objectives of this study.

3.1. Annual Publications Trend of AI in Civil Engineering toward Sustainable Development

The study included 105 publications spanning the years 1995 to 2021(April). It was revealed that the trend of publications received its high intention of researchers in 2020. However, in 2021 (April), the number of publications was 22 in the domain of AI in civil engineering, and this number is expected to increase because the currently included publications were from January–April (2021). The detailed image is depicted in Figure 2.

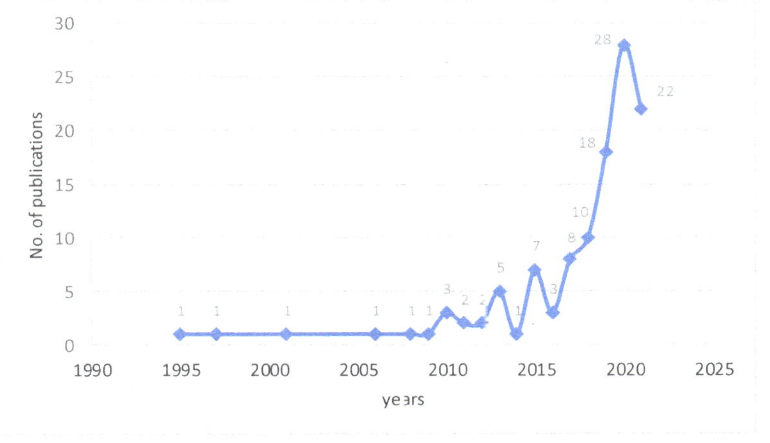

Figure 2. Annual publication trend of AI in civil engineering toward sustainable development.

3.2. Contributions of Leading Journals in Terms of AI in Civil Engineering toward Sustainable Development

Numerous studies emphasize and expound on the significance of conducting research in any field that involves the analysis of academic journals [58–60]. Such information can be useful to scholars in their search for knowledge tools, as well as to readers, who can then assist them in determining which journals are appropriate for publishing works related to AI in civil engineering [61,62]. It was revealed that the contribution of *Automation in Construction (61%)* was the greatest achievement, followed by *Advances in Civil Engineering (10%), Buildings (10%), Journal of Building Engineering (6%), Energy and Buildings (3%)*, and *International Journal of Civil Engineering (2%)*, with the remaining journals contributing 1% equally. The detailed picture is shown in Table 1.

Table 1. Contribution of journals toward sustainable development.

Research Articles	No. of Articles	Percentage
Automation in Construction	60	57%
Advances in Civil Engineering	11	10%
Buildings	11	10%
Journal of Building Engineering	7	7%
Energy and Buildings	3	3%
International Journal of Civil Engineering	2	2%
Arabian Journal for Science and Engineering	1	1%
International Journal of Design Sciences and Technology	1	1%
Journal of Architectural Engineering	1	1%
Journal of Engineering Mechanics	1	1%
Journal of Engineering Science and Technology Review	1	1%
Journal of Engineering, Design and Technology	1	1%
Journal of Infrastructure Systems	1	1%
KSCE Journal of Civil Engineering	1	1%
Malaysian Construction Research Journal	1	1%
Open Civil Engineering Journal	1	1%
Structures	1	1%

3.3. Geospatial Distribution and Comparison between Developed and Developing Countries

The United States was the most influential in terms of publications, followed by China, the United Kingdom, Singapore, Australia, the United Arab Emirates, Iran, South Korea, Denmark, and Malaysia. However, developed countries (the United States, the United Kingdom, Singapore, Australia, the United Arab Emirates, South Korea, and Denmark) had significantly more impact on publications related to AI in civil engineering than developing countries (China and Malaysia). It is recommended to collaborate with developed countries to share ideas and works that will help to increase the shortage of publications in developing countries. Figure 3 depicts a more detailed picture.

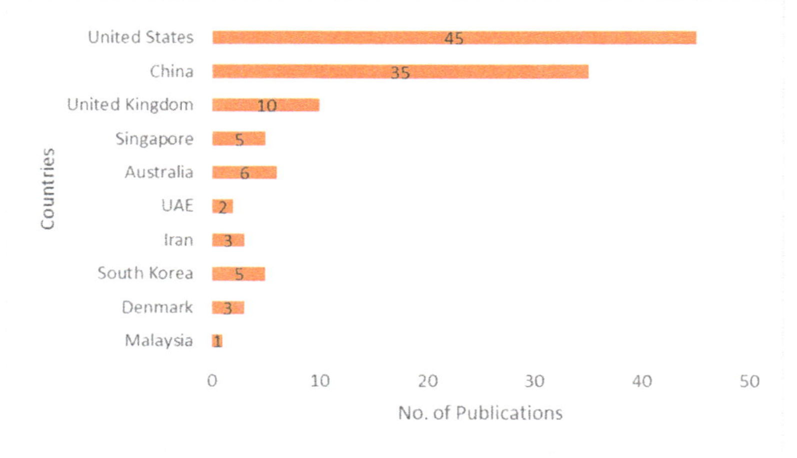

Figure 3. Geospatial distribution of research publications.

3.4. Civil Engineering Activities toward Sustainability

Civil engineering activities have an effect on construction during the project's life cycle. These impacts occur during the construction process, operating process, and final demolition as a construct reaches the end of its lifespan.

3.4.1. During Construction Process

To ensure a smooth construction phase, comprehensive plans of the project creation related to resources, schedule, cost, and others must be secured prior to the construction process [63,64]. There is a need to develop well-documented strategies that include cost reduction, duration, and scheduling of the realistic process. A suitable example is how a graphical representation of the building structures and systems will provide a complete overview during the construction process [65,66]. To ensure the effectiveness of these efforts, a comprehensive, flexible technique that serves as a framework for improvement activities is essential. This organized strategy was created as a consequence of the knowledge and experience gained from executing various sorts of improvement operations on numerous construction sites over the last six years. All of these measures have been designed to improve the construction industry's performance by decreasing waste and eliminating non-value-added activities from the building process [67].

3.4.2. Operating Process

The project will have entered a new phase when construction is completed, known as operation, and the construction team will begin the operation process when this project is up and running [68,69]. The operating method not only helps to operate and maintain a constructed building while also maintaining the safety [70] and comfort of users, but it also performs the expected functions during its life cycle [71]. To begin, maintaining a clear service flow of day-to-day operations helps track the facility's hazard and keep it safe while lowering the facility risk by ensuring that the systems work efficiently and that the number of failed operations is kept to a minimum [72,73].

3.4.3. Demolition Process

A building may be demolished using various methods or techniques, including (a) blast, (b) wrecking balls, (c) hydraulic crushers and pulverizers, and (d) the top-down process. The method of demolition chosen is determined by the project requirements, site limitations, and equipment availability [74,75]. Demolition sites are typically located in congested areas where space for large machinery such as cranes is often restricted [76]. As a result, the first three demolition methods are typically unsuitable for the majority of demolition projects, and the top-down approach is the most widely used demolition method [77,78].

3.5. Features of Civil Engineering toward Sustainability

The key features of civil engineering consist of interconnectivity, functionality, unpredictability, and individuality.

3.5.1. Interconnectivity

The interconnectivity of construction projects is most often created from the amount of work and the work combined with the degree of engagement of participation needed for the project [79,80]. Furthermore, in the construction industry, multiple workers are allocated, all of the uniform density and interconnected, which means that complex scheduling can be a problem [81,82]. The operation process should take a variety of elements into account, such as security, weather, operating hours, and time limit, and in order to deliver successful results smoothly, operations should be designed to take into consideration [83,84]. Furthermore, AI contributes to the automation of critical but time-consuming and repetitive operations, allowing humans to concentrate their time and energy on higher-value work activities. As a result, AI reveals insights that would otherwise be concealed in massive volumes of huge data sets that previously required human management and analysis, such as data generated by videos and images, as well as written reports and business papers, social media posts, and e-mail communications.

3.5.2. Functionality

The initial scope and end results of projects are decided upon when they are established. If there are delays, however, the original end-goals become hazy. Changes or circumstances that occur during the lifespan of projects create a variety of needs or for a variety of reasons [85,86]. Any shifts in size, such as expansions or decreases, must be accompanied by similar changes in the schedule and budget of the project in order for all aspects of the project's dynamic reality to remain in balance [87,88]. The development of automated environmental sustainability studies of products and regions, as well as the optimization of energy use and distribution, is underway. Robotic vehicles that optimize routes and driving styles to reduce carbon emissions are also being developed. While static environmental benefits are determined by the construction, manufacture, and transportation prior to purchase, autonomous environmental objectives are provided from post-purchase autonomous interactions between an AI-enhanced product and its environment, which include knowledge and decision-making, and are determined by the design, production, and distribution of products prior to purchase. A domestic robot, for example, may clean the house and its surroundings autonomously using tools and gadgets that it purchases and collects on its own timetable [89,90].

3.5.3. Unpredictability

In most cases, unknown risks must be detected before they manifest, which means they will invariably jeopardize the overall performance of the project. Notably, complex construction projects have a high degree of unpredictability, which is closely related to a number of variables [91,92]. Furthermore, forecasting and preparation must be carried out prior to construction under considerable uncertainty; specifically, determining the dates, times, and costs are often subject to change [93,94]. Some questions about the architecture design remain unanswered, such as if it can pass audits, if clients are happy with it, and so on. In addition, potential risks may be reduced if uncertainties are recognized and evaluated early on, raising the likelihood of a successful construction project [95–97].

3.5.4. Individuality

Construction projects vary from one another due to variations in client specifications, project size, environments, influences, and constraints, which increases the complexity of project management [98]. When applied to a new project, repeating the schedule, design scheme, finance, and strategy is insufficient; therefore, there is no point in taking the first project's budget and plans and attempting to implement them on the new one [99]. Furthermore, designers, architects, contractors, and service providers are essential in any project since a product or service can only be designed with design work and a prototype [100,101]. It means that each project is carried out by a distinct group of people, and each group has its own set of characteristics, such as ability, knowledge, experience, and communication skills [102].

3.6. AI and Sustainable Assessment

With the current popularity of AI, the widespread availability of sensors, the advent of big data, the growth of e-commerce, the rise of the information culture, the interconnection and convergence of data, knowledge with society, physical space and cyberspace have fundamentally altered the information climate for AI development, ushering in a new evolutionary process [103–106]. New technologies have also permitted the emergence of a new form of AI. These key features include the advent of data-driven perception, Internet-mediated group intelligence, and technology-based human-machine hybrids for reinforcement learning, which both agree on the prediction of what human and media data will support and where each other converge [107–109].

People comprehend with their nervous systems and respond with their bodies. This is useful in explaining AI, which is based on cognitive computation and follows a different design paradigm than that which incorporates human cognition, emotional reasoning, and realistic judgment [110,111]. This new age of AI opens the door to a multitude of civil engineering applications in daily life, enabling the construction industry to diversify its goods and products while also using AI to enter uncharted territories for long-term growth [112]. Though impressive, these technologies are highly customized to specific tasks. Each application usually necessitates years of advanced study and thoughtful, one-of-a-kind sustainable construction [113,114].

New information and communication technology can be used to usher in new methods to promote sustainable growth, such as AI and robotics, which includes product design, production, management, testing, and integration of systems in new ways of the entire lifecycle [115–118]. In addition, sensor connectivity, understanding, thinking, perception, control, material, and environmental awareness, as well as human reasoning, must occur during the manufacturing life cycle. Human-machine communication is required, life cycle decisions must be taken, and machine and data analysis and controls must be enforced [119–121]. It is recommended that with application technology, the level and capability of infrastructure construction, single applications, synergy applications, and business growth must be evaluated. The incorporation of AI into sustainable products represents an opportunity to improve these products' environmental sustainability and, as a result, increase consumers' purchasing intentions while also appealing to new consumer segments who are not attracted to conventional environmental sustainability in the traditional sense. As a result, AI-enabled environmental sustainability can help firms build new competitive advantages and sell their products more effectively to consumers. Simultaneously, the effectiveness of a product varies based on the type of user and the product. When compared to traditional environmental sustainability, AI-based environmental sustainability has the potential to attract more consumers. This may enable firms to employ environmental sustainability to more broadly engage consumers across social boundaries, resulting in increased sales while improving the environment.

In the field of civil engineering, the value of AI is extremely great. However, the value of AI toward water resource management and sustainability is discussed here.

3.6.1. Value of AI toward Water Resource Management

Water is essential for human survival. Thousands of years of human settlement and improvement decisions have been inextricably related to a reliable supply of safe and nutritious water [122,123]. One-third of the world's water is used in irrigation, with the remainder used for drinking and domestic purposes. Cities and societies all over the world work on a staggering amount of freshwater, processed water, and wastewater every day to meet the needs of human society [124,125]. Water must be treated and transported in accordance with hygiene and health standards to ensure that its material and properties meet the specifications of end-users [126,127]. Since artificial intelligence now enables managers to perform two things for water utility management: first, combine long-term growth projections with existing facilities and systems to anticipate future supply, and second, invest in new infrastructure based on results [128]. In general, the results of AI are only as useful as the data fed into it, and the performance generated by it is only as fully interpreted as the intentions of the organization that has it. As with all AI, as AI asks the questions and offers the answers, humans lose some of the information gained from solving the problems themselves [129–131]. The overall issue may be more visible, but it is not required for any particular answer to be given now. Solutions to other problems already perform several times better than human interaction, where things have a plurality of solutions, the tipping point moves from human interaction to automated responses.

3.6.2. Value of AI toward Sustainability

As discussions, conversations, awareness, and campaigns have expanded over the years, and new and emerging technologies, such as AI, have entered the civil engineering market, the emphasis has shifted away from the environment and toward AI [132,133]. Though pushing the limits of human creativity, several organizations, including Microsoft, Google, and Tesla, have made significant efforts to create "earth-friendly" AI systems [134]. Other cases have shown that it is possible, such as Google's own DeepMind AI assisting the organization in reducing data center energy usage by 40% while also lowering overall (greenhouse gas emission) GHG emissions. Advancing AI architecture has added 3% to global energy demand through data centers and has also led to ensuring sustainable energy access for people living in rural and off-grid areas while also being active in micro grids and exploring ways to integrate renewable resources [135–137]. Furthermore, in cities, AI can be used to track and control the use of electrical power distribution, with artificial grids being used to generate less electricity, making them less unreliable [138]. It is the responsibility of researchers and scientists to ensure that the data collected from AI systems are completely transparent, unbiased, reliable, and credible [139]. Greater demand for automation, as well as more stringent environmental requirements and phenomena, is accompanied by rising demand for technologies with higher levels of analysis and study-driven precision, which necessitates increased investment in R&D, development, both in multinationals and in education and government, in the latter sectors.

4. Summary of Findings

A brief summary of results is discussed in the domain of civil engineering sub-fields. It can be seen that AI has had a major effect and made impressive advances in all sub-fields of civil engineering. The details are shown in Table 2.

Table 2. Summary of findings of AI in civil engineering.

Sr No.	Applications of AI	Structure Engineering	Construction Engineering and Management	Transportation Engineering	Hydraulic Engineering	Geotechnical Engineering
1	Developing model of steel structure using AI and machine learning tools	✓				
2	Nano-material with an artificial carbonate can be used in nano-crystalline crystals	✓				
3	Artificial neural network					✓
4	To ensure irrigation and application of pesticides and herbicides are applied more effectively				✓	
5	Slope stability					✓
6	Optimization of water demand forecasting				✓	
7	AI focused on edge computing irrigation systems				✓	
8	Deep neural assessment of friction angle clay					✓
9	Forecasting daily lake level				✓	
10	Bio-inspired computational intelligence			✓		
11	Innovation management and machine learning approach		✓			
12	Forecasting to monthly discharge time series				✓	
13	Pile foundation machine learning approach					✓
14	AI approaches for management risk assessment		✓			
15	Multivariate transportation problem and its implementation			✓		
16	Vehicle traffic load prediction			✓		
17	Analysis and design of sustainable structures	✓				
18	Neural network approaches for cost estimation		✓			
19	Multi-agent system for traffic			✓		
20	Structural health monitoring	✓				
21	Investigating the soil properties					✓

5. Theoretical Framework and Future Directions

The overview of results serves as the foundation for the conceptual framework. It has been discovered that AI has a tremendous effect on construction activities such as computer vision, robotic construction, electronic augmentation, machine learning, digitalization, collaboration, and synchronization. While civil engineering includes several sub-fields such as structural engineering, construction engineering and management, transportation engineering, hydraulic engineering, geotechnical engineering, and environmental engineering, the role of AI in digitalization cannot be overstated. Machine learning techniques, in coordination with remote cameras and unmanned aerial vehicles (UAVs), provide promising non-contact solutions to civil infrastructure condition assessment in structural engineering.

Furthermore, in the transportation field, computer vision-based condition management techniques are used on railway networks. These approaches may be used to obtain data from computers for the purpose of tracking rail conditions [140]. The use of AI in the modeling of semi-arid area infiltration is important in hydraulic engineering. Infiltration is vital to stream flow, groundwater recharge, subsurface flow, and the quality and quantity of surface and subsurface water. In addition, landslides and slope failures are serious issues because they can result in the loss of life and property [141]. As a result, slope stability analysis is one of the most critical geotechnical engineering issues. Slope failures, like most geotechnical problems, are complex. The slope stability research had evolved in tandem with advances in computational geotechnical engineering [142,143]. Furthermore, reinforcement learning is a broad class of optimum control systems that increase performance in a highly dynamic, stochastic environment by estimating value functions obtained from experience, modeling, or search. The interactive environment promotes reinforcement learning, resulting in greater learning ability and adaptability. Because reinforcement learning does not need the use of a system dynamics model, it is a potential choice for ensuring long-term sustainability. The use of reinforcement learning in sustainability is certain to change the old way of energy use, and the system will gain intelligence.

Similarly, the rise of AI is expected to have an effect on global competitiveness, environmental assessment techniques, and long-term sustainability [144]. AI will produce data for more intelligent intervention targeting, minimize waste and losses in production and use, build new technologies that will change whole industries and careers, provide the requisite improvements in communication and cost savings, and bring the benefits of the rapid pace of technological growth to many people around the world. Furthermore, this theoretical framework laid a foundation for future researchers, policymakers, and practitioners to explore the limitations in future research. Figure 4 shows the findings and future directions in the theoretical framework. Furthermore, a recent study conducted by AlArjani et al. [145] built the framework for Saudi Arabia's sustainable development goals. It has made significant contributions to the body of knowledge by concentrating and spending greater resources on alternative energy sources such as solar, wind, biomass, and nuclear energy. In comparison to the current study, greater attention is placed on AI in civil engineering by emphasizing publishing trends, journal contributions, and aspects of civil engineering toward sustainability.

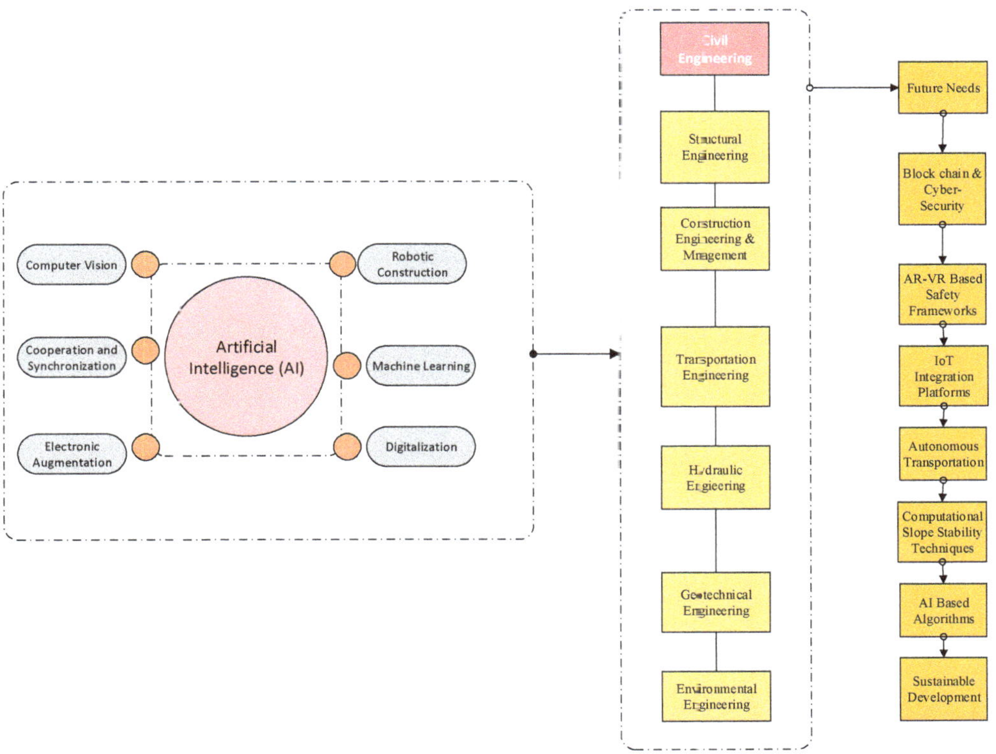

Figure 4. Theoretical framework.

6. Conclusions and Limitations

It was found that the pattern of publications in the domain of AI in civil engineering was received high attention in 2020. Furthermore, *Automation in construction* had played a significant role in AI-related publication works. However, it was discovered that developed countries paid more attention to AI research than developing countries. The contributions of this review are (i) to have a fundamental understanding of civil engineering and to demonstrate the possible importance of AI in supporting and enhancing construction work (ii) to discuss the influence of AI and sustainable assessment, which provides substantial evidence highlighting the benefits of AI techniques (iii) to provide a theoretical framework that will aid potential researchers in incorporating findings. It was revealed that AI had demonstrated the potential to accelerate the learning process, as well as to simplify and augment reasoning, which is useful in construction projects that differ in their basic characteristics. Technology and creativity have resulted in tremendous improvements in our living conditions. Technology development and innovation will also be at the forefront of the transition to a more sustainable future. Productivity-enhancing digital technologies such as AI can help to increase productivity, reduce production costs and emissions, reduce the intensity of production process resources, improve market correspondence, and enable the use of big data to make public services more accessible. From a theoretical approach, the recommendations presented in this research provide guidance on how to address any shortcomings in the process of defining additional research. From a practical aspect, this research can provide practitioners a modulated reference point that is easily accessible, as well as assist them in grasping the most recent techniques and methodology of artificial intelligence research aimed at sustainable development. This research has limitations, despite its contributions. The information is initially obtained from the Scopus database. In the future, additional data will be gathered by combining information from various databases for quantitative and qualitative studies (e.g., Google Scholar, Web of Science, and so on). Second, the focus of this study was limited to journal articles. As a matter of fact, the study's findings do not completely reflect the available AI literature. The limitations mentioned above offer excellent opportunities for further study, but they should be considered when assessing the research results.

Author Contributions: B.M.: Conceptualization, Investigation, Data creation, Writing—original draft; I.O.: Supervision, Writing—review and editing; S.D.: Writing—review and editing; S.I.: Writing—original draft; M.H.W.: Writing—review and editing. All authors have read and agreed to the published version of the manuscript.

Funding: This research received no external funding.

Institutional Review Board Statement: Not applicable.

Informed Consent Statement: Not applicable.

Conflicts of Interest: The authors declare no conflict of interest.

References

1. Tavana Amlashi, A.; Alidoust, P.; Pazhouhi, M.; Pourrostami Niavol, K.; Khabiri, S.; Ghanizadeh, A.R. AI-based formulation for mechanical and workability properties of eco-friendly concrete made by waste foundry sand. *J. Mater. Civ. Eng.* **2021**, *33*, 4021038. [CrossRef]
2. Luckey, D.; Fritz, H.; Legatiuk, D.; Dragos, K.; Smarsly, K. Artificial intelligence techniques for smart city applications. In Proceedings of the International Conference on Computing in Civil and Building Engineering, Sao Paulo, Brazil, 18–20 August 2020; Springer: New York, NY, USA, 2020; pp. 3–15.
3. Giang, D.T.H.; Pheng, L.S. Role of construction in economic development: Review of key concepts in the past 40 years. *Habitat Int.* **2011**, *35*, 118–125. [CrossRef]
4. Barbosa, F.; Woetzel, J.; Mischke, J. *Reinventing Construction: A Route of Higher Productivity*; McKinsey Global Institute: Chicago, IL, USA, 2017.
5. Donepudi, P.K.; Ahmed, A.A.A.; Saha, S. Emerging Market Economy (EME) and Artificial Intelligence (AI): Consequences for the Future of Jobs. *PalArch's J. Archaeol. Egypt/Egyptol.* **2020**, *17*, 5562–5574.

6. Stojčić, M.; Zavadskas, E.K.; Pamučar, D.; Stević, Ž.; Mardani, A. Application of MCDM methods in sustainability engineering: A literature review 2008–2018. *Symmetry* **2019**, *11*, 350. [CrossRef]
7. Shirowzhan, S.; Tan, W.; Sepasgozar, S.M.E. Digital twin and CyberGIS for improving connectivity and measuring the impact of infrastructure construction planning in smart cities. *Int. J. Geo-Inf.* **2020**, *9*, 240. [CrossRef]
8. Taeihagh, A. Governance of artificial intelligence. *Policy Soc.* **2021**, *40*, 137–157. [CrossRef]
9. Pamučar, D.; Božanić, D.; Lukovac, V.; Komazec, N. Normalized weighted geometric bonferroni mean operator of interval rough numbers–application in interval rough dematel-copras model. *Facta Univ. Ser. Mech. Eng.* **2018**, *16*, 171–191.
10. Mardani, A.; Streimikiene, D.; Zavadskas, E.K.; Cavallaro, F.; Nilashi, M.; Jusoh, A.; Zare, H. Application of Structural Equation Modeling (SEM) to solve environmental sustainability problems: A comprehensive review and meta-analysis. *Sustainability* **2017**, *9*, 1814. [CrossRef]
11. Mardani, A.; Zavadskas, E.K.; Govindan, K.; Amat Senin, A.; Jusoh, A. VIKOR technique: A systematic review of the state of the art literature on methodologies and applications. *Sustainability* **2016**, *8*, 37. [CrossRef]
12. De la Fuente, A.; Blanco, A.; Armengou, J.B.; Aguado, A. Sustainability based-approach to determine the concrete type and reinforcement configuration of TBM tunnels linings. Case study: Extension line to Barcelona Airport T1. *Tunn. Undergr. Sp. Technol.* **2017**, *61*, 179–188. [CrossRef]
13. Mardani, A.; Zavadskas, E.K.; Streimikiene, D.; Jusoh, A.; Khoshnoudi, M. A comprehensive review of data envelopment analysis (DEA) approach in energy efficiency. *Renew. Sustain. Energy Rev.* **2017**, *70*, 1298–1322. [CrossRef]
14. Adeli, H. Four decades of computing in civil engineering. In Proceedings of the CIGOS 2019: Innovation for Sustainable Infrastructure, Hanoi, Vietnam, 31 October–1 November 2019; Springer: New York, NY, USA, 2020; pp. 3–11.
15. Yucel, M.; Nigdeli, S.M.; Bekdaş, G. Artificial neural networks (anns) and solution of civil engineering problems: Anns and prediction applications. In *Artificial Intelligence and Machine Learning Applications in Civil, Mechanical, and Industrial Engineering*; IGI Global: Hershey, PA, USA, 2020; pp. 13–38.
16. Durdyev, S. Review of construction journals on causes of project cost overruns. *Eng. Constr. Archit. Manag.* **2020**, *28*. [CrossRef]
17. Goralski, M.A.; Tan, T.K. Artificial intelligence and sustainable development. *Int. J. Manag. Educ.* **2020**, *18*, 100330. [CrossRef]
18. Feroz, A.K.; Zo, H.; Chiravuri, A. Digital transformation and environmental sustainability: A review and research agenda. *Sustainability* **2021**, *13*, 1530. [CrossRef]
19. Verhoef, P.C.; Broekhuizen, T.; Bart, Y.; Bhattacharya, A.; Dong, J.Q.; Fabian, N.; Haenlein, M. Digital transformation: A multidisciplinary reflection and research agenda. *J. Bus. Res.* **2021**, *122*, 889–901. [CrossRef]
20. Kamble, S.S.; Gunasekaran, A.; Gawankar, S.A. Sustainable Industry 4.0 framework: A systematic literature review identifying the current trends and future perspectives. *Process. Saf. Environ. Prot.* **2018**, *117*, 408–425. [CrossRef]
21. Kunkel, S.; Matthess, M. Digital transformation and environmental sustainability in industry: Putting expectations in Asian and African policies into perspective. *Environ. Sci. Policy* **2020**, *112*, 318–329. [CrossRef]
22. Qureshi, A.H.; Alaloul, W.S.; Manzoor, B.; Musarat, M.A.; Saad, S.; Ammad, S. Implications of Machine Learning Integrated Technologies for Construction Progress Detection Under Industry 4.0 (IR 4.0). In Proceedings of the 2020 Second International Sustainability and Resilience Conference: Technology and Innovation in Building Designs (51154), Sakheer, Bahrain, 11–12 November 2020; IEEE: Piscataway, NJ, USA, 2020; pp. 1–6.
23. Pongiglione, M.; Calderini, C. Sustainable structural design: Comprehensive literature review. *J. Struct. Eng.* **2016**, *142*, 4016139. [CrossRef]
24. Pan, Y.; Zhang, L. Roles of artificial intelligence in construction engineering and management: A critical review and future trends. *Autom. Constr.* **2021**, *122*, 103517. [CrossRef]
25. Wang, Y.; Szeto, W.Y. Multiobjective environmentally sustainable road network design using Pareto optimization. *Comput. Civ. Infrastruct. Eng.* **2017**, *32*, 964–987. [CrossRef]
26. Ibem, E.O.; Laryea, S. Survey of digital technologies in procurement of construction projects. *Autom. Constr.* **2014**, *46*, 11–21. [CrossRef]
27. Zavadskas, E.K.; Antuchevičienė, J.; Kapliński, O. Multi-criteria decision making in civil engineering: Part I–a state-of-the-art survey. *Eng. Struct. Technol.* **2015**, *7*, 103–113. [CrossRef]
28. Yi, W.; Wang, S. Multi-objective mathematical programming approach to construction laborer assignment with equity consideration. *Comput. Civ. Infrastruct. Eng.* **2016**, *31*, 954–965. [CrossRef]
29. Di Vaio, A.; Palladino, R.; Hassan, R.; Escobar, O. Artificial intelligence and business models in the sustainable development goals perspective: A systematic literature review. *J. Bus. Res.* **2020**, *121*, 283–314. [CrossRef]
30. Mitropoulos, L.K.; Prevedouros, P.D. Incorporating sustainability assessment in transportation planning: An urban transportation vehicle-based approach. *Transp. Plan. Technol.* **2016**, *39*, 439–463. [CrossRef]
31. Zhang, L.; Pan, Y.; Wu, X.; Skibniewski, M.J. Introduction to Artificial Intelligence. In *Artificial Intelligence in Construction Engineering and Management*; Springer: New York, NY, USA, 2021; pp. 1–15.
32. Vyas, M.; Thakur, S.; Riyaz, B.; Bansal, K.K.; Tomar, B.; Mishra, V. Artificial intelligence: The beginning of a new era in pharmacy profession. *Asian J. Pharm* **2018**, *12*, 72–76.
33. Brynjolfsson, E.; Rock, D.; Syverson, C. Artificial intelligence and the modern productivity paradox: A clash of expectations and statistics. In *The Economics of Artificial Intelligence: An Agenda*; University of Chicago Press: Chicago, IL, USA, 2018; pp. 23–57.

34. Zawacki-Richter, O.; Marín, V.I.; Bond, M.; Gouverneur, F. Systematic review of research on artificial intelligence applications in higher education–where are the educators? *Int. J. Educ. Technol. High. Educ.* **2019**, *16*, 1–27. [CrossRef]
35. Nishant, R.; Kennedy, M.; Corbett, J. Artificial intelligence for sustainability: Challenges, opportunities, and a research agenda. *Int. J. Inf. Manag.* **2020**, *53*, 102104. [CrossRef]
36. Abduljabbar, R.; Dia, H.; Liyanage, S.; Bagloee, S.A. Applications of artificial intelligence in transport: An overview. *Sustainability* **2019**, *11*, 189. [CrossRef]
37. Lv, Z.; Chen, D.; Lou, R.; Alazab, A. Artificial intelligence for securing industrial-based cyber–physical systems. *Futur. Gener. Comput. Syst.* **2021**, *117*, 291–298. [CrossRef]
38. Alrukaibi, F.; Alsaleh, R.; Sayed, T. Applying Machine Learning and Statistical Approaches for Travel Time Estimation in Partial Network Coverage. *Sustainability* **2019**, *11*, 3822. [CrossRef]
39. Patrício, D.I.; Rieder, R. Computer vision and artificial intelligence in precision agriculture for grain crops: A systematic review. *Comput. Electron. Agric.* **2018**, *153*, 69–81. [CrossRef]
40. Qin, W.; Lin, Q. Construction of cultural industry development factor model based on factor analysis, artificial intelligence and big data. *Microprocess. Microsyst.* **2021**, *82*, 103880.
41. Manzoor, B.; Othman, I. Safety Management Model During Construction Focusing on Building Information Modeling (BIM). In Proceedings of the Advances in Civil Engineering Materials: Selected Articles from the International Conference on Architecture and Civil Engineering (ICACE2020), Chattogram, Bangladesh, 21–23 December 2020; Springer Nature: Basingstoke, UK, 2021; p. 31.
42. Zhu, X. Research on the reform of higher automotive engineering education under the background of artificial intelligence. *E3S Web Conf.* **2021**, *245*, 3091. [CrossRef]
43. Araszkiewicz, K. Digital Technologies in Facility Management–the state of practice and research challenges. *Procedia Eng.* **2017**, *196*, 1034–1042. [CrossRef]
44. Toorajipour, R.; Sohrabpour, V.; Nazarpour, A.; Oghazi, P.; Fischl, M. Artificial intelligence in supply chain management: A systematic literature review. *J. Bus. Res.* **2021**, *122*, 502–517. [CrossRef]
45. Núñez-Merino, M.; Maqueira-Marín, J.M.; Moyano-Fuentes, J.; Martínez-Jurado, P.J. Information and digital technologies of Industry 4.0 and Lean supply chain management: A systematic literature review. *Int. J. Prod. Res.* **2020**, *58*, 5034–5061. [CrossRef]
46. Mihic, M.; Vukomanovic, M.; Završki, I. Review of previous applications of innovative information technologies in construction health and safety. *Organ. Technol. Manag. Constr. Int. J.* **2019**, *11*, 1952–1967.
47. Bosch-Sijtsema, P.; Claeson-Jonsson, C.; Johansson, M.; Roupe, M. The hype factor of digital technologies in AEC. *Constr. Innov.* **2021**. ahead of print. [CrossRef]
48. Salehi, H.; Burgueño, R. Emerging artificial intelligence methods in structural engineering. *Eng. Struct.* **2018**, *171*, 170–189. [CrossRef]
49. Zou, Y.; Kiviniemi, A.; Jones, S.W. A review of risk management through BIM and BIM-related technologies. *Saf. Sci.* **2017**, *97*, 88–98. [CrossRef]
50. Hatami, M.; Flood, I.; Franz, B.; Zhang, X. *State-of-the-Art Review on the Applicability of AI Methods to Automated Construction Manufacturing. Computing in Civil Engineering 2019: Data, Sensing, and Analytics, Atlanta, Georgia, 17–19 June 2019*; American Society of Civil Engineers: Reston, VA, USA, 2019; pp. 368–375.
51. Kakani, V.; Nguyen, V.H.; Kumar, B.P.; Kim, H.; Pasupuleti, V.R. A critical review on computer vision and artificial intelligence in food industry. *J. Agric. Food Res.* **2020**, *2*, 100033. [CrossRef]
52. Darko, A.; Chan, A.P.C.; Adabre, M.A.; Edwards, D.J.; Hosseini, M.R.; Ameyaw, E.E. Artificial intelligence in the AEC industry: Scientometric analysis and visualization of research activities. *Autom. Constr.* **2020**, *112*, 103081. [CrossRef]
53. Yan, H.; Yang, N.; Peng, Y.; Ren, Y. Data mining in the construction industry: Present status, opportunities, and future trends. *Autom. Constr.* **2020**, *119*, 103331. [CrossRef]
54. Shukla, A.K.; Janmaijaya, M.; Abraham, A.; Muhuri, P.K. Engineering applications of artificial intelligence: A bibliometric analysis of 30 years (1988–2018). *Eng. Appl. Artif. Intell.* **2019**, *85*, 517–532. [CrossRef]
55. Mulrow, C.D. Systematic reviews: Rationale for systematic reviews. *BMJ* **1994**, *309*, 597–599. [CrossRef]
56. Tranfield, D.; Denyer, D.; Smart, P. Towards a methodology for developing evidence-informed management knowledge by means of systematic review. *Br. J. Manag.* **2003**, *14*, 207–222. [CrossRef]
57. Moher, D.; Liberati, A.; Tetzlaff, J.; Altman, D.G.; Group, P. Preferred reporting items for systematic reviews and meta-analyses: The PRISMA statement. *PLoS Med.* **2009**, *6*, e1000097. [CrossRef] [PubMed]
58. Serenko, A. The development of an AI journal ranking based on the revealed preference approach. *J. Inform.* **2010**, *4*, 447–459. [CrossRef]
59. Bobrow, D. Artificial Intelligence in perspective, a retrospective on fifty volumes of the Artificial Intelligence Journal. *Artif. Intell.* **1994**, *59*, 5–20. [CrossRef]
60. Azhar, S. Building information modeling (BIM): Trends, benefits, risks, and challenges for the AEC industry. *Leadersh. Manag. Eng.* **2011**, *11*, 241–252. [CrossRef]
61. Hosseini, M.R.; Maghrebi, M.; Akbarnezhad, A.; Martek, I.; Arashpour, M. Analysis of citation networks in building information modeling research. *J. Constr. Eng. Manag.* **2018**, *144*, 4018064. [CrossRef]

62. Badrinath, A.C.; Chang, Y.; Hsieh, S. A review of tertiary BIM education for advanced engineering communication with visualization. *Vis. Eng.* **2016**, *4*, 1–17. [CrossRef]
63. Al-Reshaid, K.; Kartam, N.; Tewari, N.; Al-Bader, H. A project control process in pre-construction phases: Focus on effective methodology. *Eng. Constr. Archit. Manag.* **2005**, *12*, 351–372. [CrossRef]
64. Osipova, E. Risk management in the different phases of a construction project: A study of actors' involvement. In Proceedings of the Nordic Conference on Construction Economics and Organsiation, Luleå, Sweden, 14–15 June 2007; Luleå Tekniska Universitet: Luleå, Sweden, 2007; pp. 307–319.
65. Noktehdan, M.; Shahbazpour, M.; Zare, M.R.; Wilkinson, S. Innovation management and construction phases in infrastructure projects. *J. Constr. Eng. Manag.* **2019**, *145*, 4018135. [CrossRef]
66. Durdyev, S.; Hosseini, M.R. Causes of delays on construction projects: A comprehensive list. *Int. J. Manag. Proj. Bus.* **2019**, *13*, 20–46. [CrossRef]
67. Yaseen, Z.M.; Ali, Z.H.; Salih, S.Q.; Al-Ansari, N. Prediction of risk delay in construction projects using a hybrid artificial intelligence model. *Sustainability* **2020**, *12*, 1514. [CrossRef]
68. Yang, J.-B.; Wei, P.-R. Causes of delay in the planning and design phases for construction projects. *J. Archit. Eng.* **2010**, *16*, 80–83. [CrossRef]
69. Chan, D.W.M.; Kumaraswamy, M.M. A comparative study of causes of time overruns in Hong Kong construction projects. *Int. J. Proj. Manag.* **1997**, *15*, 55–63. [CrossRef]
70. Manzoor, B.; Othman, I.; Manzoor, M. Evaluating the critical safety factors causing accidents in high-rise building projects. *Ain Shams Eng. J.* **2021**. [CrossRef]
71. Odeh, A.M.; Battaineh, H.T. Causes of construction delay: Traditional contracts. *Int. J Proj. Manag.* **2002**, *20*, 67–73. [CrossRef]
72. Kabirifar, K.; Mojtahedi, M. The impact of Engineering, Procurement and Construction (EPC) phases on project performance: A case of large-scale residential construction project. *Buildings* **2019**, *9*, 15. [CrossRef]
73. Demirkesen, S.; Ozorhon, B. Impact of integration management on construction project management performance. *Int. J. Proj. Manag.* **2017**, *35*, 1639–1654. [CrossRef]
74. Luciano, A.; Cutaia, L.; Cioffi, F.; Sinibaldi, C. Demolition and construction recycling unified management: The DECORUM platform for improvement of resource efficiency in the construction sector. *Environ. Sci. Pollut. Res.* **2020**, *28*, 24558–24569. [CrossRef] [PubMed]
75. Pantini, S.; Rigamonti, L. Is selective demolition always a sustainable choice? *Waste Manag.* **2020**, *103*, 169–176. [CrossRef]
76. Marzouk, M.; Elmaraghy, A.; Voordijk, H. Lean Deconstruction Approach for Buildings Demolition Processes using BIM. *Lean Constr. J.* **2019**, *9*, 147–173.
77. Alba-Rodríguez, M.D.; Martínez-Rocamora, A.; González-Vallejo, P.; Ferreira-Sánchez, A.; Marrero, M. Building rehabilitation versus demolition and new construction: Economic and environmental assessment. *Environ. Impact Assess. Rev.* **2017**, *66*, 115–126. [CrossRef]
78. Bastianoni, S.; Galli, A.; Pulselli, R.M.; Niccolucci, V. Environmental and economic evaluation of natural capital appropriation through building construction: Practical case study in the Italian context. *AMBIO A J. Hum. Environ.* **2007**, *36*, 559–565. [CrossRef]
79. Lebcir, R.M.; Choudrie, J. A dynamic model of the effects of project complexity on time to complete construction projects. *Int. J. Innov. Manag. Technol.* **2011**, *2*, 477.
80. Zhou, H.; Wang, H.; Zeng, W. Smart construction site in mega construction projects: A case study on island tunneling project of Hong Kong-Zhuhai-Macao Bridge. *Front. Eng. Manag.* **2018**, *5*, 78–87. [CrossRef]
81. Abotaleb, I.S.; El-adaway, I.H. Managing construction projects through dynamic modeling: Reviewing the existing body of knowledge and deriving future research directions. *J. Manag. Eng.* **2018**, *34*, 4018033. [CrossRef]
82. Khan, K.I.A.; Flanagan, R.; Lu, S.-L. Managing information complexity using system dynamics on construction projects. *Constr. Manag. Econ.* **2016**, *34*, 192–204. [CrossRef]
83. Teo, M.M.; Loosemore, M. Community-based protest against construction projects: A case study of movement continuity. *Constr. Manag. Econ.* **2011**, *29*, 131–144. [CrossRef]
84. Teo, M.M.M.; Loosemore, M. Community-based protest against construction projects: The social determinants of protest movement continuity. *Int. J. Manag. Proj. Bus.* **2010**, *3*, 216–235. [CrossRef]
85. Ali, A.S.; Rahmat, I. The performance measurement of construction projects managed by ISO-certified contractors in Malaysia. *J. Retail. Leis. Prop.* **2010**, *9*, 25–35. [CrossRef]
86. Nguyen, L.D.; Ogunlana, S.O. A study on project success factors in large construction projects in Vietnam. *Eng. Constr. Archit. Manag.* **2004**, *11*, 404–413. [CrossRef]
87. Shelbourn, M.; Bouchlaghem, N.M.; Anumba, C.; Carrillo, P. Planning and implementation of effective collaboration in construction projects. *Constr. Innov.* **2007**, *7*, 357–377. [CrossRef]
88. Love, P.E.D.; Edwards, D.J. Determinants of rework in building construction projects. *Eng. Constr. Archit. Manag.* **2004**, *11*, 259–274. [CrossRef]
89. Frank, B. Artificial intelligence-enabled environmental sustainability of products: Marketing benefits and their variation by consumer, location, and product types. *J. Clean. Prod.* **2021**, *285*, 125242. [CrossRef]
90. Bienvenido-Huertas, D.; Farinha, F.; Oliveira, M.J.; Silva, E.M.J.; Lança, R. Comparison of artificial intelligence algorithms to estimate sustainability indicators. *Sustain. Cities Soc.* **2020**, *63*, 102430. [CrossRef]

91. Gummesson, E. Qualitative research in marketing: Road-map for a wilderness of complexityand unpredictability. *Eur. J. Mark.* **2005**, *39*, 309–327. [CrossRef]
92. Esposito, E. The structures of uncertainty: Performativity and unpredictability in economic operations. *Econ. Soc.* **2013**, *42*, 102–129. [CrossRef]
93. You, J.; Chen, Y.; Wang, W.; Shi, C. Uncertainty, opportunistic behavior, and governance in construction projects: The efficacy of contracts. *Int. J. Proj. Manag.* **2018**, *36*, 795–807. [CrossRef]
94. Abdi, M.; Aulakh, P.S. Locus of uncertainty and the relationship between contractual and relational governance in cross-border interfirm relationships. *J. Manag.* **2017**, *43*, 771–803. [CrossRef]
95. Meardi, G.; Martin, A.; Riera, M.L. Constructing uncertainty: Unions and migrant labour in construction in Spain and the UK. *J. Ind. Relat.* **2012**, *54*, 5–21. [CrossRef]
96. Gosling, J.; Naim, M.; Towill, D. Identifying and categorizing the sources of uncertainty in construction supply chains. *J. Constr. Eng. Manag.* **2013**, *139*, 102–110. [CrossRef]
97. Arashpour, M.; Wakefield, R.; Lee, E.W.M.; Chan, R.; Hosseini, M.R. Analysis of interacting uncertainties in on-site and off-site activities: Implications for hybrid construction. *Int. J. Proj. Manag.* **2016**, *34*, 1393–1402. [CrossRef]
98. Cheung, S.O.; Wong, P.S.P.; Wu, A.W.Y. Towards an organizational culture framework in construction. *Int. J. Proj. Manag.* **2011**, *29*, 33–44. [CrossRef]
99. Adenfelt, M.; Lagerström, K. Enabling knowledge creation and sharing in transnational projects. *Int. J. Proj. Manag.* **2006**, *24*, 191–198. [CrossRef]
100. Ankrah, N.A.; Langford, D.A. Architects and contractors: A comparative study of organizational cultures. *Constr. Manag. Econ.* **2005**, *23*, 595–607. [CrossRef]
101. Cheung, S.-O. Critical factors affecting the use of alternative dispute resolution processes in construction. *Int. J. Proj. Manag.* **1999**, *17*, 189–194. [CrossRef]
102. Van Marrewijk, A. Managing project culture: The case of Environ Megaproject. *Int. J. Proj. Manag.* **2007**, *25*, 290–299. [CrossRef]
103. Ivanov, D.; Dolgui, A.; Sokolov, B.; Werner, F.; Ivanova, M. A dynamic model and an algorithm for short-term supply chain scheduling in the smart factory industry 4.0. *Int. J. Prod. Res.* **2016**, *54*, 386–402. [CrossRef]
104. Arciszewski, T.; De Jong, K.A. Evolutionary computation in civil engineering: Research frontiers. In *Civil and Structural Engineering Computing: 2001*; Saxe-Coburg Publications: Stirling, UK, 2001; pp. 161–184.
105. Lu, P.; Chen, S.; Zheng, Y. Artificial intelligence in civil engineering. *Math. Probl. Eng.* **2012**, *2012*, 145974. [CrossRef]
106. Chou, J.; Pham, A. Smart artificial firefly colony algorithm-based support vector regression for enhanced forecasting in civil engineering. *Comput. Civ. Infrastruct. Eng.* **2015**, *30*, 715–732. [CrossRef]
107. Adam, B.; Smith, I.F.C. Active tensegrity: A control framework for an adaptive civil-engineering structure. *Comput. Struct.* **2008**, *86*, 2215–2223. [CrossRef]
108. Adam, B.; Smith, I.F. Reinforcement learning for structural control. *J. Comput. Civ. Eng.* **2008**, *22*, 133–139. [CrossRef]
109. Manzoor, B.; Othman, I.; Pomares, J.C. Digital Technologies in the Architecture, Engineering and Construction (AEC) Industry—A Bibliometric—Qualitative Literature Review of Research Activities. *Int. J. Environ. Res. Public Health* **2021**, *18*, 6135. [CrossRef] [PubMed]
110. Shaheen, A.A.; Fayek, A.R.; AbouRizk, S.M. Methodology for integrating fuzzy expert systems and discrete event simulation in construction engineering. *Can. J. Civ. Eng.* **2009**, *36*, 1478–1490. [CrossRef]
111. Forcael, E.; Glagola, C.R.; González, V. Incorporation of computer simulations into teaching linear scheduling techniques. *J. Prof. Issues Eng. Educ. Pract.* **2012**, *138*, 21–30. [CrossRef]
112. Senouci, A.; Al-Derham, H.R. Genetic algorithm-based multi-objective model for scheduling of linear construction projects. *Adv. Eng. Softw.* **2008**, *39*, 1023–1028. [CrossRef]
113. Zavadskas, E.K.; Antucheviciene, J.; Vilutiene, T.; Adeli, H. Sustainable decision-making in civil engineering, construction and building technology. *Sustainability* **2018**, *10*, 14. [CrossRef]
114. Correia, A.G.; Cortez, P.; Tinoco, J.; Marques, R. Artificial intelligence applications in transportation geotechnics. *Geotech. Geol. Eng.* **2013**, *31*, 861–879. [CrossRef]
115. Cai, S.; Ma, Z.; Skibniewski, M.J.; Bao, S. Construction automation and robotics for high-rise buildings over the past decades: A comprehensive review. *Adv. Eng. Inform.* **2019**, *42*, 100989. [CrossRef]
116. Wakisaka, T.; Furuya, N.; Inoue, Y.; Shiokawa, T. Automated construction system for high-rise reinforced concrete buildings. *Autom. Constr.* **2000**, *9*, 229–250. [CrossRef]
117. Chu, B.; Jung, K.; Lim, M.-T.; Hong, D. Robot-based construction automation: An application to steel beam assembly (Part I). *Autom. Constr.* **2013**, *32*, 46–61. [CrossRef]
118. Jung, K.; Chu, B.; Park, S.; Hong, D. An implementation of a teleoperation system for robotic beam assembly in construction. *Int. J. Precis. Eng. Manuf.* **2013**, *14*, 351–358. [CrossRef]
119. Hoc, J.-M. From human–machine interaction to human–machine cooperation. *Ergonomics* **2000**, *43*, 833–843. [CrossRef] [PubMed]
120. Läubli, S.; Castilho, S.; Neubig, G.; Sennrich, R.; Shen, Q.; Toral, A. A set of recommendations for assessing human–machine parity in language translation. *J. Artif. Intell. Res.* **2020**, *67*, 653–672. [CrossRef]
121. Pacaux-Lemoine, M.-P.; Trentesaux, D.; Rey, G.Z.; Millot, P. Designing intelligent manufacturing systems through Human-Machine Cooperation principles: A human-centered approach. *Comput. Ind. Eng.* **2017**, *111*, 581–595. [CrossRef]

122. Wang, J.; Li, S.; Xiong, G.; Cang, D. Application of digital technologies about water network in steel industry. *Resour. Conserv. Recycl.* **2011**, *55*, 755–759. [CrossRef]
123. Berkhout, F.; Hertin, J. De-materialising and re-materialising: Digital technologies and the environment. *Futures* **2004**, *36*, 903–920. [CrossRef]
124. Mistry, D. Comparison of digital water marking methods. *Int. J. Comput. Sci Eng.* **2010**, *2*, 2905–2909.
125. Bressanelli, G.; Adrodegari, F.; Perona, M.; Saccani, N. Exploring how usage-focused business models enable circular economy through digital technologies. *Sustainability* **2018**, *10*, 639. [CrossRef]
126. Merchant, G. Mobile practices in everyday life: Popular digital technologies and schooling revisited. *Br. J. Educ. Technol.* **2012**, *43*, 770–782. [CrossRef]
127. Lember, V. The Increasing Role of Digital Technologies. In *Co-Production and Co-Creation*; Taylor & Francis: Abingdon, UK, 2018; p. 115.
128. Aleksandrova, E.; Vinogradova, V.; Tokunova, G. Integration of digital technologies in the field of construction in the Russian Federation. *Eng. Manag. Prod. Serv.* **2019**, *11*, 38–47. [CrossRef]
129. Chassignol, M.; Khoroshavin, A.; Klimova, A.; Bilyatdinova, A. Artificial Intelligence trends in education: A narrative overview. *Procedia Comput. Sci.* **2018**, *136*, 16–24. [CrossRef]
130. Lutz, C. Digital inequalities in the age of artificial intelligence and big data. *Hum. Behav. Emerg. Technol.* **2019**, *1*, 141–148. [CrossRef]
131. Kolbjørnsrud, V.; Amico, R.; Thomas, R.J. How artificial intelligence will redefine management. *Harv. Bus. Rev.* **2016**, *2*, 1–6.
132. Borges, A.F.S.; Laurindo, F.J.B.; Spínola, M.M.; Gonçalves, R.F.; Mattos, C.A. The strategic use of artificial intelligence in the digital era: Systematic literature review and future research directions. *Int. J. Inf. Manag.* **2020**, *57*, 102225. [CrossRef]
133. Al-Emran, M.; Mezhuyev, V.; Kamaludin, A.; Shaalan, K. The impact of knowledge management processes on information systems: A systematic review. *Int. J. Inf. Manag.* **2018**, *43*, 173–187. [CrossRef]
134. Bertone, E.; Burford, M.A.; Hamilton, D.P. Fluorescence probes for real-time remote cyanobacteria monitoring: A review of challenges and opportunities. *Water Res.* **2018**, *141*, 152–162. [CrossRef]
135. Hosseinzadeh-Bandbafha, H.; Nabavi-Pelesaraei, A.; Shamshirband, S. Investigations of energy consumption and greenhouse gas emissions of fattening farms using artificial intelligence methods. *Environ. Prog. Sustain. Energy* **2017**, *36*, 1546–1559. [CrossRef]
136. Bannerjee, G.; Sarkar, U.; Das, S.; Ghosh, I. Artificial intelligence in agriculture: A literature survey. *Int. J. Sci. Res. Comput. Sci. Appl. Manag. Stud.* **2018**, *7*, 1–6.
137. Bolandnazar, E.; Rohani, A.; Taki, M. Energy consumption forecasting in agriculture by artificial intelligence and mathematical models. *Energy Sources Part A Recover. Util. Environ. Eff.* **2020**, *42*, 1618–1632. [CrossRef]
138. Soofastaei, A. The application of artificial intelligence to reduce greenhouse gas emissions in the mining industry. In *Green Technologies to Improve the Environment on Earth*; IntechOpen: London, UK, 2018.
139. Magazzino, C.; Mele, M.; Schneider, N. The relationship between municipal solid waste and greenhouse gas emissions: Evidence from Switzerland. *Waste Manag.* **2020**, *113*, 508–520. [CrossRef]
140. Das, S.; Dey, A.; Pal, A.; Roy, N. Applications of artificial intelligence in machine learning: Review and prospect. *Int. J. Comput. Appl.* **2015**, *115*, 31–41. [CrossRef]
141. Talaviya, T.; Shah, D.; Patel, N.; Yagnik, H.; Shah, M. Implementation of artificial intelligence in agriculture for optimisation of irrigation and application of pesticides and herbicides. *Artif. Intell. Agric.* **2020**, *4*, 58–73.
142. Shahin, M.A. State-of-the-art review of some artificial intelligence applications in pile foundations. *Geosci. Front.* **2016**, *7*, 33–44. [CrossRef]
143. Shahin, M.A.; Jaksa, M.B.; Maier, H.R. Artificial neural network applications in geotechnical engineering. *Aust. Geomech.* **2001**, *36*, 49–62.
144. Manzoor, B.; Othman, I.; Gardezi, S.S.S.; Altan, H.; Abdalla, S.B. BIM-Based Research Framework for Sustainable Building Projects: A Strategy for Mitigating BIM Implementation Barriers. *Appl. Sci.* **2021**, *11*, 5397. [CrossRef]
145. AlArjani, A.; Modibbo, U.M.; Ali, I.; Sarkar, B. A new framework for the sustainable development goals of Saudi Arabia. *J. King Saud Univ.* **2021**, *33*, 101477. [CrossRef]

applied system innovation

MDPI

Article

Detection and Identification of Expansion Joint Gap of Road Bridges by Machine Learning Using Line-Scan Camera Images

In Bae Kim [1,2], Jun Sang Cho [2], Goang Seup Zi [1,*], Beom Seok Cho [3], Seon Min Lee [3] and Hyoung Uk Kim [3]

1 School of Civil, Environmental and Architectural Engineering, Korea University, 145 Anam-ro, Seongbukgu, Seoul 02841, Korea; sputnik5@naver.com
2 R&D Division, Korea Expressway Corporation, Hwaseong 18489, Korea; junsangcho@ex.co.kr
3 AI Research Division, CNSI, Seoul 05262, Korea; christpower@cnsi.kr (B.S.C.); lsm@cnsi.kr (S.M.L.); hukim@cnsi.kr (H.U.K.)
* Correspondence: g-zi@korea.ac.kr; Tel.: +82-2-3290-3324

Abstract: Recently, the lack of expansion joint gaps on highway bridges in Korea has been increasing. In particular, with the increase in the number of days during the summer heatwave, the narrowing of the expansion joint gap causes symptoms such as expansion joint damage and pavement blow-up, which threaten traffic safety and structural safety. Therefore, in this study, we developed a machine vision (M/V)-technique-based inspection system that can monitor the expansion joint gap through image analysis while driving at high speed (100 km/h), replacing the current manual method that uses an inspector to inspect the expansion joint gap. To fix the error factors of image analysis that happened during the trial application, a machine learning method was used to improve the accuracy of measuring the gap between the expansion joint device. As a result, the expansion gap identification accuracy was improved by 27.5%, from 67.5% to 95.0%, and the use of the system reduces the survey time by more than 95%, from an average of approximately 1 h/bridge (existing manual inspection method) to approximately 3 min/bridge. We assume, in the future, maintenance practitioners can contribute to preventive maintenance that prepares countermeasures before problems occur.

Keywords: bridge; expansion joint; joint gap; smart bridge maintenance equipment; sensor; structural health monitoring; line-scan camera; machine vision; machine learning

Citation: Kim, I.B.; Cho, J.S.; Zi, G.S.; Cho, B.S.; Lee, S.M.; Kim, H.U. Detection and Identification of Expansion Joint Gap of Road Bridges by Machine Learning Using Line-Scan Camera Images. *Appl. Syst. Innov.* **2021**, *4*, 94. https://doi.org/10.3390/asi4040094

Academic Editor: Luís Bragança

Received: 18 October 2021
Accepted: 16 November 2021
Published: 18 November 2021

Publisher's Note: MDPI stays neutral with regard to jurisdictional claims in published maps and institutional affiliations.

Copyright: © 2021 by the authors. Licensee MDPI, Basel, Switzerland. This article is an open access article distributed under the terms and conditions of the Creative Commons Attribution (CC BY) license (https://creativecommons.org/licenses/by/4.0/).

1. Introduction

The Korea Expressway Corporation is an institution that builds and maintains expressways in Korea, and the number of highway bridges in use reached about 9800 as of 2020. Recently, the number of narrow expansion joint gap occurrences has been increasing. It increased the most in 2018, when the number of days of summer heat waves was the greatest (see Figure 1, Table 1). If such a lack of expansion joint gap occurs, it may adversely affect the structural behavior of the bridge and can be a major threat to traffic safety [1]. Considering the characteristics of the climate of Korea, which has four distinct seasons, we can see that the importance of maintenance and the role of the bridge expansion joint responding to temperature changes between the cold and hot seasons is growing very high.

The causes of narrow expansion joint gap occurrences vary. Representatively, there are construction errors, inappropriate pre-setting, deformation of backfill, and alkaline silica reaction; recently, abnormal high temperatures have also emerged as one of the causes. The year 2018 had the highest number of heatwave days since 2013, and accordingly, the number of cases of lacking bridge expansion joint gaps and pavement blow-up damage rapidly increased. The Korea Expressway Corporation conducted a complete survey of bridges under its management and found that narrow expansion joint gap occurred in 276 bridges (2.96% of the total of 9334 bridges). Table 2 presents the main causes of the occurrence of narrow expansion joint gaps as analyzed through onsite investigations [2].

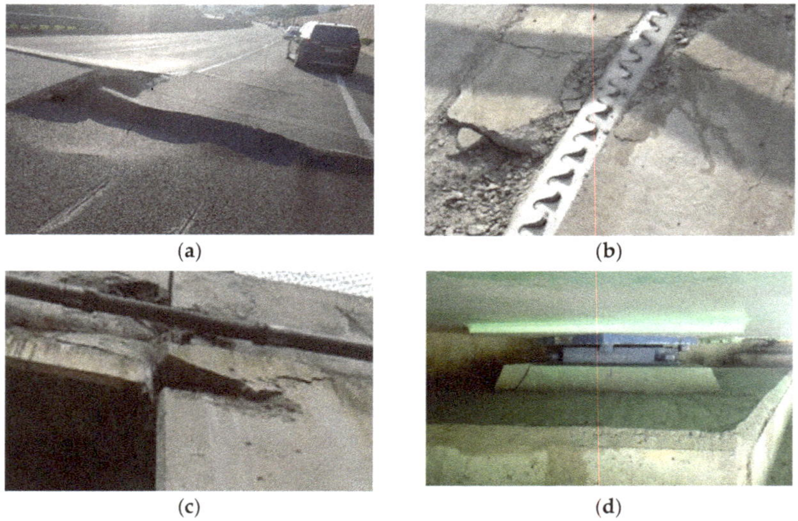

Figure 1. Examples of major damage at road bridge expansion joint: (**a**) blow-up in road pavement, (**b**) lack of joint gap in expansion joint, (**c**) damage to wing wall of bridge abutment, and (**d**) damage by excessive bridge support movement.

Table 1. Average annual temperature, average annual maximum temperature, and number of heatwave days (national).

Year	2011	2012	2013	2014	2015	2016	2017	2018	Recent Average (2011–2018)	Average Year (1981–2010)	Difference	Ratio
Average Temperature (°C)	24.0	24.7	25.4	23.6	23.7	24.8	24.5	25.4	24.5	23.6	+0.8	3.8%
Maximum Temperature (°C)	36.7	38.7	39.2	37.9	38.7	39.6	39.7	41.0	38.9	37.5	+1.4	3.7%
Number of Heatwave Days (days)	14	15	18	6	10	22	14	32	14.2	9.8	+4.4	45%

Table 2. Analysis of causes of insufficient expansion joint gap.

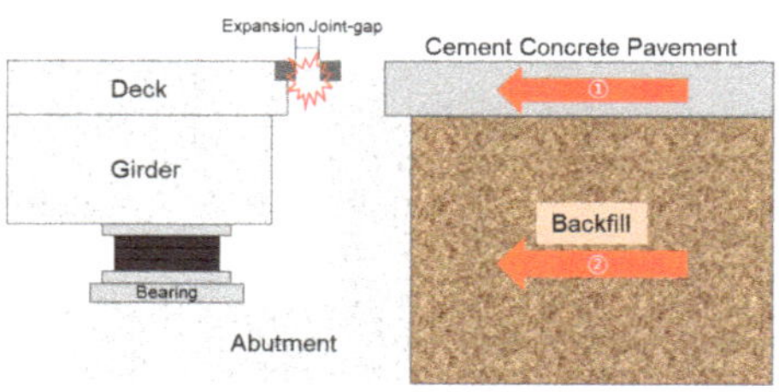

Sum (Bridges)	Major causes	
	Expansion of cement concrete pavement ①	Deformation of backfill ②
276	166 (60%)	110 (40%)

Most importantly, after bridge expansion joint damages occur, a huge budget is required for maintenance, and restoration is virtually impossible unless the bridge is remodeled. Experience shows that the life-cycle costs of a bridge expansion joint, over the life of the bridge, are many times greater than the initial costs of supply and installation, especially when consequential impacts such as traffic disruption during replacement works are considered [3].

Therefore, to solve these problems, the overall design/materials/construction/maintenance should be re-examined to supplement the relevant standards, but this has the problem of consuming a large amount of budget and time. Gaining a full understanding of the demands on the bridge's expansion joints, and how well they are performing, can enable adjustments and maintenance measures to be tailored to maximize the length of the service life with a minimum of maintenance effort [4–7].

In this study, to detect problems and minimize the damage that may occur in the bridge expansion joint at an early stage, an optimal inspection method was developed by converting a vast number of bridge expansion joint gap maintenance inspection methods from existing manual inspections to automated inspections. To this end, it was decided to incorporate an M/V technique to switch from the existing manual inspection method of the bridge expansion joint device to a precise high-speed inspection method.

M/V refers to the technology and methods used to provide imaging-based automatic inspection and analysis for such applications as automatic inspection, process control, and robot guidance, usually in industry. It refers to many technologies, software and hardware products, integrated systems, actions, methods, and expertise.

Definitions of the term M/V vary, but all include the technology and methods used to extract information from an image on an automated basis, as opposed to image processing, where the output is another image. The information can be used for such applications as automatic inspection, robot and process guidance in industry, security monitoring, and vehicle guidance [8–10]. This field encompasses a large number of technologies, software and hardware products, integrated systems, actions, methods, and expertise [10,11]. M/V is practically the only term used for these functions in industrial automation applications. It attempts to integrate existing technologies in new ways and apply them to solve real-world problems in a way that meets the requirements of industrial automation and similar application areas [10,12].

The primary uses for M/V are imaging-based automatic inspection and sorting and robot guidance [13,14]. The imaging device (e.g., camera) can either be separate from the main image processing unit or combined with it, in which case the combination is generally called a smart camera or smart sensor [15,16]. While conventional (2D visible light) imaging is most commonly used, alternatives include multispectral imaging, hyperspectral imaging, imaging various infrared bands, line scan imaging, 3D imaging of surfaces, and X-ray imaging [13]. We used the method of line scan imaging, which we think is most effective in technology that can automatically acquire images while driving at high speeds. [17–20].

One of the purposes of the research is to use a line-scan camera in M/V to record road surfaces as images with a resolution of 1.0 mm per pixel in a high-speed (100 km/h) environment. Existing line-scanning cameras are used for the purpose of identifying counterfeit bills and inspecting semiconductor wafers' surfaces on belt conveyors, so they can be said to be the best technology for introducing to fast-moving vehicles on roads [21,22]. This is suitable for this study because it is free from the problem of lens distortion that may occur in an area-scan camera [23]. According to this study, we developed a survey system equipped with line-scan cameras, Global Positioning System (GPS), and a distance measurement instrument sensor (DMI) in the vehicle, and as a result, we could obtain a safe and accurate image without blocking the road [24–28].

Recently, deep-learning-based approaches [29] have been applied to many problems in various industrial and academic fields. Visual recognition tasks, which extract information of interest from images, such as image classification [30], object detection [31], and semantic segmentation [32], have been actively studied. In particular, CNN (convolutional neural networks) have shown successful results in many visual recognition applications. In this study, models based on CNN and some deep learning techniques are applied to image analysis problems to determine the distance between expansion joints. Two CNN models are used to analyze the distance step by step. First, image classification categorizes images semantically into sub-groups. An image-classification CNN extracts important image features like texture cues and shape cues from each image, and a logistic regression [31] distinguishes them between categories based on the extracted features. Successful studies on image classification CNNs have been conducted focusing on the ImageNet benchmark dataset [33], and some design patterns have been proposed to achieve technical goals such as learning more complex patterns (ResNet) [34], light model weights (MobileNet) [35], and efficient scalability (EfficientNet) [36]. These design patterns, called CNN architectures, are applied to general image classification to reduce the need to design new models every time. Second, semantic segmentation is the process of classifying each pixel in an image belonging to a particular class. The recent success of CNN has also driven outstanding progress in semantic segmentation [37–46]. Based on CNN architecture, the semantic segmentation model extracts local contexts (a small area centered in a pixel to classify) and global contexts (overall semantics of input image) from an image and reconstructs contexts to create a class heatmap of the same size as the original image, which shows the probability of which class each pixel is. Therefore, it can be thought of as a classification problem for each pixel, considering local and global contexts.

Another purpose of the study is to develop software that measures gap width with more than 90% accuracy in the road image of the survey so that many bridges can be monitored quickly. Therefore, an algorithm was developed to measure the gap width after first finding the expansion point in the road image using machine learning. As a result, the identification accuracy was more than 95%, and the investigation time was reduced by more than 95%, from an average of about 1 h/bridge (the inspector's existing manual inspection method) to about 3 min/bridge.

The main contribution of this study is to demonstrate the possibility of developing smart maintenance techniques for road structures, in other words, the successful development of a smart maintenance system that combines machine vision and machine learning technology, for the following purposes: first, to ensure the traffic safety of vehicles on roads that are already in use; and second, to obtain the condition of road structures at the level the investigator wants with figures and accurate images.

The article is organized as follows:

- Section 2 introduces a system developed to survey the road surface while driving at high speed (100 km/h) with a line-scan camera and an M/V imaging device. It introduces operating equipment and explains the main functions and test results.
- Section 3 describes the adequacy review of pre-setting by surveying newly constructed bridge construction joints with standard computer vision methods applied to the initial system.
- Section 4 describes another detection mechanism that uses machine learning.
- Section 5 concludes the paper and proposes future work.

2. Development of Monitoring Technology for Bridge Expansion Joint Using Line-Scan Cameras

Recently, the Korean government has been actively encouraging the introduction of the latest structural safety inspection technologies through the revision of various laws and regulations. For example, Annex 10 of the Enforcement Decree of the Special Act on the Safety and Maintenance of Facilities was amended in March 2020 to newly establish a provision for "appearance investigation and image analysis using new technology or inspection robot, etc.". In other words, new technology can be applied for structure inspection. Further, the revision of Article 167 of the Occupational Safety and Health Act emphasized the safety management of internal employees by strengthening the punishment standards for employers (a maximum fine of $100,000 USD) in case of workplace accidents caused by negligence in worker safety management [1,2].

As a result, the structure management conditions of the Korea Expressway Corporation are becoming increasingly strict. The establishment of facility and performance evaluation following the revision of the Special Act on Safety and Maintenance of Facilities newly necessitated the performance evaluation of 2611 Type 3 facilities (bridges) and 4626 structures. As a result, structural inspectors performed 25,219 regular safety inspections, 2205 precise safety inspections, and 258 precise safety diagnoses in 2020, including in-house and external services. A shortage of management personnel is arising owing to this excessive workload (see Table 3).

Table 3. Number of structural inspections (2020).

Safety Management	Sum	Bridges	Tunnels	Box Culverts
Sum (EA)	27,682	15,636	2118	9928
Regular safety inspection [1]	25,219	13,648	1643	9928
Precision safety inspection [2]	2205	1783	422	-
Precision safety diagnosis	258	205	53	-

[1] Regular safety inspection: an average of approximately 450 EA/yr/branches performed by the structural staff.
[2] Precision safety inspection (2020): self-manpower performance (684EA), performing external services (1521EA).

Given that the stricter structure management conditions are imposing greater demands on manpower, it is crucial to promote inspection methods and smart technologies that can replace manpower.

The Korea Highway Corporation Road Traffic Research Institute analyzed past safety inspection and technical advice data and found that 78.3% of expansion joint devices that had to be replaced lacked a gap. Since 2016, we have been researching monitoring methods to preemptively maintain a suitable gap between expansion joints [2].

For this purpose, by integrating high-speed line-scan cameras, which are widely used in various fields such as semiconductor inspection and road pavement investigation, image-processing technology to determine the length of expansion and contraction through AI-based (i.e., machine learning) image sensing in the analysis process, and automatic control technology, we developed the **N**onstop bridge **EX**pansion joint gap measuring **U**tility **S**ystem (NEXUS) to measure the distance between expansion joints with 1.0 mm resolution while driving at a high speed of 100 km/h [1] (see Figure 2).

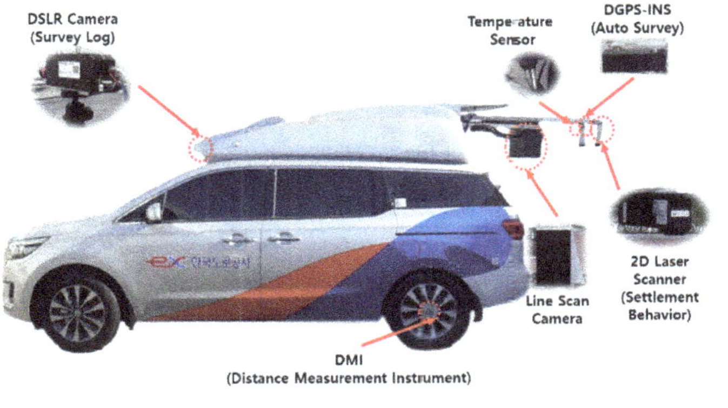

Figure 2. Development system (NEXUS) and main function configuration.

Existing methods manually measure the gap in highway bridge expansion joints through traffic control and partial blockage. By contrast, the proposed method uses a high-speed line-scan camera mounted on a vehicle driving at 100 km/h to acquire an image across a 40 cm-wide car lane. This system uses the geographical information system data of the bridge (above, longitude coordinate system) to perform automatic measurements in conjunction with the mounted GPS while driving at high speed and thereby creates a database of the expansion gaps based on accurate survey images without affecting the traffic flow. In this study, a test survey was conducted on approximately 5000 bridges along the highway, and the analysis results were used for big-data-based machine learning for developing algorithms to accurately determine the length of the expansion joint gap depending on its type and site conditions (see Figure 3) [45–62]. The NEXUS system and the on-site test survey introduction is available on our YouTube channel [63].

Figure 3. UI configuration of the analysis program: measurement point image; survey trajectory (link to GPS); front-view status; value of the joint gap.

The use of the NEXUS for performing automatic surveys of the bridge expansion joint device gap while driving at high speed reduces the survey time by more than 95%, from an average of approximately 1 h/bridge (existing manual inspection method) to approximately 3 min/bridge. In addition, if the accumulated gap is monitored and preemptive maintenance is performed before gap narrowing occurs, it can eliminate the risk factors in the future temperature expansion behavior of the bridge and contribute to traffic safety and cost reductions.

3. Initial Gap Measurement and Evaluation of New Bridge Expansion Joint Device

Previously, it was difficult to use data when necessary, as they were managed in the form of reports and fields. By contrast, by using the NEXUS device, initial data can be accumulated, and the adequacy of highway bridge expansion joints can be checked and evaluated online at any time. To confirm this, we tested the initial data for the adequacy of the pre-setting after 300 days, when the initial drying shrinkage and creep deformation roughly converge; in this regard, it can be beneficial to use the initial gap database (D/B) for the three routes completed at the end of 2016. As the pre-setting of the expansion joint device is not consistent with the daily average temperature during winter and summer at the time of construction, it is essential to adjust the gap in advance such that it has an intermediate value at the reference temperature (average daily temperature of 15 °C).

The route to be investigated has 302 expansion joint devices of three types, namely, rail type, steel finger type, and mono cell type (see Table 4). We converted the expansion gap and average temperature (obtained from the Korea Meteorological Administration) at 15 °C during the investigation and evaluated the adequacy of the pre-setting. After analyzing the initial values of the new bridges, we investigated some bridges that were not properly preset, potentially due to lacking gaps during summer (35–40 °C) or excessive gaps during winter (−15 °C) (see Figures 4 and 5).

For example, the design temperature range ΔT is from −5 °C to 35 °C in the prestressed concrete girder type expansion bridge design movement calculation standard. Therefore, the reference temperature for pre-setting can be seen as 15°C, which is the median temperature. In fact, the temperature at the time of construction of the new joint is under various temperature environments, not 15 °C, so pre-setting is performed at the time of installation to increase or decrease the spacing at the time of installation. The y-axis "Percentage of the capacity of the new joint" is to ensure that pre-setting was performed properly at the time of construction:

(1) After examining the joint gap and the average daily temperature on any day,
(2) The joint gap converted into the reference temperature of 15 °C is expressed as a percentage of the capacity of the new joint.

This value means the following:

(a) If it is close to 50%, it means that it is installed in the middle of the absolute value of the joint gap. (If it is an expansion and contraction joint with a capacity of 100 mm, it represents 50 mm when the joint gap is 15 °C, which is the reference temperature.)
(b) If it is near 10%, it means that it is installed at a small value of the absolute value of the joint gap. (In the case of an expansion joint with a capacity of 100 mm, it shows 10 mm when the joint gap is 15 °C, which is the reference temperature, so the joint gap is insufficient in summer.)
(c) If it is near 90%, it means that it is installed at a large value of the absolute value of the joint gap. (If it is a 100 mm stretchable joint, it is 90 mm when the joint gap is 15 °C, which is the reference temperature, so the gap is exceeded in winter.)

As such, if it is possible to investigate and analyze the expansion joint device gap while driving, without blocking a separate route, from the beginning, and to convert it into D/B in the bridge management system, by monitoring the changing trend, it is possible to take preemptive measures in case of emergency (such as when the gap narrows in a short time).

Table 4. Bridge expansion joint devices installation status.

Division	Total	Rail Type	Steel Finger Type	Mono Cell Type
Total	302 places	169 places	128 places	5 places
	100%	56%	42%	2%
lane A	171 places	98 places	71 places	2 places
	100%	57%	42%	1%
lane B	131 places	71 places	57 places	3 places
	100%	54%	44%	2%

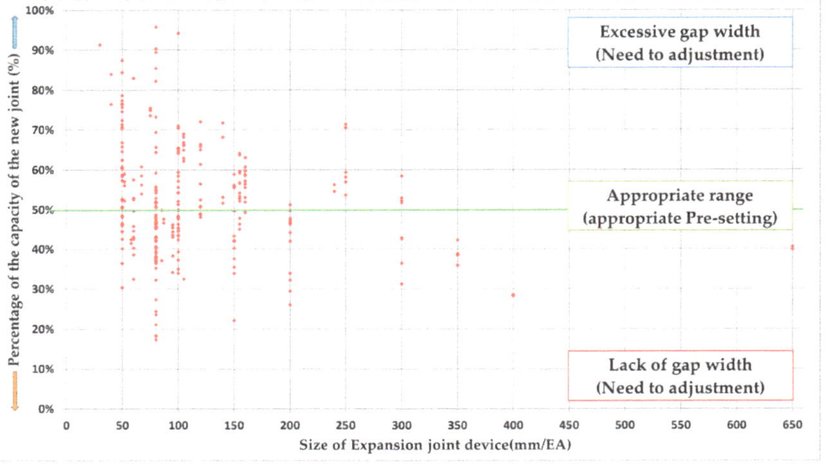

Figure 4. Pre-setting suitability evaluation of new bridge expansion joint devices.

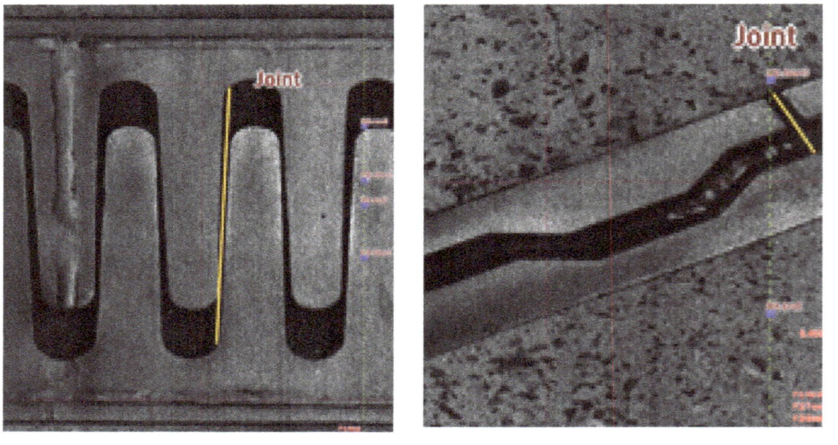

Figure 5. False-detection case of expansion joint gap.

4. Advanced Identification of Expansion Gap Using Machine Learning

Expansion joints device has different details for each development and sales company, and typical expansion joint devices constructed on road bridges in South Korea include mono cell type, finger type, and rail type. Table 5 lists the types of expansion joint devices installed on highway bridges in South Korea as of 2014.

Table 5. Expansion joint device installation status (2014).

Installation	Sum	Mono Cell Type	Finger Type	Rail Type	Others
EA	14,784	7793	1786	4228	977
Prop (%)	100	53	12	29	7

The analysis of the expansion joint gaps of 4821 bridges and 12,825 devices on the highway mainline from 2017 to 2019 using NEXUS equipment revealed various types of expansion joint devices and gap identification errors depending on the condition of the expansion joint device body installed in the field. False-detection errors include the case of erroneously identifying another gap in the expansion joint body as a gap and the case of detecting a specific part of the road surface (portholes, repair marks, cracks, etc.). On average, the gap identification accuracy was 67.5%; that for the finger type with a complex shape was the lowest at 51% (see Table 6, Figure 5). False detections were attributed to limitations of the traditionally coded image analysis algorithm. Therefore, the image identification method must be improved using machine learning, as it is superior to traditional coding.

Table 6. Gap identification accuracy by type of expansion joint (2017–2019).

Discrimination (%)	Average (%)	Mono Cell Type	Finger Type	Rail Type	Others
Accuracy	67.5	71	51	87	61
Loss	32.5	29	49	13	39

We aimed to overcome the limitations of the traditional gap identification algorithm. Machine learning can be used to learn previously investigated images as big data, and to thereby identify the expansion joint device from an input image and more accurately determine the gap value. The input image obtained using the NEXUS equipment was recorded as multiple 10,000 × 1024 pixels images, depending on the irradiation distance. The image analysis system consisted of cascaded AI vision modules, in which the work process analyzed high-resolution input images for realizing effective calculations. The AI process involves the following steps: original image input > expansion joint area extraction AI network (classification) > expansion joint area image area (cropping) input > gap extraction AI (segmentation) > gap measurement (algorithm) (see Figure 6).

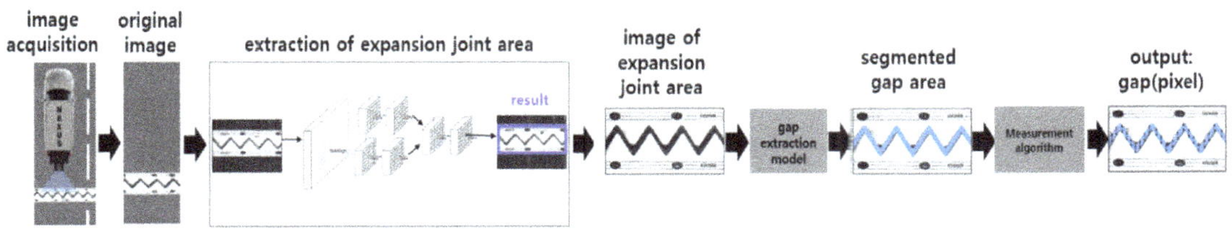

Figure 6. Expansion joint device analysis process for AI method (i.e., machine learning).

To measure the gap value, the position of the expansion joint device must be specified in the image. Further, for achieving high measurement accuracy, the expansion joint device needs to be recognized at the pixel level. This can be done using image segmentation methods used in computer vision. Further, a deep learning model with extensive data learning is used for problems with large changes in illuminance, angle, image shape, and noise generation, such as road environments.

The AI machine learning model for expansion joint device discovery was implemented to solve the classification problem using CNN as a feature extractor. The original image with a size of 10,000 × 1024 pixels was cut into an image patch with a size of 1000 × 1024 pixels to be inputted to the AI model with minimal distortion. Then, 19 image patches were generated for each line-scan image, and classification was performed to find the image patches in which the expansion joint device exists (see Figure 7).

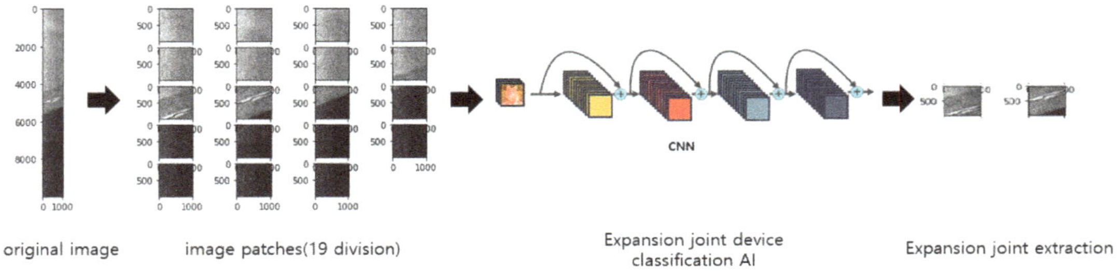

Figure 7. Image segmentation and classification to search for expansion joint device.

To find the point with the minimum gap distance, it is necessary to precisely segment the gap region from the image patch where the expansion joint device appears. The AI machine learning model for gap region extraction used U-Net, a representative segmentation model. The first AI model generated and learned training data by labeling pixels in the corresponding area to extract metal parts with characteristic textures from the line-scan images of the expansion joint device. The second AI model was trained to find the gap region between the extracted joint devices (see Figure 8).

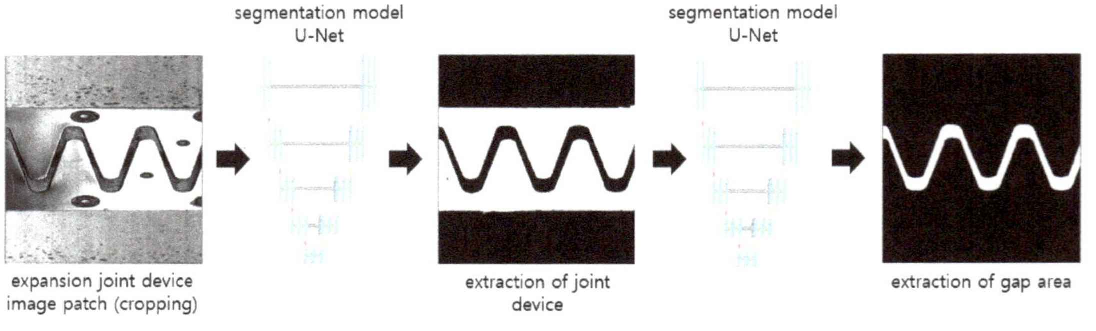

Figure 8. Segmentation process for extracting gap area.

After obtaining a binarized image with the gap region and the background, we found the point with the minimum gap distance. The minimum gap distance in pixels was converted to the final gap value in millimeters by multiplying the information in millimeters per pixel.

4.1. AI-Based Image Analysis

Line-scan images obtained using NEXUS were converted into high-resolution digital images with a length and width of 10,000 and 1024 pixels, respectively. The unit length (mm) of the pixels can be calculated by considering the vehicle speed at the time of the shooting. If the expansion joint device can be measured in pixels, the actual value of the gap distance can be inferred within the error range. Therefore, recognizing the position of the expansion joint device in the line-scan image using a program and accurately measuring the size of the gap in pixels determine the measurement accuracy.

To perform highly accurate analyses, a large number of line-scan images were trained using deep learning. To efficiently analyze high-resolution input images, cascade-type analysis procedures were configured. The analysis involved the following steps: (1) ultrafast line-scan image input, (2) extraction of small image patches in a square matrix, (3) recognition of expansion joint device among the image patches, (4) expansion gap division, and (5) gap distance analysis and actual value calculation (see Figure 9).

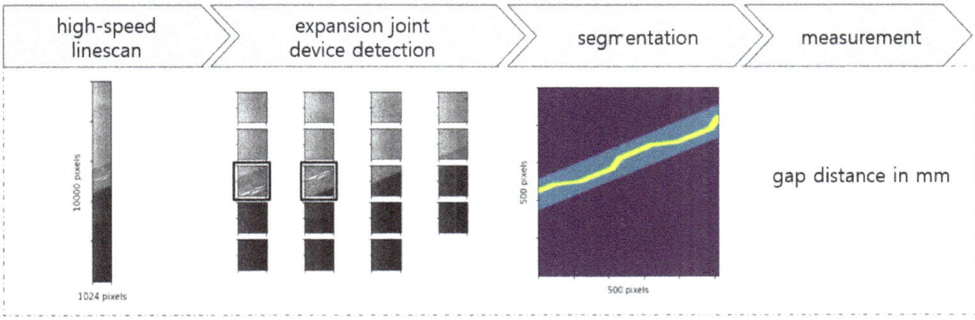

Figure 9. Spacing measurement process of expansion joint device.

4.2. Expansion Joint Device Recognition

Line-scan images have a high resolution of more than 10,000,000 pixels and require considerable memory and computational resources when being analyzed using a deep neural network. Therefore, it is necessary to divide the image into small image patches for realizing effective computations. Through the window sliding method, images were extracted every 1000 pixels in the longitudinal direction to generate a total of 19 image patches with a size of 1000 × 1024 (see Figure 10). Each image patch is again converted to a size of 224 × 224 pixels and subjected to binary classification.

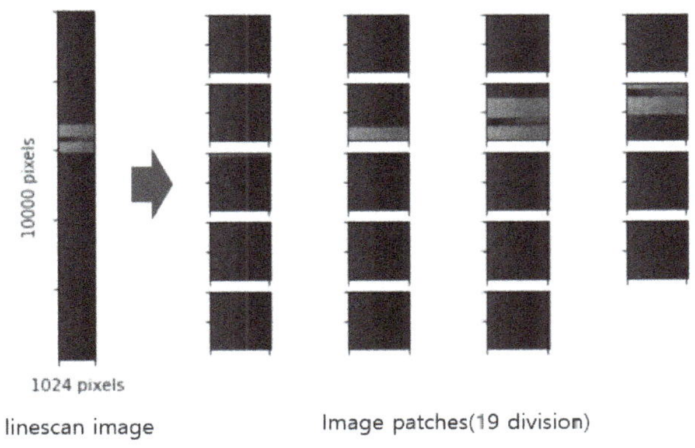

Figure 10. Line-scan image segmentation to search for expansion joint devices.

Each image patch is again converted to a size of 224 × 224 pixels for binary classification. By using a CNN, an effective neural network for image analysis, we labelled patches with the expansion joint device as y = 1 and patches without the expansion joint device as y = 0 and classified them by logistic regression. The logistic regression followed by CNN infers the probability P that there is an expansion joint in the image patch. The error of the model prediction for each image patch is calculated with binary cross entropy [64–69], which compares each of the predictions to the actual class, which can be either 0 or 1. The errors of each set of training data are summed up. In stochastic gradient methods [70–77], the cost and sum of errors is used to update current model parameters to reduce the distance from the optimal point in the parameter space. The equation of binary cross entropy is shown as follows:

$$L = -ylogP - (1-y)log(1-P) \tag{1}$$

Because class imbalance exists in a ratio of 1:9, a weight is assigned to each class.

$$L_{weighted} = -\epsilon \times ylogP - \mu \times (1-y)log(1-P) \tag{2}$$

where $\epsilon = 9.0$, $\mu = 1.0$.

The cost function J is calculated by averaging the errors for N training data and adding the L2 regularization to reduce overfitting [78–85] of the model.

$$J = \frac{1}{N}\sum_{j=1}^{N} L^{j}_{weighted} + \lambda|w|^2 \tag{3}$$

where $\lambda = 10^{-4}$.

The gradient descent method updates the model parameters in the direction of reducing the cost function J as follows:

$$w \Leftarrow w - \alpha \cdot g \tag{4}$$

where $\alpha = 10^{-4}$, $g = \frac{\partial J}{\partial w}$.

We tested ResNet, MobileNet, and EfficientNet, all of which are well-known CNN architectures [22–24]. Gradient vanishing is more likely to occur as the layers of the deep learning model deepen, and ResNet solved this problem by performing residual learning using skip connection. The structure obtained high accuracy compared to the scale of the model. MobileNet proposed a depth-wise separable convolution that reconstructed the existing convolution method to reduce the computational amount of the model. Compared to the popular models at the time of proposal, the amount of computation was significantly reduced, and the same accuracy was maintained. EfficientNet, which empirically reports a methodology to increase model complexity to improve performance, updated the state of the art for benchmark datasets.

We trained the three models to a binary classification of expansion joints and non-expansion joints (see Figure 11). Through the test, the EfficientNet model showed the highest performance, with recognition accuracy of 97.57% for the expansion joint device. Further, it was better to start learning from randomized initial parameters than through transfer learning for ImageNet. This is because the analysis image has different characteristics from the general characteristics of ImageNet. We employed EfficientNet for expansion joint detection, as it has the highest accuracy.

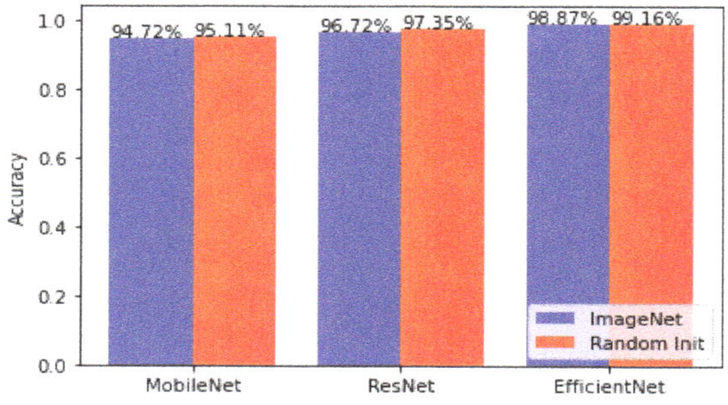

Figure 11. Recognition accuracy test results for each CNN structure.

4.3. Expansion Joint Gap Segmentation

Even if the image patch of the expansion joint is extracted from the line-scan image, image segmentation is required within the detected image to accurately measure the expansion gap. Image segmentation is a pixel-level classification that deduces the class each pixel belongs to (i.e., expansion joint device or background). A masking image representing the pixel corresponding to the expansion joint device can be generated by the image segmentation algorithm.

Figure 12 shows the image patch (left) and correct mask (right) of the expansion joint device. The neural network structure receives image patches as an input and performs binary classification of the individual pixels. The deep learning model receives image patches as the input and learns to predict a masking that is close to the correct answer. In this case, the class of a pixel should be determined by considering the global and local characteristics of the image rather than individual pixel values.

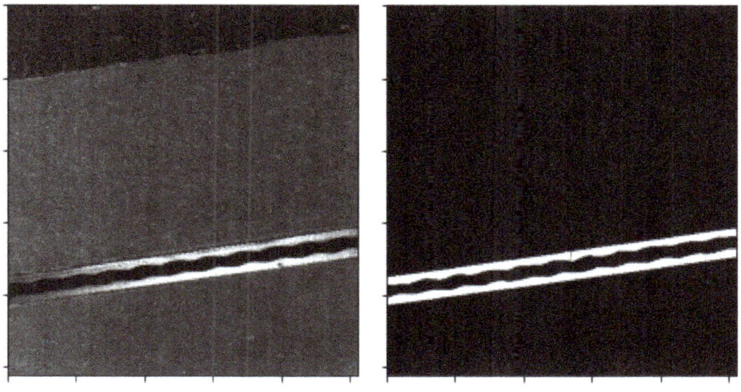

(a) (b)

Figure 12. Image of flexible expansion joint and mask of correct answer: (**a**) original image, (**b**) masked image.

This study uses a U-Net model, which was previously developed for sophisticated analyses of organ lesions in the field of biomedical science [36]. U-Net is suitable for the problem of precisely detecting the shape of the analysis target by simultaneously learning the global and local information of the image. The structure of U-Net was modified to develop EfficientNet, a neural network that extracts image information.

U-Net infers the probability that an arbitrary pixel is an expansion joint device and learns the correct answer for each pixel from the correct answer masking. Therefore, the prediction error for all pixels M of the image patch expressed as BCE is as follows:

$$L_{patch} = \sum L_{pixel} = \sum_{i=1}^{M} (-y_i \log P_i - (1 - y_i) \log(1 - P_i)),\ \ i = 1, 2, 3, \ldots, N \quad (5)$$

The final cost function J is expressed by summing the mean error for N image patches and the L2 regularization term:

$$J = \frac{1}{N} \sum_{j=1}^{N} \left(L_{patch}^{j} + \lambda |w|^2 \right) \quad (6)$$

where $\lambda = 10^{-4}$.

The gradient descent method updates model parameters in the direction of reducing the cost function J as follows:

$$w \Leftarrow w - \alpha \cdot g \quad (7)$$

where $\alpha = 10^{-4}$, $g = \frac{\partial J}{\partial w}$.

We compared the performance between U-Net's masking image and the correct masking image in the test dataset. Table 7 shows the pixel-level classification performance for the test set; the pixel precision of the expansion joint device was 96.61%, recall rate was 94.38%, and f1-score was 95.49%. The f1-score of the expansion joint device pixel detection was within 5%. In the post-processing process, by correcting the error of the predicted masking image of U-Net, the minimum gap point was detected, and the distance was measured.

Table 7. Gap identification accuracy by type of expansion joint (2017–2019).

Accuracy by Type of Expansion Joint (%)	Precision	Recall	f1-Score
Positive (pixels of expansion joints)	96.61	94.38	95.49
Negative (other pixels)	99.23	99.55	99.39

4.4. Gap Distance Analysis Algorithm

The texture of the pixels in the expansion gap area to be measured appears somewhat irregular depending on the gap, foreign matter, and type of expansion joint device. Texture irregularity is a factor that makes it difficult to distinguish pixels in the expansion gap area.

The metal surface constituting the expansion joint device has a consistent texture compared to the gap region. This means that it is easier to extract the expansion joint device than the gap area. Therefore, to analyze the gap distance, the expansion joint device is extracted first, and the gap area is extracted again from the resulting image. Image segmentation using U-Net was applied to both area extraction processes (see Figure 13).

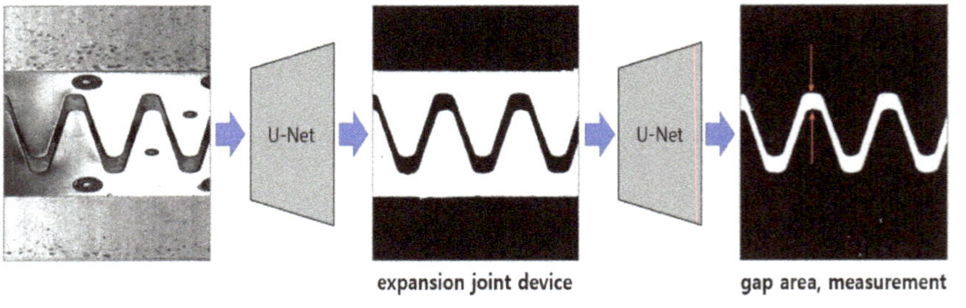

Figure 13. An approach for elaborate gap area extraction.

Because the U-Net output represents the probability that each pixel is a gap area, binarization is performed by applying a threshold value of 0.5. When defining a binarization function as a response, the formula to search the x-coordinate of the minimum gap is as given below. Note that the same applies to rail-type expansion joint devices.

$$\underset{1 \le x \le 512}{arg\ min} \sum_{y=1}^{512} response(x, y) \quad (8)$$

The input and output of the U-Net model have the size of 512 × 512. By binarizing the final output probability map, we search for the x-coordinate with the smallest number of pixels with a value of 1. The number of pixels at the coordinates is the gap distance in pixels, and the actual gap distance is obtained by multiplying the distance value per pixel.

$$(distance) = (\#\,of\ pixel) * \left(\frac{mm}{pix}\right) \quad (9)$$

Figure 14 shows the pseudocode for obtaining the minimum gap distance from the output probability map of U-Net.

```
response = threshold(output of U-Net, 0.5)
for x in range(0, width of response)
    distList[x] = # of response which value is 1
    if distList[x] is minimum
        minIdx = x
minimum distance = distList[minIdx] * mm/pixel
```

Figure 14. Pseudocode to obtain minimum gap distance from U-Net output.

4.5. Gap Identification Verification

Based on the abovementioned results of AI-based gap identification, we randomly selected 10,526 places among 12,825 expansion joint device big-data images obtained previously to determine the discrimination of the expansion joint device gap. After dividing and refining 10,526 line-scan images into 19 image patches, 289,495 sets of training data and 45,950 tests of the classification model were constructed. A total of 21,604 sets of training data and 4174 of test data of the segmentation model for measuring the expansion joint gap were refined. The results are presented below for each expansion joint device type. The result of the position where the minimum spacing was measured is indicated by a red line. For rail-type joints in which several gaps appear at once, the starting and end gaps of the part with the smallest actual gap value are indicated by red lines (see Figure 15).

We used Python 3 and TensorFlow 2 to implement and train a deep learning model using CNN, and development frameworks such as TensorFlow and PyTorch support libraries for implementing popular CNN layers and to support learning using GPUs.

We used a single NVIDIA Tesla V100 graphics card and Tensorflow to accelerate the training of the model. The EfficientNet B0 model for classification of expansion joints completed training in less than 30 epochs and took up to 4 h. A total of 259,495 training images and 30,000 validation images were used. Overall, 45,950 images for testing did not participate in the training. The U-Net model for gap region extraction completed training in less than 20 epochs and took up to 4 h, and 19,304 training images and 2300 validation images were used, while 4174 images for testing did not participate in the training.

In the environment for field application after training, one NVIDIA RTX 3070 graphics card and Tensorflow were used. In one line-scan image (1024 × 10,000), it took an average of 0.6 s to find the expansion joint and measure the gap.

After comparing the identification rate results from the existing traditional algorithm, we found that the finger type, which had the lowest discrimination accuracy, improved the most, by 41% to an accuracy of approximately 92%, and the overall average identification accuracy improved by 27.5% to 95%. Even though the identification accuracy was greatly improved, if a foreign substance existed inside the expansion gap, if a sealing agent was applied to the surface for preventing water leakage, if the body was damaged, and so on, these were included in the discriminant errors and manual correction would be required (see Table 8).

Table 8. Accuracy of expansion gap identification using machine learning algorithm.

Accuracy (%)	Average	Mono Cell Type	Finger Type	Rail Type	Others
Conventional algorithm	67.5	71	51	87	61
AI algorithm (machine learning)	95	98	92	99	91
Improvement rate	↑27.5	↑27	↑41	↑12	↑30

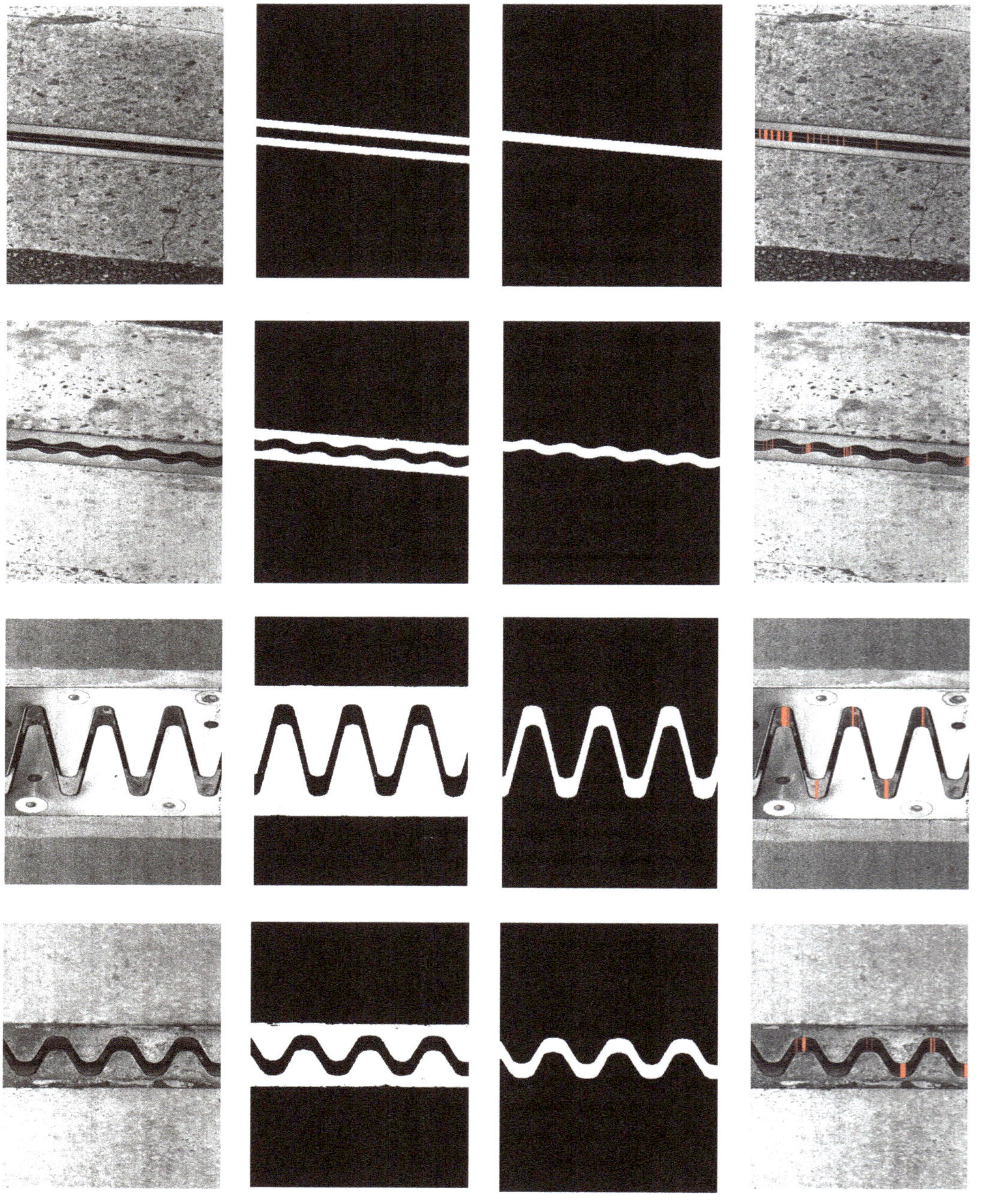

(a) (b) (c) (d)

Figure 15. Machine-learning-based verification results of expansion joint device gap identification: (**a**) original image; (**b**) extraction of expansion joint area image; (**c**) extraction of gap area image; (**d**) final measurement result image.

5. Conclusions

In this research, an automatic-image-recognition-based survey system was established on road bridges and was successfully verified through field tests. In order to ensure 1.0 mm resolution performance per pixel in a high-speed (100 km/h) environment, M/V technology using a line-scan camera was used, and accurate images could be obtained almost in real time as a result of the test. Line-scan camera technology advances rapidly, so using a better camera could further improve system performance.

Technological advances will promote the implementation of the M/V system and lower maintenance costs. Therefore, one contribution of this research is providing a solution that can apply the M/V system to road maintenance using line-scan cameras.

A survey system equipped with line-scan cameras, GPS, and DMI was designed to perform automatic investigation when accessing the bridge while driving at high speeds. The use of NEXUS for performing automatic surveys of the bridge expansion joint device gap while driving at high speed reduces the survey time by more than 95%, from an average of approximately 1 h/bridge (existing manual inspection method by an inspector) to approximately 3 min/bridge. In addition, if the accumulated gap data is monitored and preemptive maintenance is performed before gap narrowing (lack of joint gap) occurs, it can eliminate the risk factors in the future temperature expansion behavior of the bridge and contribute to traffic safety and cost reductions. This is the second contribution.

Measuring the bridge expansion joint gap through survey images has limitations in traditional algorithmic methods. Images with various objects on the road surface and shapes of various expansion joint devices are similar to big data. Therefore, by creating an algorithm by artificial intelligence technology (machine learning), more accurate survey values can be stably obtained. The machine learning model for searching for expansion joint devices in survey images used Resnet as a feature extractor, and the representative segmentation model for searching for the gap area used U-Net. These were used to solve the classification problem. Testing with a random selection of 10,526 previously acquired big data images indicated that the expansion gap identification accuracy was improved by 27.5%, from 67.5% to 95.0%. This is another contribution.

However, the main contribution of this study is to demonstrate the possibility of developing smart maintenance techniques for road structures, in other words, the successful development of a smart maintenance system that combines machine vision and machine learning technology that serves the following purposes: first, to ensure the traffic safety of vehicles on roads that are already in us; and second, to obtain the condition of road structures at the level the investigator wants with figures and accurate images. This requires satisfaction with vehicle traffic, customer safety, investigator convenience, and traffic safety, so we believe that developing road maintenance technology should consider much more complex problems than developing smart construction technology.

If this study intensifies in the future, we will be able to create a wider variety of smart road maintenance systems based on this concept and our imagination.

Author Contributions: Supervision, investigation, data creation, writing—original draft preparation, and writing—review and editing, I.B.K.; conceptualization, J.S.C. and G.S.Z.; software, formal analysis, validation, and data curation, B.S.C., S.M.L. and H.U.K. All authors have read and agreed to the published version of the manuscript.

Funding: This research received no external funding.

Institutional Review Board Statement: Not applicable.

Informed Consent Statement: Not applicable.

Conflicts of Interest: The authors declare no conflict of interest.

References

1. In Bae, K.; Byung Ju, L.; Chang Ho, P. *Development of Behavior Evaluation Method for Bridge Expansion Joints Based on Ultrafast Laser Line Scanning System*; The Korea Expressway Corporation Research Institute: Hwaseong-si, Korea, 2018; OTKCRK190185.
2. Hyun Ho, C.; In Bae, K.; Hong Sam, K.; Yu Sung, S. *A Study on Proper Construction and Management Standards for Bridge Expansion Joints to Cope with Lack of Joint-Gap*; The Korea Expressway Corporation Research Institute: Hwaseong-si, Korea, 2020; OTKCRK210798.
3. Spuler, T.; Loehrer, R.; O'Suilleabhain, C. Life-cycle considerations in the selection and design of bridge expansion joints. In Proceedings of the IABSE Congress on Innovative Infrastructures towards Human Urbanism, Seoul, Korea, 19–21 September 2012.
4. Moor, G.; Meng, N.; O'Suilleabhain, C. Remote structural health monitoring systems for bridge expansion joints and bearings. In Proceedings of the 2nd Conference on Smart Monitoring Assessment and Rehabilitation of Civil Structures, Istanbul, Turkey, 9–11 September 2013.
5. Joo, O.; Hyun Sup, S.; Sang Suk, L.; Hu Seung, K. *Bridge Expansion Joint Design and Construction*; CIR: Seoul, Korea, 2015.
6. Korea Expressway Corporation. *Technical Advisory Case Book (Bridge Support and Expansion Joint Device)*; Korea Expressway Corporation: Gimcheon-si, Korea, 2008.
7. Korea Expressway Corporation. *Expressway Construction Professional Specifications/Civil Edition*; Korea Expressway Corporation: Gimcheon-si, Korea, 2017.
8. Steger, C.; Ulrich, M.; Wiedemann, C. *Machine Vision Algorithms and Applications*, 2nd ed.; Wiley-VCH: Weinheim, Germany, 2018; p. 1, ISBN 978-3-527-41365-2.
9. Beyerer, J.; León, F.P.; Frese, C. *Machine Vision—Automated Visual Inspection: Theory, Practice and Applications*; Springer: Berlin/Heidelberg, Germany, 2016; ISBN 978-3-662-47793-9. [CrossRef]
10. Graves, M.; Batchelor, B. *Machine Vision for the Inspection of Natural Products*; Springer: Berlin/Heidelberg, Germany, 2003; p. 5, ISBN 978-1-85233-525-0.
11. Holton, W.C. By Any Other Name. *Vis. Syst. Des.* **2010**, *15*, 1089–3709.
12. Turek, F.D. Machine Vision Fundamentals, How to Make Robots See. *NASA Tech. Briefs* **2011**, *35*, 60–62.
13. Zhuang, H.; Raghavan, S. Development of a machine vision laboratory. *Age* **2003**, *8*, 1.
14. Belbachir, A.N. (Ed.) *Smart Cameras*; Springer: Berlin/Heidelberg, Germany, 2009; ISBN 978-1-4419-0952-7.
15. Dechow, D. Explore the Fundamentals of Machine Vision: Part 1. *Vis. Syst. Des.* **2013**, *18*, 14–15.
16. Wilson, A. The Infrared Choice. *Vis. Syst. Des.* **2011**, *16*, 20–23.
17. Jang, J.; Shin, M.; Lim, S.; Park, J.; Kim, J.; Paik, J. Intelligent image-based railway inspection system using deep learning-based object detection and weber contrast-based image comparison. *Sensors* **2019**, *19*, 4738. [CrossRef]
18. Li, L.; Luo, W.T.; Wang, K.C.P. Lane marking detection and reconstruction with line-scan imaging data. *Sensors* **2018**, *18*, 1635. [CrossRef]
19. Wendel, A.; Underwood, J. Extrinsic Parameter Calibration for Line Scanning Cameras on Ground Vehicles with Navigation Systems Using a Calibration Pattern. *Sensors* **2017**, *17*, 2491. [CrossRef]
20. Lopes, G.; Ribeiro, A.; Sillero, N.; Gonçalves-Seco, L.; Silva, C.; Franch, M.; Trigueiros, P. High Resolution Trichromatic Road Surface Scanning with a Line Scan Camera and Light Emitting Diode Lighting for Road-Kill Detection. *Sensors* **2016**, *16*, 558. [CrossRef]
21. Chien, J.-C.; Wu, M.-T.; Lee, J.-D. Inspection and Classification of Semiconductor Wafer Surface Defects Using CNN Deep Learning Networks. *Appl. Sci.* **2020**, *10*, 5340. [CrossRef]
22. Wang, J.; Lee, S. Data Augmentation Methods Applying Grayscale Images for Convolutional Neural Networks in Machine Vision. *Appl. Sci.* **2021**, *11*, 6721. [CrossRef]
23. Chen, A.; Orlov-Levin, V.; Meron, M. Applying High-Resolution Visible-Channel Aerial Scan of Crop Canopy to Precision Irrigation Management. *Proceedings* **2018**, *2*, 335. [CrossRef]
24. Amziane, A.; Losson, O.; Mathon, B.; Dumenil, A.; Macaire, L. Reflectance Estimation from Multispectral Linescan Acquisitions under Varying Illumination—Application to Outdoor Weed Identification. *Sensors* **2021**, *21*, 3601. [CrossRef] [PubMed]
25. Wu, N.; Haruyama, S. The 20k Samples-Per-Second Real Time Detection of Acoustic Vibration Based on Displacement Estimation of One-Dimensional Laser Speckle Images. *Sensors* **2021**, *21*, 2938. [CrossRef] [PubMed]
26. Tzu, F.-M.; Chen, J.-S.; Hsu, S.-H. Light Emitted Diode on Detecting Thin-Film Transistor through Line-Scan Photosensor. *Micromachines* **2021**, *12*, 434. [CrossRef] [PubMed]
27. Kim, H.; Choi, Y. Autonomous Driving Robot That Drives and Returns along a Planned Route in Underground Mines by Recognizing Road Signs. *Appl. Sci.* **2021**, *11*, 10235. [CrossRef]
28. Xu, D.; Qi, X.; Li, C.; Sheng, Z.; Huang, H. Wise Information Technology of Med: Human Pose Recognition in Elderly Care. *Sensors* **2021**, *21*, 7130. [CrossRef]
29. Goodfellow, I.; Bengio, Y.; Courville, A. *Deep Learning*; MIT Press: Cambridge, MA, USA, 2016.
30. Tamuly, S.; Jyotsna, C.; Amudha, J. Deep learning model for image classification. In Proceedings of the International Conference on Computational Vision and Bio-Inspired Computing, Coimbatore, India, 25–26 September 2019; Springer: Cham, Switzerland, 2019.
31. Liu, L.; Ouyang, W.; Wang, X.; Fieguth, P.; Chen, J.; Liu, X.; Pietikäinen, M. Deep Learning for Generic Object Detection: A Survey. *Int. J. Comput. Vis.* **2020**, *128*, 261–318. [CrossRef]

32. Lateef, F.; Ruichek, Y. Survey on semantic segmentation using deep learning techniques. *Neurocomputing* **2019**, *338*, 321–348. [CrossRef]
33. Russakovsky, O.; Deng, J.; Su, H.; Krause, J.; Satheesh, S.; Ma, S.; Huang, Z.; Karpathy, A.; Khosla, A.; Bernstein, M.; et al. Imagenet large scale visual recognition challenge. *Int. J. Comput. Vis.* **2015**, *115*, 211–252. [CrossRef]
34. Howard, A.G.; Zhu, M.; Chen, B.; Kalenichenko, D.; Wang, W.; Weyand, T.; Andreetto, M.; Adam, H. Mobilenets: Efficient convolutional neural networks for mobile vision applications. *arXiv* **2017**, arXiv:1704.04861.
35. Tan, M.; Le, Q. Efficientnet: Rethinking model scaling for convolutional neural networks. In Proceedings of the International Conference on Machine Learning, PMLR, Long Beach, CA, USA, 9–15 June 2019.
36. Ronneberger, O.; Fischer, P.; Brox, T. U-net: Convolutional networks for biomedical image segmentation. In Proceedings of the International Conference on Medical Image Computing and Computer-Assisted Intervention, Munich, Germany, 5–9 October 2015; Springer: Cham, Switzerland, 2015.
37. Dreiseitl, S.; Ohno-Machado, L. Logistic regression and artificial neural network classification models: A methodology review. *J. Biomed. Inform.* **2002**, *35*, 352–359. [CrossRef]
38. Long, J.; Shelhamer, E.; Darrell, T. Fully convolutional networks for semantic segmentation. In Proceedings of the IEEE Conference on Computer Vision and Pattern Recognition, Boston, MA, USA, 7–12 June 2015.
39. Noh, H.; Hong, S.; Han, B. Learning deconvolution network for semantic segmentation. In Proceedings of the IEEE International Conference on Computer Vision, Santiago, Chille, 7–13 December 2015.
40. Badrinarayanan, V.; Kendall, A.; Cipolla, R. SegNet: A Deep Convolutional Encoder-Decoder Architecture for Image Segmentation. *IEEE Trans. Pattern Anal. Mach. Intell.* **2017**, *39*, 2481–2495. [CrossRef]
41. Chen, L.-C.; Papandreou, G.; Kokkinos, I.; Murphy, K.; Yuille, A.L. DeepLab: Semantic Image Segmentation with Deep Convolutional Nets, Atrous Convolution, and Fully Connected CRFs. *IEEE Trans. Pattern Anal. Mach. Intell.* **2018**, *40*, 834–848. [CrossRef] [PubMed]
42. Hartung, J.; Jahn, A.; Bocksrocker, O.; Heizmann, M. Camera-Based In-Process Quality Measurement of Hairpin Welding. *Appl. Sci.* **2021**, *11*, 10375. [CrossRef]
43. Martins, J.; Nogueira, K.; Osco, L.; Gomes, F.; Furuya, D.; Gonçalves, W.; Sant'Ana, D.; Ramos, A.; Liesenberg, V.; dos Santos, J.; et al. Semantic Segmentation of Tree-Canopy in Urban Environment with Pixel-Wise Deep Learning. *Remote. Sens.* **2021**, *13*, 3054. [CrossRef]
44. Mohajerani, Y.; Wood, M.; Velicogna, I.; Rignot, E. Detection of Glacier Calving Margins with Convolutional Neural Networks: A Case Study. *Remote. Sens.* **2019**, *11*, 74. [CrossRef]
45. Hirahara, K.; Ikeuchi, K. Detection of street-parking vehicles using line scan camera and scanning laser range sensor. In Proceedings of the EEE IV2003 Intelligent Vehicles Symposium. Proceedings (Cat. No. 03TH8683), Columbus, OH, USA, 9–11 June 2003; IEEE: Piscataway, NJ, USA, 2003.
46. Dvorák, M.; Kanich, O.; Drahanský, M. Scalable Imaging Device using Line Scan Camera for Use in Biometric Recognition and Medical Imaging. In Proceedings of the 14th International Joint Conference on Biomedical Engineering Systems and Technologies (BIOSTEC 2021)-Volume 1: BIODEVICES, Online Streaming, 11–13 February 2021; pp. 160–168.
47. Zhang, L.; Yang, F.; Zhang, Y.D.; Zhu, Y.J. Road crack detection using deep convolutional neural network. In Proceedings of the 2016 IEEE International Conference on Image Processing (ICIP), Phoenix, AZ, USA, 25–28 September 2016; IEEE: Piscataway, NJ, USA, 2016.
48. Shi, Y.; Cui, L.; Qi, Z.; Meng, F.; Chen, Z. Automatic road crack detection using random structured forests. *IEEE Trans. Intell. Transp. Syst.* **2016**, *17*, 3434–3445. [CrossRef]
49. Bang, S.; Park, S.; Kim, H.; Kim, H. Encoder–decoder network for pixel-level road crack detection in black-box images. *Comput. Aided Civ. Infrastruct. Eng.* **2019**, *34*, 713–727. [CrossRef]
50. Kruachottikul, P.; Cooharojananone, N.; Phanomchoeng, G.; Chavarnakul, T.; Kovitanggoon, K.; Trakulwaranont, D.; Atchariyachanvanich, K. Bridge sub structure defect inspection assistance by using deep learning. In Proceedings of the 2019 IEEE 10th International Conference on Awareness Science and Technology (iCAST), Morioka, Japan, 23–25 October 2019; IEEE: Piscataway, NJ, USA, 2019.
51. Pereira, V.; Tamura, S.; Hayamizu, S.; Fukai, H. Classification of paved and unpaved road image using convolutional neural network for road condition inspection system. In Proceedings of the 2018 5th International Conference on Advanced Informatics: Concept Theory and Applications (ICAICTA), Krabi, Thailand, 14–17 August 2018; IEEE: Piscataway, NJ, USA, 2018.
52. Guo, W.; Wang, N.; Fang, H. Design of airport road surface inspection system based on machine vision and deep learning. *J. Phys. Conf. Ser.* **2021**, *1885*, 052046. [CrossRef]
53. Mei, Q.; Gül, M. A cost effective solution for pavement crack inspection using cameras and deep neural networks. *Constr. Build. Mater.* **2020**, *256*, 119397. [CrossRef]
54. Maeda, H.; Sekimoto, Y.; Seto, T.; Kashiyama, T.; Omata, H. Road Damage Detection and Classification Using Deep Neural Networks with Smartphone Images. *Comput. Civ. Infrastruct. Eng.* **2018**, *33*, 1127–1141. [CrossRef]
55. Cha, Y.-J.; Choi, W.; Suh, G.; Mahmoudkhani, S.; Büyüköztürk, O. Autonomous Structural Visual Inspection Using Region-Based Deep Learning for Detecting Multiple Damage Types. *Comput.-Aided Civ. Infrastruct. Eng.* **2018**, *33*, 731–747. [CrossRef]
56. Chehri, A.; Saeidi, A. IoT and Deep Learning Solutions for an Automated Crack Detection for the Inspection of Concrete Bridge Structures. In *International Conference on Human-Centered Intelligent Systems*; Springer: Singapore, 2021.

57. Wang, D.; Zhang, Y.; Pan, Y.; Peng, B.; Liu, H.; Ma, R. An Automated Inspection Method for the Steel Box Girder Bottom of Long-Span Bridges Based on Deep Learning. *IEEE Access* **2020**, *8*, 94010–94023. [CrossRef]
58. Mukherjee, R.; Iqbal, H.; Marzban, S.; Badar, A.; Brouns, T.; Gowda, S.; Arani, E.; Zonooz, B. AI Driven Road Maintenance Inspection. *arXiv* **2021**, arXiv:2106.02567.
59. Maeda, H.; Sekimoto, Y.; Seto, T.; Kashiyama, T.; Omata, H. Road damage detection using deep neural networks with images captured through a smartphone. *arXiv* **2018**, arXiv:1801.09454.
60. Siriborvornratanakul, T. An automatic road distress visual inspection system using an onboard in-car camera. *Adv. Multimed.* **2018**, *2018*, 2561953. [CrossRef]
61. Abdellatif, M.; Peel, H.; Cohn, A.G.; Fuentes, R. Hyperspectral imaging for autonomous inspection of road pavement defects. In Proceedings of the 36th International Symposium on Automation and Robotics in Construction (ISARC), Banff, AB, Canada, 21–24 May 2019.
62. Zhao, X.; Li, S.; Su, H.; Zhou, L.; Loh, K.J. Image-based comprehensive maintenance and inspection method for bridges using deep learning. In *Smart Materials, Adaptive Structures and Intelligent Systems*; American Society of Mechanical Engineers: New York, NY, USA, 2018; Volume 51951.
63. The NEXUS System and the On-Site Test Survey Introduction. Available online: https://www.youtube.com/watch?v=S7p6P3VG-40 (accessed on 27 July 2018).
64. Cox, D.R. The regression analysis of binary sequences. *J. R. Stat. Soc. Ser. B (Methodol.)* **1959**, *21*, 238. [CrossRef]
65. He, K.; Zhang, X.; Ren, S.; Sun, J. Deep Residual Learning for Image Recognition. In Proceedings of the 2016 IEEE Conference on Computer Vision and Pattern Recognition (CVPR), Las Vegas, NV, USA, 27–30 June 2016; pp. 770–778. [CrossRef]
66. Schwing, A.G.; Urtasun, R. Fully connected deep structured networks. *arXiv* **2015**, arXiv:1503.02351.
67. Manzoor, B.; Othman, I.; Durdyev, S.; Ismail, S.; Wahab, M.H. Influence of Artificial Intelligence in Civil Engineering toward Sustainable Development—A Systematic Literature Review. *Appl. Syst. Innov.* **2021**, *4*, 52. [CrossRef]
68. Grác, Š.; Beňo, P.; Duchoň, F.; Dekan, M.; Tölgyessy, M. Automated Detection of Multi-Rotor UAVs Using a Machine-Learning Approach. *Appl. Syst. Innov.* **2020**, *3*, 29. [CrossRef]
69. Xue, D.; Wang, X.; Zhu, J.; Davis, D.N.; Wang, B.; Zhao, W.; Peng, Y.; Cheng, Y. An adaptive ensemble approach to ambient intelligence assisted people search. *Appl. Syst. Innov.* **2018**, *1*, 33. [CrossRef]
70. Robbins, H.; Monro, S. A stochastic approximation method. *Ann. Math. Stat.* **1951**, *22*, 400–407. [CrossRef]
71. Zhao, W.; Meng, Z.; Wang, K.; Zhang, J.; Lu, S. Hierarchical Active Tracking Control for UAVs via Deep Reinforcement Learning. *Appl. Sci.* **2021**, *11*, 10595. [CrossRef]
72. Pantho, M.J.H.; Bhowmik, P.; Bobda, C. Towards an Efficient CNN Inference Architecture Enabling In-Sensor Processing. *Sensors* **2021**, *21*, 1955. [CrossRef] [PubMed]
73. Hennessy, P.J.; Esau, T.J.; Farooque, A.A.; Schumann, A.W.; Zaman, Q.U.; Corscadden, K.W. Hair Fescue and Sheep Sorrel Identification Using Deep Learning in Wild Blueberry Production. *Remote Sens.* **2021**, *13*, 943. [CrossRef]
74. Minnetti, E.; Chiariotti, P.; Paone, N.; Garcia, G.; Vicente, H.; Violini, L.; Castellini, P. A Smartphone Integrated Hand-Held Gap and Flush Measurement System for in Line Quality Control of Car Body Assembly. *Sensors* **2020**, *20*, 3300. [CrossRef]
75. Pham, T.-A.; Yoo, M. Nighttime Vehicle Detection and Tracking with Occlusion Handling by Pairing Headlights and Taillights. *Appl. Sci.* **2020**, *10*, 3986. [CrossRef]
76. Zhang, T.; Hu, X.; Xiao, J.; Zhang, G. A Machine Learning Method for Vision-Based Unmanned Aerial Vehicle Systems to Understand Unknown Environments. *Sensors* **2020**, *20*, 3245. [CrossRef]
77. Guo, Y.; Chai, L.; Aggrey, S.E.; Oladeinde, A.; Johnson, J.; Zock, G. A Machine Vision-Based Method for Monitoring Broiler Chicken Floor Distribution. *Sensors* **2020**, *20*, 3179. [CrossRef]
78. Hawkins, D.M. The Problem of Overfitting. *J. Chem. Inf. Comput. Sci.* **2004**, *44*, 1–12. [CrossRef] [PubMed]
79. Brehar, R.; Mitrea, D.-A.; Vancea, F.; Marita, T.; Nedevschi, S.; Lupsor-Platon, M.; Rotaru, M.; Badea, R.I. Comparison of Deep-Learning and Conventional Machine-Learning Methods for the Automatic Recognition of the Hepatocellular Carcinoma Areas from Ultrasound Images. *Sensors* **2020**, *20*, 3085. [CrossRef]
80. Azimi, M.; Eslamlou, A.D.; Pekcan, G. Data-Driven Structural Health Monitoring and Damage Detection through Deep Learning: State-of-the-Art Review. *Sensors* **2020**, *20*, 2778. [CrossRef] [PubMed]
81. Zhou, J.; Pan, L.; Li, Y.; Liu, P.; Liu, L. Real-Time Stripe Width Computation Using Back Propagation Neural Network for Adaptive Control of Line Structured Light Sensors. *Sensors* **2020**, *20*, 2618. [CrossRef] [PubMed]
82. Huang, Y.; Qiu, C.; Wang, X.; Wang, S.; Yuan, K. A Compact Convolutional Neural Network for Surface Defect Inspection. *Sensors* **2020**, *20*, 1974. [CrossRef] [PubMed]
83. Zhang, Q.; Liu, Y.; Gong, C.; Chen, Y.; Yu, H. Applications of Deep Learning for Dense Scenes Analysis in Agriculture: A Review. *Sensors* **2020**, *20*, 1520. [CrossRef] [PubMed]
84. Guo, Y.; He, D.; Chai, L. A Machine Vision-Based Method for Monitoring Scene-Interactive Behaviors of Dairy Calf. *Animals* **2020**, *10*, 190. [CrossRef] [PubMed]
85. Hochreiter, S. The vanishing gradient problem during learning recurrent neural nets and problem solutions. *Int. J. Uncertain. Fuzziness Knowl.-Based Syst.* **1998**, *6*, 107–116. [CrossRef]

applied system innovation

MDPI

Article

6G Enabled Industrial Internet of Everything: Towards a Theoretical Framework

Prafulla Kumar Padhi [1,*] and Feranando Charrua-Santos [1,2]

[1] Eletromechanical Department, University of Beira Interior, 6201-001 Covilha, Portugal; bigares@ubi.pt
[2] C-MAST—Centre for Mechanical and Aerospace Science and Technologies, University of Beira Interior, 6201-001 Covilha, Portugal
* Correspondence: prafulla.k.padhi@ubi.pt

Abstract: Currently, the deficiencies of the 5G mobile system as an enabler of Internet of Everything (IoE) applications are stimulating global research activities to focus on the sixth generation (6G) wireless system. The potential of IoE is enormous and growing exponentially. With the dawn of the fifth industrial revolution, IoE is transposing into industrial Internet of Everything (IIoE) projects which are complex and are eventuating to become a prominent technology for all industries offering new opportunities. This study embodies a synthesis of 6G, IoT, IoE, IIoE exhaustive literature review advancing knowledge to facilitate theory development. For the first time, a novel theoretical framework for the 6G-enabled IIoE (henceforth referred to as 6GIIoE) system was developed. Judiciously, sequential methodology is best suited for this emerging discipline research to create significant new knowledge in the literature contributing eternal insights to expound valuable contexts to ruminate significant findings. The theoretical framework developed recognizes 6GIIoE priority areas, challenges, and applications bestowing a guide for 6G-enabled IIoE initiatives divulging opportunities for future research activities.

Keywords: 6GIIoE priorities; 6GIIoE challenges; 6GIIoE applications; information system; sequential methodology; 6GIIoE theoretical framework

Citation: Padhi, P.K.; Charrua-Santos, F. 6G Enabled Industrial Internet of Everything: Towards a Theoretical Framework. *Appl. Syst. Innov.* **2021**, *4*, 11. https://doi.org/10.3390/asi4010011

Received: 25 December 2020
Accepted: 5 February 2021
Published: 11 February 2021

Publisher's Note: MDPI stays neutral with regard to jurisdictional claims in published maps and institutional affiliations.

Copyright: © 2021 by the authors. Licensee MDPI, Basel, Switzerland. This article is an open access article distributed under the terms and conditions of the Creative Commons Attribution (CC BY) license (https://creativecommons.org/licenses/by/4.0/).

1. Introduction

The connection of embedded devices, sensors and machines via the Internet has revolutionized the way people live, work and play. In 1999, Kevin Ashton of Massachusetts Institute of Technology (MIT—Auto ID Center) used the term "Internet of Things—IoT" [1]. The IoT connects all the devices, sensors, or machines with each other in a physical system with the Internet or Intranet or wireless communication such as 6G [2] to communicate with each other for user friendly services. The IoT era is being reinstated by the IoE [3] era, where ubiquitous connectivity is possible. IoE connectivity is a broad topic that provides different solutions for massive data to be transferred to the edge of the network. With the emergence of the IoE era, industrial control systems (ICSs) [4] (p. 1) are shifting into IIoE [5] and includes all use cases, challenges, and applications for all industries. The evolution of mobile communications has been driven primarily by continual incessant demand for mobile communication capacity for higher data rates. During the last decade, a cardinal paradigm transformation happening gradually from the rate-centric enhanced mobile broadband (eMBB) services towards ultra-low latency communications (URLLC) [6] (p. 57065) and high reliability related to IIoE system that connects billions of devices, censors, machines, and people.

The fifth generation (5G) wireless communication standardization is over, and deployment has started in many countries globally. The development of 5G [7] mobile systems reveals the immanent limitations relative to its original engagement as an enabler for internet of everything (IoE) applications. A smattering of technologies will be developed to play a pivotal role towards the end of the 6G research process to attain comprehensive specification towards a 6G global standard. One such eye-catching technology is quantum computing and communications (QCC) [6] (p. 57067) that can provide long distance networking and security. At present, major research pursuits are focused on the quantum technologies realm which will intersect with 6G deployment. Starting in the 2030s, to achieve sustainable competitive edge of wireless communications applications, some industries and academia have commenced to conceptualize the sixth generation (6G) foundation for the emerging communication needs. Even though there is 6G enthusiasm, the basic architectural and performance components of the 6G mobile system remain largely undefined. Hence, one of the objectives of this study is to outline a novel holistic, efficacious, cogent vision that construes the conception of a 6G enabling IIoE system.

6G, in our opinion, is not a sheer expedition of additional high-frequency bands wireless spectrum, but it will be an authentic concurrence of breathtaking latent services and emerging technological trends. The role of 6G communication will revolutionize IIoE with mobile Teraband reliable low latency communication (MTRLLC) [6] (p. 57069) and massive network capacity accelerating the rate of data download, enabling the of use real-time data in the industrial operations so that data can be easily shared remotely. A massive volume of data can be shared between the machines via edge (cloud) platform and can be downloaded as well as uploaded at the edge of the 6G network to support connected intelligent industrial applications. The information communication technology (ICT), operational technology (OT) and emerging 6G communications are going to converge to form the 6G IIoE system. The 6GIIoE system is an emerging high-performance growth engine that has inspired global research and development (R & D) enthusiasm and opened application opportunities for the "connected intelligent industrial world" (CIIW) [8] (p. 3), [9], [10] (p. 102), [11] for the emerging fifth industrial revolution.

The literature review bestowed a synthesis of 6G, IoT/IoE priority areas, challenges, applications, architectures, and their potential influence in industries. Drawing from the exhaustive review, we developed a novel theoretical framework for the adoption and execution of 6GIIoE system.

1.1. Organization and Roadmap

The rest of the paper is organized as follows: In Section 2, we present an exhaustive literature review and synthesis. In Section 3, the theoretical foundation outlines the 6GIIoE boundaries, relevant theories that support the theoretical framework. The sequential methodology found to be the best-suited for this study is described in the Section 4. The 6GIIoE paradigm is presented outlining 6G technologies enabler, new service classes and IIoE overview in Section 5. A 5-layer 6GIIoE modular architecture is proposed in Section 6. A list of 6GIIoE priority areas, applications and challenges is described in Section 7. In the Section 8, a novel theoretical framework and contributions of the framework components are discussed. Consolidated lessons learned are explained in Section 9. We outline the future research agenda and research gaps in Section 10. Recommendations are offered for the academia and practitioners in the Section 11. Section 12, finally, concludes that the developed theoretical framework falls under a theory's phenomenon that has not been covered by prior theories.

1.2. Background and Problem Formulation

There has been a scattering of theories proposed and developed in the IoT domain. Our exhaustive literature review demonstrates that theoretical framework on the 6GIIoE system approbation and execution is non-existent in the literature.

Researchers place an eminent value on theoretical frameworks. There is prevailing concern that the information systems (IS) [12] discipline on developing theories does not exist relevant to the discipline that deals with 6GIIoE system, operating system (OS) software solutions and topics, namely philosophy, anthropology, socioeconomic, psychology and ethical dimensions. The IS community of researchers and practitioners does not even share a common definition of the term 'information system (IS) and does not have an appropriate influential theory to comprehend information handling activities in human structures. To date, the IS discipline is not well understood by academia in other fields, and even in fields related to IS. Furthermore, the development of new theories and the depuration of prevailing theories have been relatively ignored within the IS field [12] (pp. 2–4, 20). Confirming the lack of research publications on theory advancement in IS discipline related to 6G and IIoE, an extensive review of literature produced no research papers that deals with 6GIIoE theoretical framework in a multi-dimensional value co-creation approach. The importance of the 6GIIoE system topic is gaining traction in the industrial applications bestowing enormous value on co-creation. A pressing issue in this regard is how industrial organizations can adapt their traditional structures and execute the innovation in the 6GIIoE field to create value.

To fill the above-mentioned void, this study bestows a comprehensive understanding to offer holistic guidance to develop a theoretical framework.

1.3. Key Contributions

- A comprehensive contribution of our research work is the creation of a novel theoretical framework on the 6GIIoE system approbation and execution which is non-existent in the literature.
- Sequential methodology is employed, for the first time, to exploit adaptive approaches of various applications that identifies the future directions in this field of 6GIIoE system research and development.

2. Literature Review and Synthesis

An exhaustive review of relevant literature is paramount in any academic research endeavor. A persuasive review generates a sound foundation for propelling new knowledge. It provides theory development and identifies where the research gap exists. Despite a great deal having been written regarding IoT, there is a captious deficit of papers on IoE research-based studies and IoE literature reviews. A detailed review of IS innovation, theoretical framework development [10] (pp. 97–108), [13] (pp. 10–12), [14] related to IoT/IoE was done. A void in the literature exist particularly regarding the 6G-enabled IIoE studies. In the Table 1 (summary of the literature review), we demonstrate the research paper classification of the relevant research papers reviewed that confirms our view that 6G-enabled IIoE research studies are non-existent. The literature review bestows an analysis of 6G, IoT/IoE and the relevant priority areas to focus, applications, challenges, and architectures and their influence in industries. Drawing from the exhaustive literature review, we developed a novel theoretical framework for the adoption and execution of 6GIIoE system.

Even though the focus was on the 6G-enabled IIoE system, literature related to 6G and IIoE priority areas, challenges, applications, and architectural framework strengthened through the literature review was utilized for reinforcing research and to help in the development of the theoretical framework. We utilized a strategic approach to explore literature relevant to 6G, IoE, and IIoE, selecting these academic research works included in the literature review. The following steps were taken to determine relevant literature (i) using the key terms 6G, IoT, IoE, IIoE, Google Scholar (database-nature and timeliness of the topic), (ii) Google and Firefox was also used for industry white papers related to 6G, IoE, IIoE and various industrial applications, (iii) exploration of the leading journals such as IEEE, Science Direct (Elsevier), Emerald Insight was done, (iv) a search of academic research databases was pursued for relevant articles with the keywords IS, 5G, 6G, IoT IoE, IIoT, IIoE, theories, (iv) articles and conference proceedings listed in the references are related closely to the theme of the research paper, (v) summarized version of the articles and industry papers related to the theoretical framework and security/privacy matter of the topic.

Table 1. Summary of the Literature Review.

Reference	Publication Year	Research Direction	Main Focus
Alsharif et al. [2]	2020	6G Research Activities	6G Networks
CISCO [3]	2020	IoE	IoE Economy
Singh et al. [5]	2020	IoE/ Blockchain	IoE/Fog Computing
HUAWEI [8]	2020	Digital solutions	IoT
Siemens [9]	2020	IoT/2050	Industrial IoT
Nord et al. [10]	2019	IoT/Theoretical Framework (TF)	Systematic Review
Falkenberg et al. [12]	1995	Information System (IS) Theory	IS Limitation
Aquilani et al. [13]	2020	Theoretical Framework (TF)	TF
Giordani et al. [15]	2020	Few 6G use cases & requirements	6G
Dang et al. [16]	2020	6G Vision	Security & Privacy
Huang et al. [17]	2020	Survey Towards 6G	Wireless Evolution
Gui et al. [18]	2020	6G New Horizon	6G Security
Chowdhury et al. [19]	2020	6G Applications	6G Research Directions
Mahmood et al. [20]	2020	6G Networking	Machine Communications
Janbi et al. [21]	2020	6G/ Smarter IoE	IoE/6G/AI
Sekaran et al. [22]	2020	6G/IoT	6G/IoT Automation.
Wang et al. [23]	2020	Industrial Control System/IIoT	IIoT
Yang et al. [24]	2019	Potential 6G Techniques	6G Techniques
Saad et al. [25]	2019	6G Vision & Trends	6G Services
Clazzer et al. [26]	2019	IoT Applications in 6G proposed	6G/IoT
Tariq et al. [27]	2019	Speculative 6G study	6G
Zhang et al. [28]	2019	6G/Super IoT	6G/IoT Aspects
Eppi Center [29]	2019	IoE	IoE
Dai et al. [30]	2019	Blockchain/IoT	IoT

This review identified a substantial number of research publications that bestowed deep insights relevant to this research paper. Despite several research publications on 6G, IoT/IoE, there is a severe scarcity of papers on 6G-enabled IIoE system research-based studies. To our surprise, theory as related to 6G-enabled IIoE system adoption and execution is non-existent in the literature. We have outlined the review of the research papers as indicated by the citations below for various relevant topics and the selection of the literature criteria. A synthesis of our literature review identified that there were no papers considered by authors in the development of a theoretical framework on 6G enabled IIoE system adoption, execution, and recommendation research agenda. In the next section the study reflects on the exhaustive review of the recent literature on the 6GIIoE system including emerging 6G networking and technologies, existing IoT/IoE, information system (IS) discipline related to various theories as the theoretical foundation, sequential methodology, CIIW, and the contributions to the theories.

2.1. Selection of Literature Criteria

The exhaustive literature review for this study was done according to the following criteria:

- Well established databases (Scopus, IEEE Xplore, MDPI), peer-reviewed double blind reputable international journals and books from Emerald Insight, Elsevier, Wiley, Science Direct, MIT Technical Review publications, Nature, Springer, and Google, were the basis of this research.
- EPPI-Reviewer includes systematic reviews, meta-analyses, 'narrative' reviews and meta-ethnographies containing over a million items—version 4.11.5.2 (16 November 2020) [31].
- Peer-reviewed international journals and books published from the years 1980 to 2020 were consulted.
- Focus and Emphasis was given to recent books and journals to review and analyze the relevant information.
- The web search engine (Google Scholar) that indexes the full text scholarly articles.
- International Conferences proceedings.

- PhD Thesis (Published Ones).
- Market and Scientific White Papers on connected intelligent industrial trends, perspectives, and applications.

2.2. Essential Insights and Relationships between Relevant Topics

A comprehensive review of the literature helped in identifying the key dimensions of the topics closely relevant to 6GIIoE systems' thinking. The literature review was divided into the following segments that bestows essential insights and the relationships between the relevant topics:

2.2.1. Theoretical Framework Development/Information System (IS) Discipline

Information systems (IS) are complex multi-dimensional phenomena [12] (p. 449) and signifies the use of various disciplines to study the development of theory based on the diversity of contexts within the relevant technologies of information. Although diversity in information systems practice, research has evolved and gained importance, but no well-accepted common theoretical basis for information system. Furthermore, there is no central corpus of a well-accepted theory.

Based on the content outlined in the sub-Section 1.2 to reaffirm our perspectives, the following scholars' points of view, as a validity, on the IS discipline are as follows: (i) Swanson [32] (p. 6) concludes that current theories describe little about IS innovation, (ii) Avgerou [32] (p. 2) contends that information system is a relatively unexplored subject, and (iii) Fichman [32] (p. 6) explains signs of exhaustion in the current research.

2.2.2. 6GIIoE System

1. 1.6G system requirements and technologies are going to bestow the following new theme: (i) novel man-machine interfaces; (ii) ubiquitous distributed computing and the edge (cloud) services; (iii) multi-sensory data fusion to create eXtended reality (XR-augmented reality-AR/mixed reality-MR/virtual reality-VR) [25] (p. 1) immersed experiences; and (iv) the physical world will be driven by precision sensing and actuated control. The IoE is an organic, complex, evolving ecosystem with limitless boundaries. 6GIIoE architectures in the literature is non-existent. This study argues for the support of 5-layer architecture for 6GIIoE ecosystem and the trends provides the formulation of design for the next generation infrastructure [6] (pp. 57063–57070).
2. CIIW is the fusion of 6G, IoE and other emerging relevant technologies promises new possibilities, opportunities, services and immersive user experiences ubiquitously offering connected intelligent industrial applications such as connected intelligent factories (CIF), connected intelligent transportation (CIT), connected intelligent cities (CIC), connected intelligent robots and drones (CIRD), connected intelligent food and beverage (CIFB), connected intelligent retail (CIR) and other connected intelligent applications. Based on the 6G connectivity vision, as outlined below in the Section 5, and with 6GIIoE applications, the CIIW is going to revolutionize endowing digitization towards personalization for every industry to provide enormous benefits.

3. Theoretical Foundation and Approach

3.1. 6GIIoE Foundations and Boundaries

Even though the 5G mobile system can support various IoT services, it may not be able to completely fulfill the requirements of IoE novel applications. Hence, 6G mobile systems are conceptualized to vanquish limitations of the 5G wireless system. IoE-based smart services raises the need for 6G wireless networks and comprises four pillars: data, things, people, and processes that are intelligently connected and describes a world where billions of objects have sensors to distinguish measurement and appraise their situation. IOE is also a concept that broadens the IoT prominence on machine-to-machine (M2M) communications to define an augmented complex system that also encircles people and processes connecting over private or public networks utilizing proprietary and/or standard protocols. IoE applications are categorized from sensor tools, smarter mobile devices, machine learning (ML) systems, interfaces used for remote appliances to machine learning systems to various types of distributed intelligent automated hardware. The Industrial Control Systems (ICS), which are closely related to people's lives, play a vital role in the development of the IIoE whose security affects the entire IIoE. ICS is connected to the Internet and unprotected in the cyberspace. To protect these precious assets, intrusion detection systems (IDS) have drawn much attention.

3.1.1. Four Pillars of IOE

Currently, IoT and 5G have started their honeymoon and are growing shinier applications such as remote surgery, smart factories, and unmanned vehicles. The 6G's requirements will redefine the mobile system architecture and will empower use cases. Distinguishing that connectivity as the foundation, 6G will make future technological breakthroughs redefining the business models around data consumption because it will be not just a physical standard but also a virtual one providing new opportunities for novel networks services. In the quest to metamorphose industries for the digital era, IoE is one of the most significant components of ICT foundation.

IoE is the intelligent connection of people, process, data, and things with billions of objects to detect capacity and appraise their standing using various protocols. The IoE concept comprises of communication between machine to machine (M2M) communications, machine-to-people (M2P), and people to people (P2P) interactions. IoE is also regarding all the information collected from the four primary pillars (people, process, data, and things).

Without "people", everything else would be pointless and "people" themselves become the nodes on the Internet. "People", in the IoE, are to connect to the Internet in numerous ways through a myriad of devices or sensors. The evolution of is Internet toward the IoE, where everyone will be connected in relevant ways.

In the connected world of IoE, "process" plays a key role in how each of these pillars—people, data, and things—interact with each other to deliver value for bestowing immersive user experience. With the exact process, connectivity becomes accordant and value is added due to the precise information being delivered to the right person at the right time in the appropriate manner.

The IoT is a vital pillar of the IoE. "Data" is still the foundation to be created, stored, then transmitted from one place to another and will continue to be at the epicenter of the internet paradigm. Currently, the number of new things connected to the Internet exceeds the number of new users connected to the Internet. Hence, we contend that the third generation "Internet for the people" became the fourth generation "Internet about things for the benefit of the people".

"Things" are composed of physical items like devices, sensors, and machines that are connected to each other and to the Internet. "Things" sense data evolve into being context-aware and bestow information to help machines and people in the IoE endowing relevant valuable decisions.

In summary, we have articulated about how the four pillars of IoE foster innovation in the new technology to empower user experience. As the internet progresses toward IoE, people will be connected in valuable ways. We believe that every challenge is an opportunity in dissimulation. Very often, huge opportunities emerge inscribing from colossal challenges. IoE will change industries and our daily lives. When the history of IoE is penned, its success or lack of success will be obstinate by: what is the benediction of IoE to humanity?

3.1.2. 6G Challenges

6G is not necessarily a single technology but a collection of mobile system technologies. A fundamental concept of the 6G future opportunity is the principle of 'New Services', from spectrum to infrastructure. Tiny cell network densification is the dominant theme of 6G wireless evolution. Densifying networks, while avoiding increasing inter-cell interference, is a key challenge. Additional key technical aspect of 6G networks will be the use of high-frequency spectrum, specifically the use of millimeter wave (mmW) spectrum with extreme low latency.

6G cost modelling is a challenge that allows one to compare the difference between data traffic demand and network costs for different network scenarios. The costs of 6G infrastructure densification depend heavily on the required throughput density, periodic interest rate, and tiny highly dense base station infrastructure investment.

6G considers the spatial rollout of telecommunications technologies over time. Often researchers focus on just one key aspect, such as modelling spatial viability. In mobile systems, there has been limited spatial modelling of the potential rollout of new technologies.

Network densification and additional spectrum will play a key role in delivering 6G, which will have a significant influence on the delivery economics. There has been limited modelling of both the spatial and temporal dynamics of the rollout for 6G. While the current research on 6G networks provides insight into the vague technical specifications of various technologies, little emphasis has been placed on the rollout implications for different global infrastructure strategies.

3.2. Discussion on Theories Related to Information Systems/IoT

Corroborating the spasmodic research publications on theory advancement in IS/IoT discipline, some of the of research papers in this domain are the following:

2006—Leidner and Kayworth promulgated a survey and review of theories on Information Technology and culture conflict [33].
2012—Evaluating and developing theories in the IS discipline was written with riche in content and published by Weber [34].
2016—Wiener et al. developed an expanded theoretical framework for Information Systems Projects [35].
2018—Chaudhuri and Cavoukian published a framework for IoT privacy [35].
2018—Hsu and Lin conducted a study examining factors seen as contributing to IoT service adoption using a research model based on the value-based adoption model [36].

3.3. Relevant Theories Supporting 6GIIoE

- The communication theory studies the process of information, interdisciplinary disciplines of interpersonal communications, psychological paradigm, philosophical and social dimension [37].
- IoT interface theory defines the interaction between the thing, service, and an accessor for which it is formalized as an interface automaton. For 6GIIoE applications, where "everything" and services interact with the physical world [38].
- The systems theory applied to 6GIIoE systems analysis applications. One of the vital mechanisms of systems analysis is systems thinking, enabling the contouring of systems from a broad perspective rather than specific events in the system [39].
- The Theory of Transparency where perpetual experience is often said to be transparent and be aware of the properties of the objects around us. This is an introspective fact. A transparency theory bridges the gap between them. The fundamental of transparency requires that the actions and decisions of industries be open to inquiry [40].

4. Methodology

Notwithstanding 6G still being nascent in its research and development, a quest to support 6G R&D decision-making in government and industry has emerged. Despite significant technological and economic uncertainty, exploration of how the potential 6G rollout can be pursued both spatially and temporally is in policy formulation. Therefore, the coverage, cost, and rollout consequences of 6G networks globally are explored by extrapolating 5G characteristics. In this study, we focus on an appropriate methodology that covers 6G vision connectivity and its impact of annual capital intensity, infrastructure sharing and the end-user speed. Furthermore, varying annual capital intensity or deployment of a shared tiny high density cell network that can greatly influence the span of time to reach the 90% threshold benefiting ubiquitous connectivity globally is explored. Additionally, by integrating new and existing spectra, a 6G network capable of achieving Tera Bit speed with massive network capacity and extreme low latency is considered for universal service globally.

Based on the 6G challenges outlined in the sub-Section 3.1.2, the following are some of the research opportunities based on the sequential methodology that should be investigated: (i) cost modeling, (ii) ways to reduce the costs of 6G infrastructure rollout is via infrastructure sharing, (iii) spectrum and coverage obligations—frequency bands that are yet to be defined and allocated, therefore spectrum costs for 6G are unknown at present, (iv) sensitivity analysis to identify the cost boundaries which will have influence on 6G coverage and capacity, and (v) 6G policy implications.

A sequential methodology [41], Ref. [42] (p. 639) process enables the analysis team to concisely define the requirements, then planning for design prior to execution and ultimately focus on network densification and spectrum. A prospective study to evaluate 6GIIoE system is an integral part of developing and launching 6G network services and IIoE system technologies. In this study, we have proposed sequential approach to conduct analysis, design, implementation, and test to allow adaptations or modifications to aspects of a 6GIIoE system to validate the integrity of the trial. The application of sequential methodology in 6GIIoE system trials has the possibility of significantly improvement of the flexibility, and efficiency. To further advance the performance of 6GIIoE system trials, the sequential method can exploit novel adaptive approaches of various applications identifying the future directions in this field of research and development. The sequential methodology for the 6GIIoE system, as shown in Figure 1a,b, is adopted phase by phase: (i) the analysis, (ii) the design, (iii) implementation, and (iv) the testing.

(a)

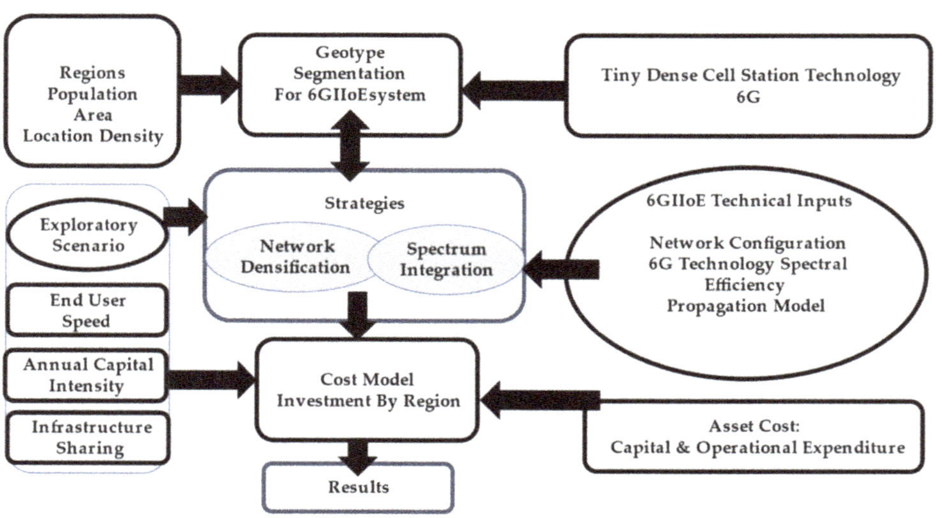

(b)

Figure 1. (a) Phase by phase sequential methodology process. (b) Sequential Methodology Process.

Each part of the sequential methodology, step by step, is explained below to deal with the sources of uncertainty:

4.1. Exploratory Scenario

A set of exploratory scenarios should be utilized to demonstrate the temporal dynamics of the rollout and potential costs of research and deployment with respect to the 6GIIoE system. The exploratory scenario captures the uncertainty associated with the key technological, economic, and regulatory changes that focus on the relevant areas such as capital intensity, infrastructure sharing, and end-user speed.

4.2. Strategies

Two primary strategies need to be considered for network dimensioning and end user speed required in the exploratory scenario. The first option is to integrate spectrum into existing high-density tiny cells for achieving total geographic coverage in 6G to provide more bandwidth in areas of high demand. The second option is where network densification consists of 6G tiny high-density cell station rollout to be deployed using specific THz spectrum.

4.3. Geotype Segmentation

Depending on the geographical region, geotypes to be created to represent the key supply-side variables that affect rollout costs. Geotypes are classified based on population density as per the geographical region. The forecasted population coverage is derived from assumptions to interpolate the population coverage for a specific annual capital intensity.

4.4. Infrastructure Sharing

An overview of the cost model should be considered for a non-virtualized 6GIIoE infrastructure. The capital expenditure is considered based on annual capital intensity in relation to the pace of 6GIIoE system roll out. When 6GIIoE equipment costs are unknown, future generations of network equipment with enhanced performance tend to be a similar estimate in price to those of 4LTE advanced or 5G systems.

4.5. Dimension of Networks

Each exploratory scenario was to be done using a model to calculate the minimum number of tiny high-density cell stations required.

To attain the results, the rollout of a 6GIIoE system under business-as-usual conditions needs to be illustrated. Then exploration of the temporal dynamics under exploratory variables including annual capital intensity, degree of infrastructure sharing in deployments, and required end-user speed executed by using the cumulative population covered to represent the spatial dimension for a period.

5. 6GIIoE System Paradigm

5.1. 6G Connectivity Vision

6G connectivity vision ought to be those needs that 5G cannot satisfy and can be defined with the following: (i) ubiquitous, (ii) intelligent, (iii) deep, and holographic. 6G mobile system [2] is expected to provide a massive coverage that will enable users to communicate ubiquitously with Tera bit per second speed that will re-shape the wireless communication evolution. 6G mobile systems will evolve to offer space/air/ground/sea integrated communication and the wireless tactile network as shown in Figure 2. 6G spectrum, technology, and features goal are shown in Table 2.

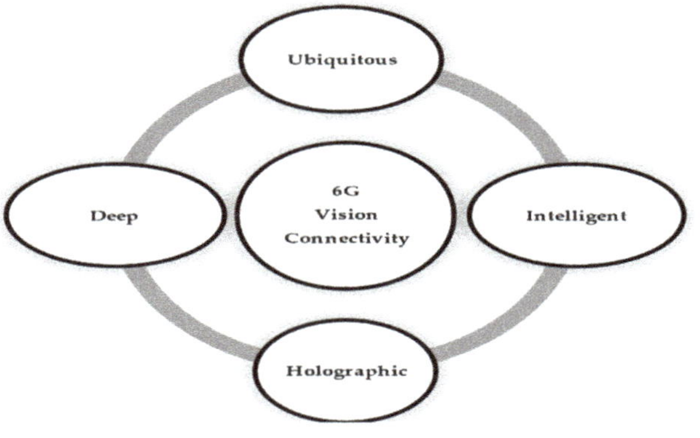

Figure 2. 6G Connectivity Vision.

Table 2. 6G Spectrum, Technology, and Features Goals.

Spectrum	Technologies	Features
Tera Hz	AI Based Wireless Communication	Space/Air/Ground/Sea
Visible Light Communication (VLC)	Advanced Signal Processing	Integrated Communication
Flexible Spectrum	Compressed Sensing	Wireless Tactile Network
(Free Full Duplex and Spectrum Sharing)	Very Large-Scale Antenna Processing for THz, New Channel Coding.	

1. From deep coverage to "Deep connectivity" with the following characteristics: (i) Deep Sensing: Tactile Internet; (ii) Deep Learning/AI, and (iii) Deep Data Mining; Deep Mind: Telepathy, Mind-to-Mind Communication.
2. High fidelity AR/MR/VR will be ubiquitous, and holographic communication and display can also be carried out anytime and anywhere to enjoy a fully immersed holographic interactive experience. This communication vision of called "Holographic connectivity" with the following features: (i) holographic communication, (ii) high fidelity AR/MR/VR, and (iii) AR/MR/VR with seamless coverage anytime, anywhere.
3. The following are the characteristic of "Ubiquitous Connectivity": 3-dimensional coverage and connection, meaning the connection of anytime and anywhere, including Integration of Space-Air-Ground-Sea communication. Comparing "Deep Connectivity" and "Ubiquitous Connectivity" versus "Deep Connectivity" the latter emphasizes the depth of the connected object, while the former affirm the breadth of the distributed area.
4. "Intelligent Connectivity" is the nucleus of the 6G network, while "Deep Connectivity", "Holographic Connectivity" and "Ubiquitous Connectivity" establish the 6G network trunk.

The above four characteristics combined make the future 6G network's organic "soul".

5.2. 6G New Service Classes

6G new services [2,25], applications with performance indicators should be introduced to offer mobile Teraband reliable low latency connectivity (MTRLLC), massive ultra-reliable, low latency communications (mURLLC), human-centric services (HCS), and multi-purpose services (MPS) as shown in the Table 3.

Table 3. 6G New Services, Performance Indicators and Applications.

6G New Services	Performance Indicators	Applications
MTRLLC	Rate Reliability Latency Requirements including Mobile	IoT/IoE/IIoT/IIoE
mURLLC	Environment Energy Efficiency, Reliability & Scalability Massive Network Capacity	XR (AR/MR/VR) CIIW Legacy eMBB/URLLC
HCS	QoPE (Quality of Physical Experience)	wireless computer-brain interactions (WCBI)
MPS	Stability, Sensing, Latency, Energy	Haptics, Telemedicine & Mapping

5.3. IoE/IIoE Brief Overview

CISCO [3] have used the term IoE and the foundation is built upon the "four pillars" of people, data, process, and things, as shown in the Figure 3. IoE extends industrial processes to enrich the lives of people and the intelligent connection of people, process, data, and things and describes a world where billions of objects have sensors to detect, measure and assess their status using standard and proprietary protocols.

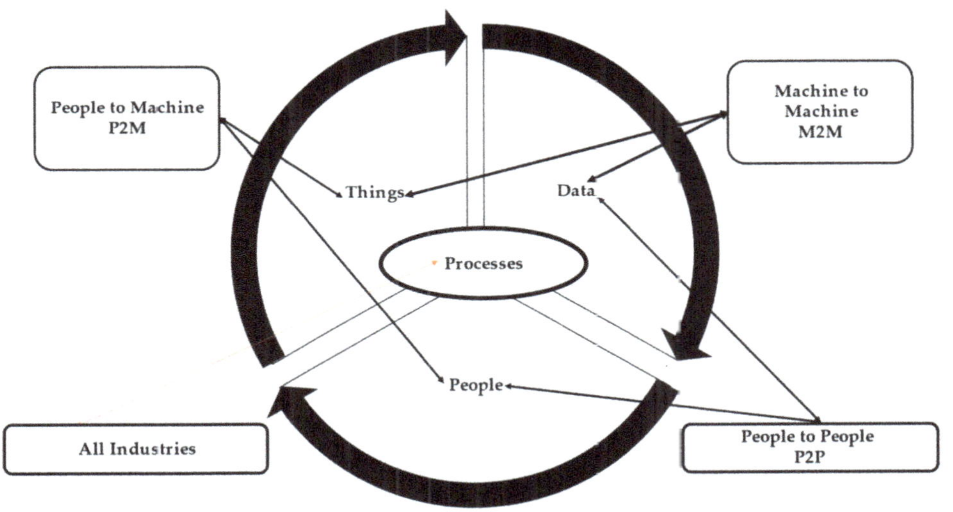

Figure 3. Industrial Internet of Everything.

Globally, people from different socio-economic backgrounds, religions and diversity using increasingly computing devices and the Internet because of the successful deployment of various innovative and killer applications based on IoE. This scenario has created a phenomenon of pervasive ubiquitous computing, universal usability or user-friendly interface design, and cross-cultural usability. Exploiting the cross-cultural universal usability and information system (IS) issues focusing on the world wide web (WWW) and mobile interactive dialogue using IoE becoming the trends open Innovations and requires further research in depth. This is a significant trend by users in how the 'IS' is utilized currently. Hence, the success of IoE/IIoE must consider the impact of cross-cultural usability by deep research requiring additional valuable research time and effort.

In this study, IIoE is defined: Smart devices or sensors are embedded into the intelligent systems for the product life cycle to build an information infrastructure empowering industry to expand their existing applications and even develop new ways to operate production. The enveloping of people, processes, data, and things is at the heart of IoE that creates the 'value' with respect to user experience. The IoT growth waves are leading to the eventual complete IoE. The advantages involve the use of IoE in supporting the attainment of socioeconomic and environmental goals. With each successive wave of added features and greater network connectedness, this leads to the IoE with many novel opportunities as well as risks. The 6GIIoE system has the potential to extract and analyze real-time massive data from the billions of devices or sensors or machines connected via 6G network and then to apply to aid all industrial applications. A study by CISCO predicted that $14.4 trillion may be exploited within ten years (2013–2023) by implementing IoE with M2M, M2P and P2P [3].

IIoE projects involve propagation of sensors, devices, complex processing events, and APIs. Project implementers overinflate the built-in integration of IoE platforms without due deliberation to the intricacies involved. The authors suggest the following strategic areas for industries to customize their mobile re-architecture: (i) Mobile App development services; (ii) Social Media Integration; (iii) IoE Integration Services. IIoE projects are complex and are emerging fast to become a popular technology for all industries.

Concurrently, intelligent offices are seeking to leverage smooth and streamlined operations for all industries. Predictive servicing is becoming the gold standard in customer service, therefore IIoE remote monitoring for the predictive servicing provides peace of mind and a meaningful value proposition:

- Sensors need to be installed at the customer's premises to be monitored visually via relevant applications.
- Publication of user data to the Cloud to build an overall picture of the user's consumption habits, and trends.
- Remote monitoring app to alert customers when the consumable falls below a set level and improve customer experience.
- IIoE remote monitoring with the consumables supplier's ERP (Enterprise Resource Planning) system for scheduled delivery.
- Specialize the situation for each specific industrial application, product access and purchase.

5.4. IIoE Trend: Massive Data Analytics and Predictive Maintenance

IoE sensors, devices and endpoints are growing exponentially in size, form, function and "massive data" remains the foundation and mettle of IoE applications and software. This scenario empowers to identify novel ways of improving and leveraging the interoperability of people, data, processes, and assets. Looking into the future, by 2030 over hundreds of trillions of gigabytes of data will be generated by IIoE that enables all types of applications including deep learning of manufacturing operations globally. Massive data analytics (MDA) is a growing trend phenomenon. IIoE analytics of massive data is fundamentally different from current data management paradigms to assemble the massive data in a unique manner providing meaningful useful insights and optimization of the connected intelligent factory (CIF) operation. The efficiency of an IIoE system efficiency grows if data analysists can analyze in a timely fashion the incoming data and operationalize those insights, and massive data sharing concepts among competitors will be one of the ideal ways to make sense of this massive volume of information. With the disposition of AI and/or machine learning (ML)-based predictive maintenance capabilities, remote massive data collection IIoE platforms via 6G communications can quickly update machinery and devices enabling minimizing down time substantially and reducing cost significantly.

5.5. Role of 6G Revolutionizing IIoE

6G will become the "eyes and ears" of all IIoE systems since it provides real-time massive data collection and analysis. The pivotal role of 6G communications in IIoE, referred to as 6GIIoEsystem, will empower and precipitate fifth industrial revolution mission and offers the following advantages to create significant value:

- The extremely low latency, massive network capacity, high-speed (>100 Gbps) for massive data exchange hastens the realization of digitization towards personalization.
- Enables reducing operation time due to extremely low latency and increase in massive data throughput leading to enhance process productivity with quality control and leverage optimization process for expanded connected intelligent automation.
- Can attenuate the complexity of supply chain networks, and warehouse management, helping businesses to efficiently operate in dynamic industrial environments.
- Endows the next era of AI driven enterprises powered by 6G connected edge (cloud) computing percussing into massive volume of data to unleash insights that would otherwise not be attainable.
- Entrusting the sustainability of all industries managing the vast volume traffic to hyper e-commerce by enhancing 6G network accessibility at a faster pace accelerating online purchases, order placements, customer delivery and have the potential to disrupt on-site job functions through virtual meetings as well as remote working environment.

5.6. Why 6GIIoE Is So Appealing?

An unprecedented proliferation of new IoE services for industrial applications are emerging such as extended reality (XR) services (augmented, mixed, and virtual reality (AR/MR/VR)) to telemedicine, flying vehicles, wireless computer-brain interfaces (WCBI), and connected intelligent systems for all industries. These emerging IIoE services will need computing functionalities and end-to-end co-design of communication which has been overlooked. To exploit these unique challenges for the new breed of services require massive bandwidth frequencies beyond sub-6 GHz to transform wireless systems to meet the requisite IIoE scenario. To overcome these future challenges to catalyze the deployment of new IoE services require 6G) wireless system. The drivers of 6G will be a convergence of past trends and emerging trends that include new services and the recent revolution devices such as smart wearables, implants, AI, 3D environmental mapping and XR devices.

There is a growing need to connect exponentially more devices, more sensors and install things like real-time cameras to perform massive data analytics (MDA). It is becoming vitally important to achieve better reliability, greater scalability, performance, easier accessibility, broader reach with a lower cost of ownership and need to increase connectivity with ultra-low latency. 6G is an ideal solution and can make a real difference for industries seeking augmented reality (AR) and virtual reality (VR) to help operators service equipment. To interact with robots is change and there is a need for real-time video to do spatial analytics for man and machines interactions to unlock a new level of intelligence with a new layer of connectivity with 6G that helps drive better business outcomes. 6GIIoE is a realm that will dramatically shape the manufacturing landscape, the way factories operate, and workforce needs. While the following financial rewarding possibilities of CIF certainly appeal to the dreamers and innovators, the clear payoff ensures that it is a pragmatic path that industry decision-makers will keenly follow:

- Maximizes revenue growth.
- Reduces operating cost.
- Increases asset efficiency.

6. 6GIIoE Architectural Framework

Based on the literature review standards, specifications, and reference architectures, (specifically) of 6G-enabled IIoE system is non-existent. Scholars are working diligently to develop 6G standards and specifications. We are of the opinion that global 6G standards and specification by the ITU (International Telecommunication Union) will emerge prior to the year 2030. 6G communication architecture should enable IIoE devices, sensors, machines to not only connect to the Internet but also to communicate with one another autonomously. The current literature describes several 6G and IoT/IoE architectures [43,44] (p. 134010). In contrast to the 3 and 4-layers architectures nevertheless, there is no proposed global 6G and IIoE standard architecture yet and drawing from the authors' extensive experience in 5G-enabled IoT/IoE architecture design as well as from the literature review of relevant research work, we believe multiple layers are essential to capture the complexity of the 6GIIoE system architecture. Hence, we have proposed and discussed below a simplified 5-layer 6GIIoE modular architecture as shown in Figures 4 and 5:

- The perception layer comprises the physical components: all sensors, devices, and machines.
- The transmission layer includes the information communication.
- The middleware layer consists of communication protocols, 6G, Wi-Fi, edge (cloud) computing and software to analyze data.
- The application layer deals with IIoE intelligent applications.
- The business layer defines the business model.

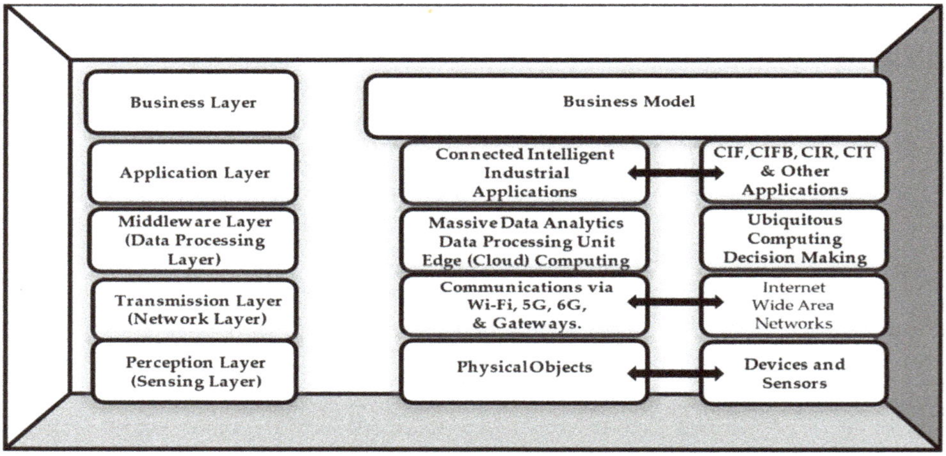

Figure 4. Five Layer 6GIIoE Architectural Framework.

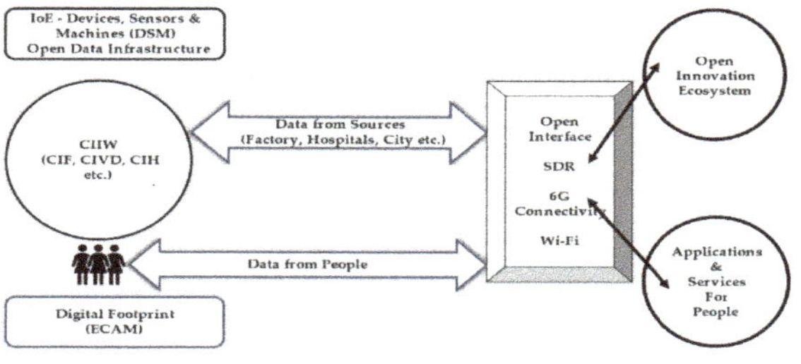

Figure 5. 6GIIoE System Architectural Framework.

The operating system (OS) is the virtual application platform that aggregates open innovations for various industrial applications and people (e.g., Apps can be developed using intelligence from the Digital Edge-Cloud on which it operates) and should serve as the 'operations center' for the public services domain. The Digital Edge-Cloud (DEC) [44] denotes the edge-cloud platform (with embedded hardware and software) that aggregates intelligence spanning multiple use cases, and more widely across IIoE applications, namely CIIW etc. that could benefit from composite information.

We cite one use case example: smart cities with 6GIIoE system. We propose significant pilot efforts around the globe that can be applied for Smart Cities with 6GIIoE

System. 'Realizing the vision of smart cities in the USA' initiatives can be focused on environmental aspects with underlying IoE technologies beyond Edge-Cloud-Analytics-Mobile (ECAM). Having a universal digital footprint for various IoE applications could help tap into the enormous benefits. A wireless communications infrastructure, such as 6G, is built on fiber-optic and/or Wi-Fi (6G-based) backbone network to serve the platform that moves massive data from devices/sensors/machines (DSM) for aggregation in the DEC. DSM infrastructure can be formed of existing physical infrastructure equipped with wired/wireless P2P, P2M and M2M devices/sensors/machines to enable detection and notification of events to the higher layers in the 6GIIoE architecture. Our view is that ECAM (Edge/Cloud/Analytics/Mobility) technologies enable powering up the 'Digital Footprint' in the 6GIIoE system architectural framework, as shown in the Figures 4 and 5, though the complexities of a large-scale CIIW deployment are yet to be understood. The major challenge would be in identifying scalable, cost-effective, and interoperable technology solutions to build the device/sensor/machine (DSM) infrastructure ensuring raw/filtered massive data flows through to the 'Digital Footprint' Considerable R& D work needs to be pursued to actualize the operations of various connected intelligent industrial applications.

Previously security involved a simplistic approach followed the layered architecture for the IoE. This ignores both the complexity of the 6GIIoE system and the need to provide security functionality in different places and in different forms. For example, authentication is needed both for applications and all individual devices. Therefore, in this study, a new model is proposed which is layered in vertical and horizontal planes to inter-operate. The horizontal planes cover the architectural components of the 6GIIoE system, from devices to service composition up through applications to the end-user. The vertical planes cover security services required at various system architecture levels and mechanisms.

The proposed 6GIIoE architecture is composed of many elements from gateways, sensors, device management, connectivity, and application platforms. It is also important to understand that the functional components of a 6GIIoE architecture that encapsulate the diverse security requirements and issues, therefore, a five-layer, three-dimensional security architecture is essential. It becomes daunting for an enterprise to take the initiative to construct the 6GIIoE architecture at the outset with these elements. For CIIW, the infrastructure layer requires the following technology, unique characteristics and challenges that should be addressed effectively, as described below:

- interoperable that keeps up with the industry's dynamic needs. Sensor/Machine Infrastructure Layer—Technology, Characteristics, and Challenges:
- Inexpensive, light weight, miniaturized IoE—devices, sensors, and machines with almost no physical security and memory capability.
- 6G network connectivity with low power communication protocols.
- Support traditional security algorithms.
- Embedded in physical structures so that wireless devices that operate autonomously with secure remote management.
- Ultra-low power circuits and communications.
- Devices/sensors/machines that can run on a single battery.
- Ambient and regenerative energy harvesting capabilities.
- Adaptive to real-time analysis of sensor/device/machine massive data.

As the 6GIIoE ecosystem continues to evolve, the following trends should be considered formulating architecture and design for the infrastructure.

6.1. Prototyping Hardware

For the industrial applications, the challenge of 6GIIoE hardware architecture and design depends on the confounding array of use case requirements. Example: Taking sensor temperature. Based on the criteria or requirements like temperature range, stability, response time and accuracy, the selection of sensors from hundreds available is key. A typical wireless sensor may not be available that fully meets the needs. That is where 6GIIoE hardware rapid prototyping helps. Hardware prototyping enables to build a customized hardware prototype in a short span. A ready-to-use portfolio, 6Gwireless modules, compatible sensor, interface, and compilers as well as prototype boards can create the optimal hardware mix-and-match that may match with specific industrial use case. Through quick hardware prototyping, enterprises can ratify the 6GIIoE technical and business viability of their target solution in a cost-effective and agile manner creating a foundation for a successful outcome.

6.2. Wireless Connectivity

Generally, a factory uses and operates with legacy industrial systems that are being rarely connected by wide area networks or latest wireless communications such as 5G or emerging 6G. Furthermore, the legacy industrial systems employ proprietary communication protocols for automation purposes and the lack of interoperability among these protocols hampers the execution of a monitoring and control network for the entire operations of the factory. Typically, data is captive within discrete control loops generating numerous data silos on the manufacturing floor.

6.3. Software-Defined Radio

At present many industrial facilities have implemented wired networks (LAN) and wireless networks such as Wi-Fi. A legacy IIoT architecture incorporates multiple radio protocols and standards. Most industries are exploring the deployment of new types of connectivity such as 5G or the emerging 6G to exploit high-value use cases. Hence, it is important to create backward compatible and an efficient 6GIIoE architecture that can accommodate software-defined radio (SDR)—a radio communication [44] (p. 134011) method where most signal processing is done using software. IIoE gateways leveraging SDR or 6G can implement and decode different protocols to save infrastructure cost and complexity. With 6G wireless connectivity, the above architecture can achieve simple software updates offering the possibility to enterprises to dynamically adapt to future operational and technological changes.

6.4. Portable Container-Based Platform Design

A portable, container-based design 6GIIoE platform provides industrial users with full flexibility and enables easy migration to another server as required without compromising the functionality of the application. Customized applications can be packaged and delivered in a portable standardized container. Users can decide which specific platform functions and applications with the modular architecture.

7. 6GIIoE Priority Areas, Applications, and Challenges

7.1. Priority Areas

Based on the exhaustive literature review, survey (mentioned in the Section 3.2) [33] and white paper published by Siemens, Huawei, Cisco [3,8,9], findings divulged that IoT/IoE affects various industries, and the topmost priority areas are customer experience, asset management and financial decision making.

First on the list of priority area is to enhance customer experience. The following ways 6G enabled IIoE system can enhance the customer experience:

- Perpetual monitoring of customer experience with company offerings.
- Personalizing the situation for each customer.
- Updating products and services based on learning over time with customer interactions.
- Continuous innovation product access and purchasing experience.

The second and third priority areas are asset management and financial decision making, respectively. The 6G enabled IIoE system can play a significant role in financial decision making by facilitating asset tracking and managing real time visibility. Top priority areas may differ based on industrial applications. Customer experience and financial decisions (asset management tracking) facilitate a sound foundation of functional areas for practitioners and academic researchers to appraise.

Furthermore, the following specific priority areas of 6G and IIoE needed attention:

- 6G should entrust MTRLLC for mobility management at high frequency mmWave bands and THz beyond.
- 6G requires a move towards radio centric system design and end-to-end co-design 3CLS [25] (p. 3) (Convergence of Communications, Computing, Control, Localization, and Sensing) driven by artificial intelligence (AI).
- 6G vision should be driven by a diverse portfolio of applications, technologies, and techniques.
- 6G high density tiny cell base station paradigm should embark upon a new era of smart surfaces communicating with human-embedded implants.
- Performance target analysis and optimization of 6G should operate in 3D space.
- IIoE: customer focused perspectives—the term "IIoE" is about the ways IoE can be used for industrial applications. IIoE focuses on improving connectivity between devices, time/cost savings and achieving efficiency optimization for industries. IIoE is amid a fundamental transformation with connected devices growing, up to 75 billion by 2025 [45], more and more enterprises are entering the industrial IoE arena. In the past, IoE interconnectivity was exclusively for enterprises that had the resources to build an IoE system. Nowadays, companies opt for a pre-built and full-stack managed IoE service that can offer the necessary hardware, software, and edge (cloud) infrastructure to handle billions of concurrent device or sensors, or machine connections to edge (cloud) network infrastructure such as 6G.
- IIoE: predictive maintenance and analytics (PMA)—Industries are using IoE-enabled systems to send out notifications to maintenance crews to predict when equipment is about to fail. This reduces the amount of maintenance time and allows maintenance crews to spend time to focus on other tasks. IoE-enabled systems collect massive data on how often these notification failures are sent out. Industries can use and communicate this massive data via 6G to create maintenance timelines and save significant cost and time, allowing teams to service equipment before it even fails. Predictive maintenance and analytics (PMA) [46] are becoming one of the most important use cases for industrial analytics.

7.2. Applications

The motivation of this section is to discuss various industrial application areas of IoE. Numerous industrial applications and services can be employed via 6GIIoE system. Here we briefly outline some of the example application areas. The CIIW comprises of industries related to transportation, vehicles, offices, buildings, manufacturing (factories), retail stores, E-commerce, food and beverage, health, port/harbor, financial services, banks, insurance entities, logistics/supply chains, energy sector and other applications becoming priorities areas to be connected intelligently. The CIIW is going to be a pervasive phenomenon in which objects (devices, sensors, machines, and wearables) can process, store, and exchange information seamlessly using IoE technologies seeking to be intelligently connected. Below are some of the 6GIIoE applications in the CIIW that can leverage the growth opportunities [8] (p. 3), [9,10] (p. 102), [11]:

- CIF—Connected Intelligent Factories.
- CIFB—Connected Intelligent Food & Beverage.
- CIFSBI—Connected Intelligent Financial Services, Banks, and Insurance.
- CIT—Connected Intelligent Transportation.
- CIC—Connected Intelligent City.
- CIE—Connected Intelligent Energy system.
- CIR—Connected Intelligent Retail
- CIH—Connected Intelligent Healthcare
- CIHP—Connected Intelligent Harbor and Port.

This study specifically describes two specific applications (CIF and CIFB) related to connected intelligent industrial world (CIIW):

- CIF-connected intelligent factories: At present, factories are connected by wired connections with fieldbus or industrial Ethernet, for connecting equipment to control systems to empower the factories with a secure and structured way for massive data access and control. 6G can target factory automation with extremely low latency communication with massive network capacity realizing the mission-critical applications in an industrial environment enabling devices or sensors to "offload" some processing capacity to the network. The IIoE needs a variety of technologies complementing each other to deliver the enhancements in agility, efficiency, and automation and requires real-time streaming for actionable intelligence, machine learning capabilities for identifying patterns and predicting behavior, XR (AR/MR/VR) to complement manual guidance and training. The 6GIIoE factory platform [22] (p. 143460), [47] (pp. 2–4) analytics software can achieve the following benefits for many industries: (i) makes IoE simple and easy to deploy without writing code, (ii) connects and manages any asset and analyzes any amount of data automatically in real time, turning data into actionable insights to reduce costs and improve efficiencies to build innovative and differentiating applications, and (iii) continuously improve production processes benefiting to visualize analytics for faster decisions, analyzing to solve problems, and monitoring operational performance. The 6GIIoE factory platform, as shown in the Figure 7, is an edge (cloud) platform using open-source-software via application programming interface (API) in the automotive industry. It stores massive data collected via data connection interface (DCI) from a variety of equipment's and autonomous logistic systems in a single cloud network. Thus, the 6GIIoE system realm will dramatically shape the 21st century manufacturing landscape transforming the way factories operate, and workforce needs. While the following financial rewarding possibilities of CIF appeals to the dreamers and innovators, the clear payoff ensures that it is a pragmatic path for industry decision-makers to follow keenly: (i) maximizes revenue growth, (ii) reduces operating cost, and (iii) increases asset efficiency. The development of a 6GIIoE factory platform should include the following: (i) the ability to develop a software platform (in-house and keep improving as well as evolving the platform even after the software platform has been put into operation, (ii) enhancing the firm's ability to leverage KLAS (Kanban, Lean, Agile and Scrum) [48–51] development techniques, (iii) acquiescing to share massive amount of data with partners as an open source platform to develop and enhance applications linking various devices to deliver various products and to notify the operator of equipment abnormalities contributing to the realization of a safe and secure system, and (iv) software engineers of an industrial firm can work swiftly and collaboratively by hastening improvements in 6G, ICT, OT, and IoE technologies attaining efficiency, quality, and cost savings.
- CIFB-connected intelligent food and beverage: The traction of food consumption is exponential globally as countries develop food and beverage supply chains and increase throughput to meet the demand. The primary aspect of empowering IoE (availability of 5G/6G connectivity, processing and cost of devices or sensors) have reached at a point to wide scale execution of IoE solutions across the food and beverage (F & B) value chain [22] (p. 143461). One key IoE use case for the F & B industry is advance tracking—cargo level tracking to complement asset level tracking by filling in the massive critical data visibility gaps at each pallet level and enabling the entire supply chain visibility and managing risk for the F& B stakeholders. The Internet of F & B can facilitate an unparalleled level of data visibility and traceability for all the stakeholders of the F & B industry leveraging 6GIIoE system technologies in connectivity and security as shown in Figure 6.

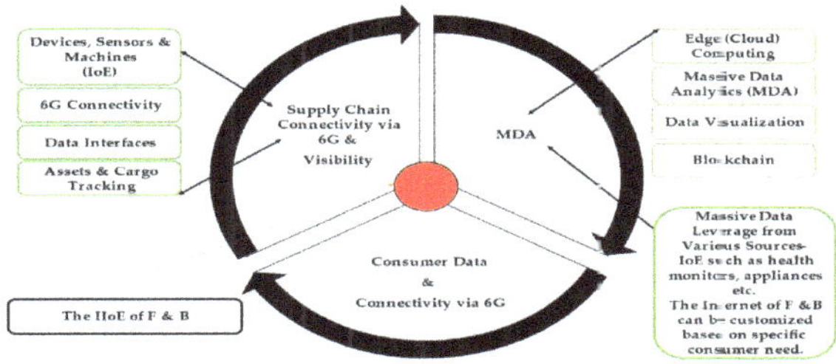

Figure 6. The 6GIIoE System for Food & Beverage—The future of Food and Beverage Traceability.

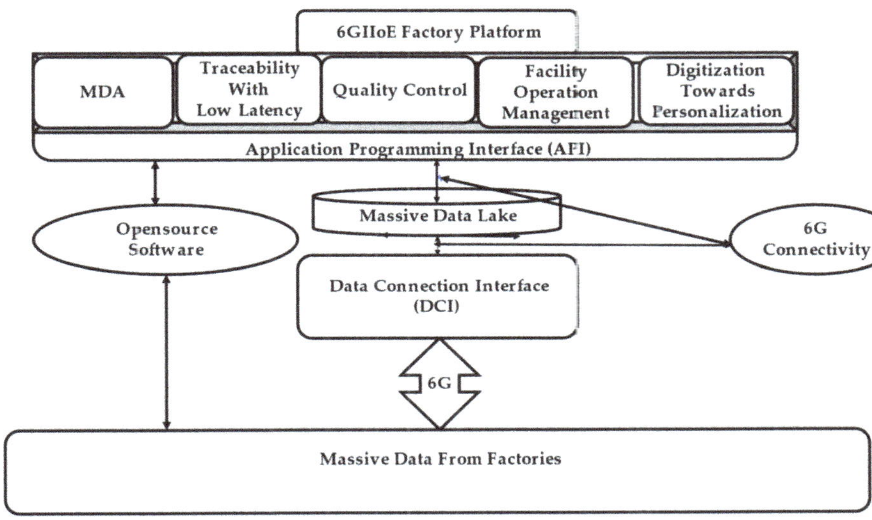

Figure 7. Connected Intelligent Factories.

7.3. Challenges

Every generation of technology approbation and execution has its daunting challenges. 6G-enabled IIoE system-based solutions involve many technologies, generating a complex environment. The following are the challenges [22]:

- Integration of new technology solutions into prevailing technologies.
- Management of a complex system
- 6G Networking issues, specifically cloud/edge data analytics and management.
- Security and privacy

6G-enabled IIoE system services is perceived as an integral part of most industries' future requirements in a ubiquitous manner. The privacy, security and trust are of utmost concerns, specifically when 6G-enabled IIoE technologies dependent on sensors, devices, machines, wireless communication networks, cloud storage, and the relevant software solutions. As per the survey of the USA technology leaders, the topmost obstacle to IoE growth are privacy and security worry. Furthermore, the following are the primary security challenges IoT: mobile system security, authentication, trust, secured software, and privacy. We suggest the following standards to relevant stakeholders for the ways to address the IoE security issues:

- Security measures must be given priority attention.
- At the design phase the security issues need to be addressed.

- Proven security practices need to be adopted during the building phase.
- Transparency is vital across all phases of IoE adoption and execution.

While the advantages of 6G enabled IIoE services are irrefutable, security and privacy aspects are not keeping pace with technology innovation. Hence the authors contend that 6GIIoE system security and privacy concerns must be taken care of prior to user trust in offering the applications. Industries face daunting challenges [22] (pp. 143455–143459) to balance the benefits of new technology and the risk associated with data breaches. Furthermore, the pressure to keep everything "intelligent" is increasing. IIoE can provide great benefits only when smart machinery is reliable. While there are many IIoE technical challenges, such as scaling the solution, it is important to focus on the two aspects: (i) how to generate recurring revenue stream and profit, and (ii) how to find a product-market fit. Essentially, the value proposition and the business model. While value propositions are the benefits for the IIoT system, a compelling business model ought to demonstrate to attain that value to market.

Additional challenges to adopt intelligent infrastructure for the 6GIIoE systems deployment include the following:

- Connectivity outage—constant need for uninterrupted connectivity if an enterprise
- Reliability in performance and delivering value to the customer.
- Data storage, and data management performance.
- Processing capacity and maintaining visibility of "things".
- Building legacy and IIoE infrastructure.
- Moving traditional OT to the edge of the network.
- Selecting the right tools.
- Accessibility anytime and anywhere,
- Interoperability in a heterogeneous environment.

8. Towards a Novel Theoretical Framework

The literature review shows that there have been few theories developed in IoT domain. This is the first time an integrated 6G-enabled IIoE system of challenges, priority and application areas in an enactment and execution of novel theoretical framework was developed. The primary contribution of our research work is the creation of a novel theoretical framework on the 6GIIoE system approbation and execution which is non-existent in the literature. In our view, a theory makes a novel contribution to a discipline in the following ways:

- A theory's nucleus phenomenon might not have been covered by previous theories.
- A theory's originality might emerge due to essential changes it causes to extant theory outlining the bounds of the theory concisely.
- A theory could be acknowledged as novel due to the fact that its construct bestows well-known central phenomenon in new ways.

The theoretical framework developed in this paper falls under a theory's focal phenomena that has not been covered by prior theories.

The proposed novel theoretical framework is based on a set of relevant theories, models, concepts, 6GIIoE paradigm mentioned above in the Sections 3–7. The developed theoretical framework is classified into cyberspace and physical space and has the following aspects, (i) technology innovation and adoption of 6GIIoE, (ii) 6GIIoE stakeholders—as people influenced by the interest in the 6GIIoE system, (iii) usage consideration of 6GIIoE priority areas, challenges, specifically privacy and security, and (iv) usage change as people gain confidence and trust in the 6GIIoE system.

It is imperative to understand the enabling technologies through the intervention of 6G and IIoE from a virtual environment to every day live environment. In this context, 6G and IIoE represents the actual link between virtual space and physical space [13] (pp. 11–12), specifically it functions as cyber-physical systems (CPS) domain. CPS is an evolution of cyberspace that provides the IIoE objects positioned at the epicenter of the framework as a link between physical space and cyberspace. Hence, 6GIIoE system offers one of the central enabling technologies of this theoretical framework.

The 6GIIoE priority areas will vary from industry to industry depending upon customer experience, asset management, and financial decision making. Stakeholders (partnerships) play a vital role in the formulation of a theory; therefore, researchers need to pay attention and decide which stakeholders (partnerships) should be included in a theory. Most theoretical frameworks in the IS discipline are process frameworks or variance [52]. Scholars suggest that theoretical development should be from a process perspective due to the technology adoption. 6GIIoE challenges (security, privacy, confidence, trust) are considered as an important aspect of this study, hence security measures are paramount to attain confidence and trust in the 6G-enabled IIoE services. Confidence creation and conviction management plays a prominent role in the 6GIIoE system for reliable data fusion, improved customer privacy, services with context-aware intelligence, and information security [10] (p. 103). Additionally, in this study, additional contributions, based on the extensive literature [10] (p. 103), on confidence and conviction management was established for the development of a 6GIIoE system theoretical framework. In our view, confidence and conviction creation and sustainability ought to be a proactive approach in all phases of 6GIIoE system adoption and execution. Furthermore, scholars have conducted extensive surveys of trust analysis and computation models for IoT concluding with guidance for trust computation research. Based on such prior surveys and the insights, this study gave credence using those survey judgements while developing the novel theoretical framework. While confidence and conviction management, security and privacy are essential to the 6GIIoE system adoption and execution success, privacy and security are the harbinger to the creation and sustainability of trust management [10] (p. 103). Thus, the above insights regarding the developed framework help to investigate 6G-enabled IIoE system.

The development of the proposed theoretical framework is also focused on the building blocks and practices as shown in the Figure 8 are: (i) the design of 6GIIoE architecture to exploit the priorities, challenges, and opportunities, (ii) cooperation and participation among the stakeholders in the entire 6GIIoE value chain, (iii) interactive actions between the 6G, and IIoE applications to achieve optimum performance in CIIW. Based on the building blocks mentioned below, the 6GIIoEtheoretical framework is constructed and the framework embraces the stakeholders' interactive work that includes 6G, IIoE applications. This framework can be used as an education model and guide for all industry stakeholders including suppliers, providers, customers, and decision makers who are seeking 6GIIoE systems.

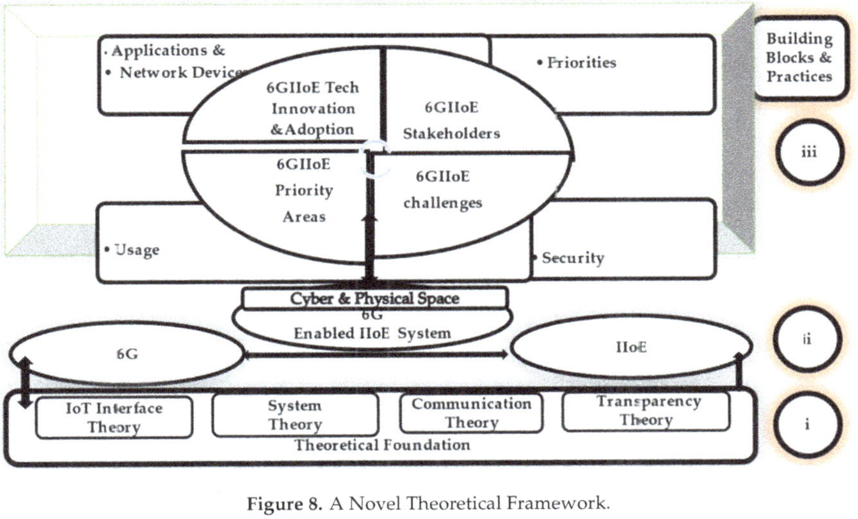

Figure 8. A Novel Theoretical Framework.

The following are additional contributions of the framework components related to industrial applications, and challenges research:

- 6GIIoE system involves people to machines (P2M), machines communicating with machines (M2M) and people to people (P2P). All stakeholders are paramount for the ultimate success of the 6GIIoE system. Keeping this in mind, cogitation of challenges and priority areas governing to 6GIIoE system adoption are cavillous and must be addressed cautiously prior to the 6GIIoE execution. This theoretical framework contributes to 6G, and IoE/IIoE knowledge guiding practitioners specifically imparting attention to the challenges of security and privacy to attain trust and successful use of the 6GIIoE system in industrial applications. In view of this contention, we accentuate that as the number of connected devices, sensors, machines increase, at a decisive level, industry professionals need to evaluate the opportunities and threats which the emergence of 6G enabled IIoE systems might affect to their enterprises indicating that the business models may need to be redefined.
- The theoretical framework outlined in this study aids this aim. The theoretical framework bequeaths to the body of knowledge in the discipline of 6G, IoE, IIoE, and CIIW offering researchers additional explanation and prediction on the theory constituents, thus, crevices opportunities for diversified research. An ample scope for testing this theory and constituents of the theory should be alternatively investigated as well.

9. Consolidated Lessons Learned

The following sub-sections present a consolidated lesson learned from this study.

9.1. Theory Development for IS Innovation

The study of IS innovation is a complex, multi-dimensional phenomenon with dynamic interactive characteristics that invites a novel theoretical framework. Our primary investigation is to identify a theoretical framework that can accommodate the complex and multidimensional landscape of IS innovation. The prevailing theories related to IS discipline do not adequately account for the dynamic relationship between the innovator and environment. An ecological systems theory can provide an appropriate framework for researchers to approach the topic of innovation and to examine its relationship with information systems. The concepts derived from the literature review can be used to map the layers based on ecological systems theory, which is not within the scope of this study, therefore not covered. It is vitally important to note that IS studies must place the internal organizational processes within the wider philosophical, socioeconomic, and ethical context.

9.2. 6G Future Benefits

We opine that a new set of service classes and IIoE applications will emerge due to 6G mobile systems. Several cutting-edge technologies will mature along the same time of 6G deployment to potentially play an important role towards the end of the 6G global standardization. One such prominent technology is quantum computing and communications (QCC) which can provide highly secured long-distance networking in Tera bit per second speed with massive network capacity and extremely low latency. Other 6G technologies will include integration of non-RF and RF links. Industries are motivated to make investments to attain substantial value proposition [53] (p. 1) from 6GIIoE deployment enabling new business models. In addition to the benefits mentioned above, the value drivers of 6G connectivity for the industries offer the following unique value proposition: (i) high reliability, (ii) greater bandwidth, and (v) operational efficiency. The DNA of 6G value proposition is as follows:

- Human centric services with improved decision making.
- Beyond 6GHz with tiny high-density cell stations for data bountiful environments.
- Communication with large intelligent surfaces.
- MTRLLC with automation agility.

- Intelligence efficiency.
- mMTRLLC connectivity latency and reliance

9.3. IIoE Value Proposition

IoE are not conventional computing devices. They are physical devices connected to "5G & Beyond" wireless communication or Wi-Fi or Bluetooth. To operate IoE devices, digital services are required to ensure its systems resiliency. IIoE connectivity unlocks the massive data, enabling enterprises to create new revenue streams, bestow operational efficiencies, cost savings, and enhance customer engagement. To exploit the power of IIoE data one must first tackle connectivity. There are many ways to select the connectivity options and it is daunting to make the right connectivity choices. Multi-National Operators (MNOs) offer different geographical reach; hence collaboration is essential to achieve ubiquitous coverage. A flexible, simple, and cost-efficient service is required to manage the connectivity complexity for enterprises. As the fraction of IoE deployments increases exponentially, efficient IoE subscriber management becomes imperative to deploy IIoE projects. Value proposition design, proven IIoE new business models [53] (pp. 1–4) and a simple (user-friendly plug-play) framework are the triumphant combination to create a successful IIoE strategy. The promise of IIoE is endless and the most valuable ones are enhanced customer experience, increased sales revenue, exceptional customer support, recurring revenue stream, and outstanding quality.

In this study, we suggest two tools, as lessons learned, to discover IoE value for industrial applications: (i) the IoE evolution canvas and (ii) the IoE value guide. The following three steps can be executed for industrial applications: (i) collect the data, (ii) make analysis and decision, and (iii) act. The execution can be done in three ways: (i) manually, (ii) assisted, and (iii) automation—the benefit of IoE can be achieved through automation.

IoE value guide is recommended to attain distill opportunities and insights. First, identify and describe IoE value proposition by reviewing the data collection, the analysis and decision making, the appropriate actions (human and machine), and results (customer competitive advantages). Customer benefit primarily depends on data collection, analysis and gathering the insights. The information obtained will create the canvas model and the algorithm required to act. It is vitally important to understand the customer need and determination of return on investment (ROI) in any IIoE project. It is paramount to understand how edge input/output (I/O) generates a simpler way to meet the needs of IIoE applications. IIoE projects demand special skills and involve many phases to execute them. I/O for the 6GIIoE projects will pose many challenges and are expensive to execute. To achieve simplicity, security, and cost-effectiveness for your 6GIIoE projects, this study proposes three solutions for improving I/O connectivity: (i) real time (on-time) measurement of equipment and (ii) controlling sensors and devices. It is important to apply the following edge I/O the ideal problem-solver: (i) Web-based configuration, (ii) embedded software, (iii) built-in security.

9.4. 6GIIoE System Applications for the CIIW Spectrum

Each industrial vertical sector of the CIIW will require a different set of 6GIIoE system applications to support its growth with their own unique set of demands. The 6GIIoE technology enablers briefed earlier focused to improve the efficiency of data transmission with low latency and massive network capacity. The exploitation of tiny high-density cell stations, network coding and beamforming will bring key advantages to the data plane. The benefits of a 6GIIoE ecosystem are yet to be well defined for network carriers, smartphone companies, software developers and edge (cloud) infrastructure providers.

10. Future Research Agenda & Research Gaps

Some of the future research areas and/or research gaps regarding IS discipline related to developing theories and 6GIIoE system are grouped into the following categories.

10.1. 6G Enabled IIoE System Theory Development

A void exists in the literature establishing the theory development on 6G-enabled IoE/IIoE system adoption and execution considering the priority areas, challenges, applications, and enormous opportunities on the horizon for connected intelligent industrial applications. Hence, we suggest researchers enhance knowledge for further studies in the technology areas pertaining to 6G, IoE/IIoE, CIIW to broaden and cultivate a thorough review and theoretical framework constituents where research void or gap prevail.

10.2. IS Innovations

IS discipline is well positioned at the intersection of technical, business, and social applications of information communication technology (ICT), operational technology (OT) and its essential dimensions of IIoE. 6GIIoE is emerging as a digital innovation to unleash opportunities in personalized customer experience. The emergence of 6GIIoE heralds a new dimension of a digital destiny with influences that are not yet fully known. This scenario invites the opening of valuable opportunities for scholarly inquiries. Hence, we propose that relevant future research work is essential to develop theories on 6G and IIoE supporting philosophical, socioeconomic, and ethical dimension specifically for connected intelligent industrial applications to create authentic value for the society at large. Thus, we highlight future research agenda in the following manner:

- Multi-level exploration of IS and heterogeneity.
- Thematic influence domains: (i) impact on organizations; (ii) impact on technology; (iii) impact on individuals; and (iv) impact on society.

10.3. Smart Service Contracts (SSC)

While applications are examined related to 6GIIoE ecosystem, there is much more significant potential for using smart service contracts (SSC), distributed ledgers, and other blockchain-related technologies. The research gaps in this domain are the following: (i) increasing the transaction rate per second; (ii) providing architectural means for IoE devices or sensors to utilize the relevant technologies, (iii) allowing secured ways to correct unintended errors in SSC; and (iv) clarification of hidden costs when using blockchains to allow objective comparisons with traditional alternatives [54] (p. 7).

11. Recommendations

The following recommendations are based on extensive literature review related to IS, 6G, IoE/IIoE, and CIIW:

11.1. Theory Development on 6G enabled IIoE system

Research scholars place a significant value on theoretical frameworks. Even with such importance, the development of novel theories and depuration of existing theories have been comparatively slighted in the areas of information system (IS) regimen. Validating the scant numbers of research publications on the IS discipline theory development, an exhaustive review of literature presented no research papers found in the literature related to 6GIIoE system. Theories bestow the perception of real-world phenomena and can be perceived as specific ontologies; such as a theory is the nature of real work. Since a void exists in the literature with regard to a theoretical framework for 6GIIoE adoption and execution considering the priority areas, challenges, and applications, we recommend the development of theories related to 6G-enabled IIoE system to enhance research scholars' knowledge for further studies.

11.2. 6G Network Specification Requirements

We suggest the following recommendations for 6G specification requirements for the connectivity (deep, intelligent, ubiquitous, holographic) vision as outlined below in Section 5 of this study:

- 6G network application ought to be human centric.

- High security, and privacy should be key features of 6G.
- Application Types: (i) MTRLLC, (ii) mURLLC, (iii) HCS, and (iv) MPS
- Device Types: (i) Sensors and Devices related CIIW, (ii) Connected Robotics and Autonomous Systems (CRAS), (iii) XR and CBI Equipment, and (iv) Intelligent Plants.
- Architecture: (i) Cell free smart surfaces at High Frequency supported by $\leq$100 m tiny and dense mmWave cell access, (ii) Hotspots served by Drone carried Base Stations, (iii) Sub-6GHZ cells, and (iv) 5-Layer Platform.
- Frequency Bands: (i)Sub—6 GHZ, (ii) mmWave for mobile access, (iii) Exploration of THz bands (above 140 GHz), and (iv) Non-RF (e.g., Optical, Visible Light Communication (VLC) etc.).
- End to End Delay: Less than 1 millisecond
- Processing and Radio only Delay: 10 Nano Seconds
- Rate Requirements: One (1) Tera Bit Per Second.
- Spectral and Energy Efficiency Gains: 1000$\times$ in bps/Hz/m^3 (volumetric)
- End to End Reliability: Seven 9 Seconds

11.3. IIoE/Industrial Control Systems (ICS)

In the IoE era, ICSs [4] (p. 1) are no longer a closed network and suffer from various threats and/or attacks. With the emergence of the fifth industrial revolution, IoE is shifting into IIoE. ICS drives the supervision and management of industrial processes performed by critical infrastructures for industries. The increase in attacks and/or threats influencing ICS drawing attention to the use of Intrusion Detection Systems (IDS) [55]. The malware used to attack critical infrastructure has become increasingly sophisticated and has mimicked normal network traffic. Hence, research pursuit to investigate to protect the threats and/or attacks through machine learning and deep learning with novel detection technique for detecting threats and/or attacks is essential.

11.4. Securing the 6GIIoE Infrastructure

The fusion of 6G, IoE, and other emerging digital technologies posing immense new opportunities, and ubiquity experiences. There are many exciting trends that are going to shape the future, inspiring a new age of digital destiny. The insatiable, end-user hunger for network capacity and speed such as 6G is a key differentiator in a digital journey. Network technologies will ultimately deliver the IIoE. The following are the recommendations for the industries to achieve safe and secure connectivity: (i) manage security at every level of the 6GIIoE system, (ii) protect the identity of the objects and users, (iii) execute multifactor authentication, and (iv) protect identities and not gateways.

11.5. Cyber Physical System (CPS)

Most industries are moving into an era of autonomous IIoE and its security must be considered a crucial element. To maintain the emerging growth rate in IIoE, future threats or attacks need the utmost attention. Furthermore, interconnectivity and new application scenarios such as Machine to Machine (M2M) need to be considered. This drives the growth of devices and sensors towards IIoE. The cyber security threats via 6G wireless communications are poised to emerge. Hence, there is a growing demand for the design of new communications systems for Industrial applications. Thus, the industrial world with its fusion of technologies requires Cyber Physical Systems [56]. Hence, research pursuit of resource saving, low cost, and effective alternative to conventional cryptographic methods is imperative to empower confidence in IIoE systems.

11.6. Security of IIoEvia ICMETRIC Technology

IIoE holds significant promise for enhanced supply chain efficiencies, quality control, communications, and business productivity for most industries. 6G is a game changer network technology paradigm that is emerging to tackle the enormous potential of IIoE data associated with the industrial applications. However, the use of IIoE generates new vulnerabilities and potential threats. To successfully adopt and execute the 6GIIoE system, it is imperative to improve and enhance the security of 6G commutations in IIoE. For future industrial manufacturing, the security must be incorporated to deal with the threats by designing it into the IIoE ecosystem at an early stage of the product design life cycle to achieve security in a cost-effective manner. Thus, security of the 6GIIoE applications can be executed by design, utilizing the ICMetric technology [57] that uses hardware and software features as well as the specification of devices or sensors in the IIoE setting.

12. Conclusions

Despite inadequate prior research encompassing the IS innovation spectrum to develop theoretical framework, this study developed a novel theoretical framework falls under a theory's focal phenomenon that has not been covered by prior theories. Furthermore, the 6GIIoE priority areas, challenges, applications were identified which offers significant benefits for the connected intelligent industrial applications, and a novel theoretical framework was developed based on the findings and insights. This study has some limitations, specifically the lack of an empirical test of the theoretical model proposed and provides further possible investigation.

Author Contributions: Conceptualization, P.K.P.; Investigation, P.K.P.; Methodology, P.K.P.; Supervision, F.C.-S.; Writing—original draft, P.K.P.; Writing—review & editing, P.K.P. All authors have read and agreed to the published version of the manuscript.

Funding: This research received no external funding.

Data Availability Statement: MDPI Research Data Policies at https://www.mdpi.com/ethics (accessed on 24 December 2020).

Acknowledgments: The authors would like to greatly thank Fundação para a Ciência e Tecnologia (FCT) and C-MAST (Center for Mechanical and Aerospace Science and Technologies), under project UIDB/00151/2020 for their reviewer support.

Conflicts of Interest: The authors declare no conflict of interest.

References

1. Ashton, K. That 'Internet of Things' Thing. *RFID* **2009**, *22*, 97–114.
2. Alsharif, M.H.; Kelechi, A.H.; Albreem, M.A.; Chaudhry, S.A.; Zia, M.S.; Kim, S. Sixth Generation (6G) wireless networks: Vision, research activities, challenges and potential solutions. *Symmetry* **2020**, *12*, 676. [CrossRef]
3. CISCO-Internet of Everything (IoE): Top 10 Insights from Cisco's IoE Value at Stake Analysis for the Public Sector. Available online: http://www.cisco.com/web/about/ac79/docs/IoE/IoE-VAS_Public-Sector_Top-10-$\backslash$Insights.pdf; https://www.cisco.com/web/about/ac79/docs/innov/IoE.pdf (accessed on 17 October 2020).
4. Wang, C.; Wang, B.; Liu, H.; Qu, H. Anomaly detection for industrial control system based on autoencoder neural network. *Wirel. Commun. Mob. Comput.* **2020**, *2020*, 1–10. [CrossRef]
5. Singh, P.; Nayyar, A.; Kaur, A.; Ghosh, U. Blockchain and fog based architecture for internet of everything in smart cities. *Future Internet* **2020**, *12*, 61. [CrossRef]
6. Viswanathan, H.; Mogensen, P.E. Communications in the 6G Era. *IEEE Access* **2020**, *8*, 57063–57074. [CrossRef]
7. Padhi, P.K. 5G & Beyond: Governance, global policy perspectives, and societal benefits. *Int. J. Eng. Technol. Res. Manag.* **2020**, *4*, 87–114.
8. Touching an Intelligent World. Huawei Global Industry Vision (GIV) White Paper. 2018, pp. 1–41. Available online: https://www.huawei.com/minisite/giv/en/index.html (accessed on 7 December 2020).
9. Siemens Digital Industries Software. 2020. Available online: https://www.plm.automation.siemens.com/global/en/webinar/track-and-trace-farm-to-table-iot/52777; https://www.plm.automation.siemens.com/global/en/industries (accessed on 8 December 2020).
10. Nord, J.H.; Koohang, A.; Paliszkiewicz, J. The Internet of Things: Review and theoretical framework. *Expert Syst. Appl.* **2019**, *133*, 97–108. [CrossRef]
11. Lueth, K.L. The 10 Most Popular Internet of Things Applications Right Now. IoT Analytics. 2015. Available online: https://iot-analytics.com/10-internet-of-thingsapplications/ (accessed on 21 December 2020).
12. Falkenberg, E.D.; Lyytinen, K.; Verrin-Stuart, A.A. Limitations of information systems theory and practice: A case for plural-ism. In *Information System Concepts. IFIP Advances in Information and Communication Technology*; Springer: Boston, MA, USA, 1995; pp. 14–15. Available online: http://eprints.lse.ac.uk/2579/1/ISwhatsortofscience.pdf (accessed on 11 December 2020).

13. Aquilani, B.; Piccarozzi, M.; Abbate, T.; Codini, A.P. The role of open innovation and value co-creation in the challenging transition from industry 4.0 to society 5.0: Toward a theoretical framework. *Sustainability* **2020**, *12*, 8943. [CrossRef]
14. Costello, G.J.; Donnellan, B.; Curley, M. A theoretical framework to develop a research agenda for information systems innovation. *Commun. Assoc. Inf. Syst.* **2012**, *33*, 26. [CrossRef]
15. Giordani, M.; Polese, M.; Mezzavilla, M.; Rangan, S.; Zorzi, M. Toward 6G networks: Use cases and technologies. *IEEE Commun. Mag.* **2020**, *58*, 55–61. [CrossRef]
16. Dang, S.; Amin, O.; Shihada, B. From a Human-Centric Perspective: What Might 6G Be? 2020. Available online: https://arxiv.org/pdf/1906.00741.pdf (accessed on 12 December 2020).
17. Huang, C.; Zappone, A.; Alexandropoulos, G.C.; Debbah, M.; Yuen, C. Holographic MIMO Surfaces for 6G Wireless Net-Works: Opportunities, Challenges, and Trends. 2020. Available online: https://arxiv.org/pdf/1911.12296.pdf (accessed on 12 December 2020).
18. Gui, G.; Liu, M.; Tang, F.; Kato, N.; Adachi, F. 6G: Opening new horizons for integration of comfort, security, and intelligence. *IEEE Wirel. Commun.* **2020**, *27*, 126–132. [CrossRef]
19. Chowdhury, M.Z.; Shahjalal, M.; Ahmed, S.; Jang, Y.M. 6G wireless communication systems: Applications, requirements, technologies, challenges, and research directions. *arXiv* **2019**, arXiv:1909.11315v1.
20. Mahmood, N.H.; Alves, H.; López, O.A.; Shehab, M.; Osorio, D.P.M.; Latva-Aho, M. Six key enablers for machine type communication in 6G. *arXiv* **2019**, arXiv:1903.05406.
21. Janbi, N.; Katib, I.; Albeshri, A.; Mehmood, R. Distributed Artificial Intelligence-as-a-Service (DAIaaS) for smarter IoE and 6G environments. *Sensors* **2020**, *20*, 5796. [CrossRef] [PubMed]
22. Sekaran, R.; Patan, R.; Raveendran, A.; Al-Turjman, F.; Ramachandran, M.; Mostarda, L. Survival study on blockchain based 6g-enabled mobile edge computation for IoT automation. *IEEE Access* **2020**, *8*, 143453–143463. [CrossRef]
23. Wang, W.; Liu, F.; Zhi, X.; Zhang, T.; Huang, C. An Integrated deep learning algorithm for detecting lung nodules with low-dose CT and its application in 6G-enabled internet of medical things. *IEEE Internet Things J.* **2020**, 1. [CrossRef]
24. Yang, P.; Xiao, Y.; Xiao, M.; Li, S. 6G wireless communications: Vision and potential techniques. *IEEE Netw.* **2019**, *33*, 70–75. [CrossRef]
25. Saad, W.; Bennis, M.; Chen, M. A Vision of 6G wireless systems: Applications, trends, technologies, and open research problems. *IEEE Netw.* **2019**, *34*, 134–142. [CrossRef]
26. Clazzer, F.; Munari, A.; Liva, G.; Lazaro, F.; Stefanovic, C.; Popovsk, P. From 5G to 6G: Has the time for modern random access come? *arXiv* **2019**, arXiv:1903.03063.
27. Tariq, F.; Khandaker, M.R.A.; Wong, K.-K.; Imran, M.A.; Bennis, M.; Debbah, M. A speculative study on 6G. *IEEE Wirel. Commun.* **2020**, *27*, 118–125. [CrossRef]
28. Zhang, L.; Liang, Y.-C.; Niyato, D. 6G Visions: Mobile ultra-broadband, super internet-of-things, and artificial intelligence. *China Commun.* **2019**, *16*, 1–14. [CrossRef]
29. Internet of Everything (IoE). *Internet Everything*; IoE: London, UK, 2019; Available online: https://ioe.org/Eppicenter (accessed on 15 December 2020).
30. Dai, H.-N.; Zheng, Z.; Zhang, Y. Blockchain for Internet of Things: A Survey. *IEEE Internet Things J.* **2019**, *6*, 8076–8094. [CrossRef]
31. EPPI-Reviewer. Available online: https://eppi.ioe.ac.uk/CMS/Default.aspx?alias=eppi.ioe.ac.uk/cms/er4& (accessed on 5 December 2020).
32. Swanson, E.B. Information systems innovation among organizations. *Manag. Sci.* **1994**, *40*, 1069–1092. [CrossRef]
33. Leidner, D.E.; Kayworth, T. Review: A review of culture in information systems research: Toward a theory of information technology culture conflict. *MIS Q.* **2006**, *30*, 357. [CrossRef]
34. Weber, R. Evaluating and developing theories in the information systems discipline. *J. Assoc. Inf. Syst.* **2012**, *13*, 1–30. [CrossRef]
35. Chaudhuri, A.; Cavoukian, A. The Proactive and Preventive Privacy (3P) framework for IoT privacy by design. *EDPACS* **2018**, *57*, 1–16. [CrossRef]
36. Hsu, C.-L.; Lin, J.C.-C. Exploring factors affecting the adoption of internet of things services. *J. Comput. Inf. Syst.* **2018**, *58*, 49–57. [CrossRef]
37. Slepian, D.; Wyner, A.S.O. Rice's contributions to Shannon theory. *IEEE Trans. Inf. Theory* **1988**, *34*, 1374. [CrossRef]
38. Lohstroh, M.; Lee, E.A. An Interface Theory for the Internet of Things. *Min. Data Financ. Appl.* **2015**, 20–34. [CrossRef]
39. Cacciapuoti, A.S.; Caleffi, M.; Tafuri, F.; Cataliotti, F.S.; Gherardini, S.; Bianchi, G. Quantum internet: Networking challenges in distributed quantum computing. *IEEE Netw.* **2019**, *34*, 137–143. [CrossRef]
40. Averill, E.W.; Gottlieb, J. *Two Theories of Transparency*; Springer: New York, NY, USA, 2019; pp. 1–21. [CrossRef]
41. Todd, S. A 25-year review of sequential methodology in clinical studies. *Stat. Med.* **2006**, *26*, 237–252. [CrossRef] [PubMed]
42. Oughton, E.J.; Frias, Z. The cost, coverage and rollout implications of 5G infrastructure in Britain. *Telecommun. Policy* **2018**, *42*, 636–652. [CrossRef]
43. Khan, L.U.; Yaqoob, I.; Imran, M.; Han, Z.; Hong, C.S. 6G wireless systems: A vision, architectural elements, and future directions. *IEEE Access* **2020**, *8*, 147029–147044. [CrossRef]
44. Akyildiz, I.F.; Kak, A.; Nie, S. 6G and beyond: The future of wireless communications systems. *IEEE Access* **2020**, *8*, 133995–134030. [CrossRef]

45. Statista Research Department, Internet of Things—Number of Connected Devices Worldwide 2015–2025. Available online: https://www.statista.com/statistics/471264/iot-number-of-connected-devices-worldwide/ (accessed on 17 December 2020).
46. AVANET/ABACUS: Predictive Maintenance with IoT: The Road to Real Returns. Available online: https://www.avnet.com/wps/portal/abacus/solutions/markets/industrial/predictive-maintenance-iot/ (accessed on 12 December 2020).
47. Kalsoom, T.; Ramzan, N.; Ahmed, S.; Ur-Rehman, M. Advances in sensor technologies in the era of smart factory and industry 4.0. *Sensors* **2020**, *20*, 6783. [CrossRef] [PubMed]
48. Riaz, M.N. Implementation of Kanban techniques in software development process: An empirical study based on benefits and challenges. *Sukkur IBA J. Comput. Math. Sci.* **2019**, *3*, 25–36. [CrossRef]
49. Roth, N.; Deuse, J.; Biedermann, H. A framework for System Excellence assessment of production systems, based on lean thinking, business excellence, and factory physics. *Int. J. Prod. Res.* **2019**, *58*, 1074–1091. [CrossRef]
50. Gochhait, S.; Butt, S.A.; Jamal, T.; Ali, A. *Cloud Enhances Agile Software Development*; IGI Global: Hershey, PA, USA, 2020; pp. 28–49.
51. Khalid, A.; Butt, S.A.; Jamal, T.; Gochhait, S. Agile scrum issues at large-scale distributed projects. *Int. J. Softw. Innov.* **2020**, *8*, 85–94. [CrossRef]
52. Rivard, S. The ions of theory construction. *MIS Q.* **2014**, *38*, 3–13.
53. Hudson, D. Value propositions for the internet of things: Guidance for entrepreneurs selling to enterprises. *Technol. Innov. Manag. Rev.* **2017**, *7*, 5–11. [CrossRef]
54. Hang, L.; Kim, D.-H. Reliable task management based on a smart contract for runtime verification of sensing and actuating tasks in IoT environments. *Sensors* **2020**, *20*, 1207. [CrossRef] [PubMed]
55. Kim, S.; Jo, W.; Shon, T. APAD: Autoencoder-based Payload Anomaly Detection for industrial IoE. *Appl. Soft Comput.* **2020**, *88*, 106017. [CrossRef]
56. Chen, B.; Yang, Z.; Huang, S.; Du, X.; Cui, Z.; Bhimani, J.; Xie, X.; Mi, N. Cyber-physical system enabled nearby traffic flow modelling for autonomous vehicles. In Proceedings of the 2017 IEEE 36th International Performance Computing and Communications Conference (IPCCC), San Diego, CA, USA, 10–12 December 2017; pp. 1–6.
57. Tahir, R.; McDonald-Maier, K. ICMetrics based Industrial Internet of Things (IIoT) Security in the Post Quantum World. In Proceedings of the World Congress on Engineering (WCE 2018), London, UK, 4–6 July 2018; Available online: http://www.iaeng.org/publication/WCE2018/WCE2018_pp347-348.pdf (accessed on 18 December 2020).

applied system innovation

MDPI

Article

6G Enabled Tactile Internet and Cognitive Internet of Healthcare Everything: Towards a Theoretical Framework

Prafulla Kumar Padhi [1,*] and Fernando Charrua-Santos [2]

[1] Department of Industrial Engineering & Management, University of Beira Interior, 6201-001 Covilha, Portugal
[2] Electromechanical Department, University of Beira Interior, 6201-001 Covilha, Portugal; bigares@ubi.pt
* Correspondence: prafulla.k.padhi@ubi.pt

Abstract: Digital era deficiencies traditionally exist in healthcare applications because of the unbalanced distribution of medical resources, especially in rural areas globally. Cognitive data intelligence, which constitute the integration of cognitive computing, massive data analytics, and tiny artificial intelligence, especially tiny machine learning, can be used to palpate a patient's health status, physiologically and psychologically transforming the current healthcare system. To remotely detect patients' emotional state of diagnosing diseases, the integration of 6G enabled Tactile Internet, cognitive data intelligence, and Internet of Healthcare Everything is proposed to form the 6GCIoHE system that aims at achieving global ubiquitous accessibility, extremely low latency, high reliability, and elevated performance in cognitive healthcare in real time to ensure patients receive prompt treatment, especially for the haptic actions. Judiciously, a model-driven methodology is proffered to facilitate the 6GCIoHE system design and development that adopts different refinement levels to incorporate the cognitive healthcare requirements through the interactions of semantic management, process management, cognitive intelligence capabilities, and knowledge sources. Based on the 6GCIoHE system architecture, applications, and challenges, the aim of this study was accomplished by developing a novel theoretical framework to captivate further research within the cognitive healthcare field.

Keywords: cognitive data intelligence; cognitive healthcare; tiny machine learning; 6GCIoHE theoretical framework

Citation: Padhi, P.K.; Charrua-Santos, F. 6G Enabled Tactile Internet and Cognitive Internet of Healthcare Everything: Towards a Theoretical Framework. *Appl. Syst. Innov.* **2021**, *4*, 66. https://doi.org/10.3390/asi4030066

Received: 25 July 2021
Accepted: 6 September 2021
Published: 10 September 2021

Publisher's Note: MDPI stays neutral with regard to jurisdictional claims in published maps and institutional affiliations.

Copyright: © 2021 by the authors. Licensee MDPI, Basel, Switzerland. This article is an open access article distributed under the terms and conditions of the Creative Commons Attribution (CC BY) license (https://creativecommons.org/licenses/by/4.0/).

1. Introduction

The shortcomings of the 5G [1] (p. 1) mobile system as a key enabler of the Internet of Everything (IoE) [1] (p. 6) are addressed. To brace the cognitive development of a global ubiquitous smart healthcare system, research activities are being pursued on the next-generation 6G communication system [1] (p. 2). Appendix A provides an acronym list.

The 6G revolution is envisioned to connect everything and control trillions of devices—macro to micro to nano—for the digitization future. Time-sensitive healthcare applications such as haptic (involving touch, sight, and sound) actions and holographic connections displaying three dimensional images assist healthcare professionals using emotion-sensing wearable devices to monitor mental health, heartbeats, oxygen level, glucose, blood pressure, and much more, as shown in Figure 1.

The Tactile Internet [2,3] bestows a new paradigm in people to machine (P2M) interaction featuring extremely low latency and short transit, high availability, great reliability, and an enormous level of security to create systems with real time interactive actions. Currently, based on the 5G advancements and deployment, the stage is being set for the emergence of 6G enabled Tactile Internet [1–3], a new paradigm, to provide the integration of mobile edge-clouds and eXtended Reality [1] (p. 5) (eXR—augmented reality—AR, virtual reality—VR, and mixed reality—MR) for sensory and haptic controls to address the future smart healthcare systems with less than one millisecond reaction time, offering physical tactile experiences remotely.

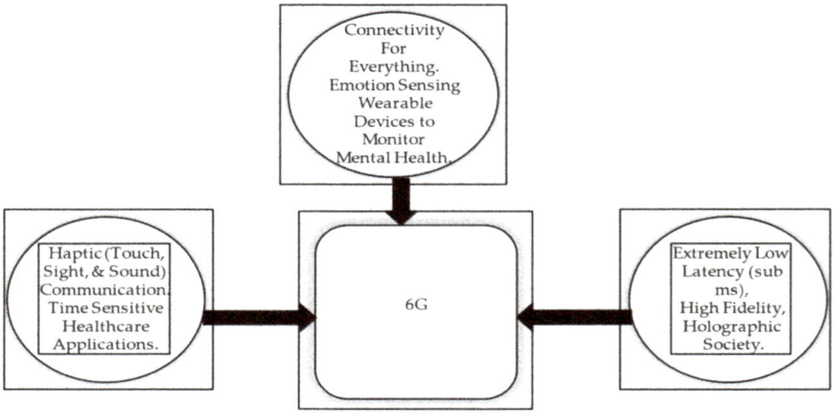

Figure 1. Examples of 6G for healthcare applications.

Cognitive technology (CT) [4] is a discipline of computer science that mimics human brain functions through various means such as cognitive computing (CC) [4] and tiny machine learning (tiny ML) [5]. CC is a propitious tool for analyzing massive data for better holistic ideas to help healthcare businesses accomplish faster and smarter decisions. Tiny AI is an emerging avant-garde discipline to revolutionize the healthcare industry that intersects between tiny ML and embedded IoE devices and that requires intimate knowledge of hardware, software, algorithms, and applications. The integration of CC, massive data analytics (MDA) [6], and tiny ML constitutes cognitive data intelligence (CDI) [4–6] to offer better healthcare solutions by providing relevant information to assist stakeholders in decision-making.

The Internet of things (IoT) [1] is an online, dynamic, rapidly changing world that empowers industries such as healthcare. As IoT contains billions of devices, the spectrum usage efficiency faces a daunting challenge in harboring that many devices within the sparse spectrum. The IoE is a superset of IoT and is replaced by the IoE era, where ubiquitous connectivity is essential. The IoE connectivity is a discipline that provides various solutions for massive data to be exchanged to the edge of the cloud infrastructure. With the emergence of the IoE era, industrial control systems (ICSs) [1] are switching into the Industrial Internet of Everything (IIoE) [1] for various industries, especially for healthcare. The Internet of Healthcare Everything (IoHE) [1] is a sub-classification of IoE/IIoE which explains uniquely identifiable devices connected to the 6G enabled Tactile Internet that are able to communicate with each other for smart healthcare applications, enabling localization and real-time information about healthcare stakeholder assets. In this study, IoT/IoE and IoHE terms are used interchangeably.

To satisfy ever-expanding bandwidth demand, the CT-based IoHE, namely cognitive IoHE (CIoHE) [1,4], is an assuring capability and approach for healthcare professionals to collect the real time physiological data of the patients. Thus, the integration of 6G enabled Tactile Internet, CDI, and IoHE formalizes the 6GCIoHE system [1–6], which plays a vital role in cognitive healthcare applications.

The primary purpose of this study was to develop a novel theoretical framework based on the 6GCIoHE system architecture, applications, and challenges that lie at the heart of the cognitive healthcare domain.

1.1. Background and Problem Formulation

At present, the healthcare sector is facing numerous challenges. The deficiencies of the 5G mobile system as an enabler of IoE have inspired global research activities to focus on the 6G wireless system. The advancement of the 6G Tactile Internet, CDI, and IoHE can play a pivotal role in providing interoperability for healthcare applications to alleviate the challenges of the present healthcare system. Hence, we propose a holistic approach to form the 6GCIoHE system for improving the healthcare domain. Academic researchers assign an esteemed value to theory development. Although there is an increasing interest in comprehending theory, the process of theory advancement is rarely addressed with contributions from other subjects or topics to harmonize them meaningfully. To date, in the context of the 6GCIoHE system, there is neither a central corpus of a well-accepted theory nor a theoretical framework in the literature. Because of the theory development void in the literature related to the 6GCIoHE system for healthcare applications, this study offers comprehensive new knowledge to develop a novel theoretical framework with a significant contribution of value.

Although the issues of privacy, security, and trust were extensively discussed in the literature, a gap exists in establishing the CIoHE global standards. The CIoHE ought to be part of a global healthcare ecosystem to evaluate security, privacy, and trust considerations and not splinter into inconsistent sets of rules or standards. One must engage with healthcare stakeholders to support the development of global standards to foster innovation and promote security and privacy solutions.

1.2. Significant Contributions

The contributions and the significance of this study are as follows: (i) the study discusses the foundations of the CIoHE framework, which helps improve services and experiences, reduce costs, and promote next-gen healthcare facilities and automates the workflow of patient care; (ii) for the first time, the enabling technologies that constitute the 6GCIoHE system architecture demonstrate machine to machine communication, haptic actions, interoperability, data movement, and information exchange that allows the healthcare industry to deliver efficiency; (iii) we describe the opportunities and the challenges that emerge in executing the 6GCIoHE system, including the CDI scalability requirement and the concerns of data security, privacy, as well as risk perceptions. Thus, the primary contribution of this study was achieved by developing the novel 6GCIoHE theoretical framework and bestowing new knowledge in the literature.

1.3. Organization Summary

The structure of the paper comprises of the following sections and the related topics:

Sections	Topics
Literature review	6G enabled Tactile Internet, CDI, and IoHE.
Theoretical foundation	Research boundary, related frameworks, and relevant theories.
Methodology	Model-driven methodology mapping 6GCIoHE system.
6GCIoHE system paradigm	Enabling technologies.
6GCIoHE architecture	A 5-layer modular architecture is proposed.
Applications and challenges	6GCIoHE system for smart healthcare.
Theoretical framework	Building blocks of the 6GCIoHE system theoretical framework.
Consolidated lessons learned	6GCIoHE system for next-gen healthcare.
Recommendations	For academia and practitioners.
Conclusion	Theory's prodigy not covered by prior theories.

2. Literature Review

The literature review and the theoretical framework are intrinsically linked for logically understanding and developing the disparate yet interconnected essential parts of the literature review.

A thorough review of relevant literature in any study undertaking bestows a sound foundation for actuating new knowledge and creates theory development as well as identifies where the research gap exists. The goal was to enhance the comprehension of the 6GCIoHE system technologies that support software, hardware development, testing, and evidence of its use in the healthcare sector. The investigation of topics, research directions, and insights, as shown in Tables 1 and 2 included the following: (i) the theoretical foundation of technologies related to 6G, Tactile Internet, CDI, IoHE; (ii) the 6G Tactile Internet and the IoE systems ascendancy leveraged by software defined networking (SDN) and network function virtualization (NFV) technologies [7], strengthening their comprehensive security, resilience, and flexibility, thus providing substantial added value; (iii) the Tactile Internet for smart healthcare in the 6G domain for ultra-reliability and fast response time; (iv) the 6G enabled Tactile Internet as the next revolution for the haptic communication system; (v) the challenges and the standards for the Tactile Internet in the 6G domain; (vi) the future of population health management (PHM) [8] (p. 10) trussed to cognitive computing, which converts unstructured data into structured data by utilizing massively parallel processing and tiny ML; and (vii) cognitive applications where clinicians can collaborate with cognitive computing systems to improve healthcare.

Table 1. Summary of the 6G/Tactile Internet/CDI/IoHE literature review.

References	Research Directions/Insights
Alsharif et al. [9]	6G Networks Research Activities
Dang et al. [10]	6G Human Centric Perspective
Gui et al. [11]	6G New Horizon and Security
Chowdhury et al. [12]	6G Applications and Research Directions
Janbi et al. [13]	6G/Smarter IoE/AI
Zhang et al. [14]	6G/Super IoT
Ateya et al. [15]	Towards Tactile Internet
Simsek et al. [16]	5G Enabled Tactile Internet
Antonakoglou et al. [17]	Haptic actions over the 5G Tactile Internet
Dai et al. [18]	IoMT to Combat COVID-19
Atlam et al. [19]	IoT and AI
Cisotto et al. [20]	Healthcare Services over Cellular Systems
CISCO [21]	IoE Economy
Siemens [22]	Industrial IoT 2050

Table 2. CC, AI, and ML theoretical framework (TF) review.

References	Research Directions/Insights
Wan et al. [23]	CC and wireless communications for healthcare
Bini S.A. [24]	AI, ML, Deep Learning, and CC
Norden et al. [25]	Promise of CC: Ushering in a New Era
Khalil et al. [26]	CC and the Future Healthcare: IBM Watson
Polson et al. [27]	Cognitive theory-based of user interfaces
Nord et al. [28]	IoT/TF/Systematic Review
Falkenberg et al. [29]	Information System (IS) Theory/Limitation
Aquilani et al. [30]	Value Creation TF

The following steps were taken to investigate the relevant literature: (i) usage of the key terms 6G, CC, MDA, tiny ML, and IoHE; (ii) review of Google, Firefox, and leading journals such as IEEE, MDPI, Science Direct (Elsevier), and Emerald Insight was used related to 6G, Tactile Internet, CDI, IoHE, and healthcare applications as well as challenges; (iii) search of academic research databases related to 5G, 6G, Tactile Internet, CC, tiny ML, CDI, IoHE, and associated theories; (iv) industry white papers were used related to the research paper themes; and (v) security, risk perception, and privacy matters concerning trust in the IoHE were investigated.

3. Theoretical Foundation

3.1. 6GCIoHE System Boundaries and Research Approach

Reflecting on the history of technology accepting theories, scholars have understood the evolution and the development of such theories. Although there has been an increase in comprehending theory, the process of theory advancement is addressed rarely with contributions from other disciplines to harmonize them meaningfully. The reason behind accepting or rejecting any new technology investigation remains the most censorious area in the information communication technology (ICT) discipline. Technology acceptance theories aim to express the concept of the user's knowledge of how one would use such technology. Most technology theories were introduced to measure the degree of satisfaction and acceptance toward any technology of ICT system. Therefore, for any novel technology execution, there would be many variables that influence the execution of a specific technology.

Theories of technology describe the factors that shape technological innovation and the impact of technology on society. From the perspective of ICT, there are four theories that have frequently been utilized in the advent of behaviorism, cognitivism, constructivism, and connectivism as learning theories for the digital age. The aims of ICT theory are threefold: (i) to provide a comprehensive and compelling overview of cutting-edge ICT; (ii) to identify and discuss the fundamental principles underlying technologies such as CDI, 6G Tactile Internet, and IoHE; and (iii) to investigate the reciprocal relationship between these technologies and societies [31].

It is critical to lay higher foundations for theory building from a greater cumulative and coherent research base. Consequently, this research considered the underlying challenges in research of ICT and reflected especially on the 6GCIoHE system to reflect the principles of the disciplined inquiry consisting of cognitive healthcare applications.

3.2. 6G Enabled Tactile Internet Framework for Healthcare

The theoretical foundation of wireless communication is based on fundamental principles from communication and information theory, signal processing, detection, and estimation. The 6G framework is an emerging network technology, and distinct devices characteristically evolve to the subsequent paradigm value-added services and are achieved through the development of a distributed service 6G enabled Tactile Internet framework, as shown in Figure 2, that encourages tactile apps for the healthcare industry.

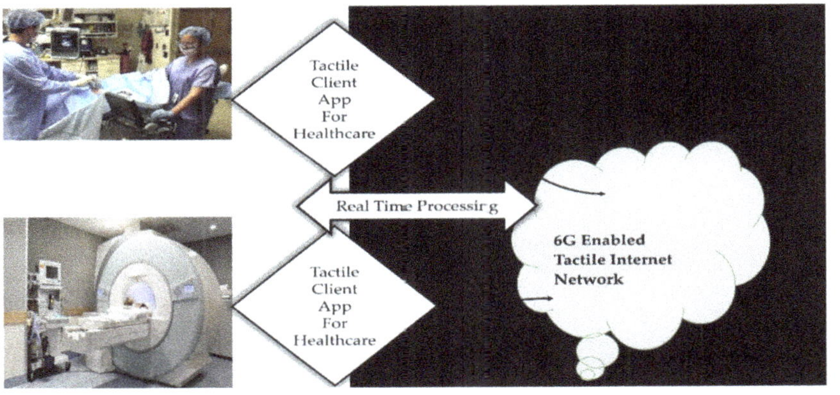

Figure 2. 6G enabled Tactile Internet framework for healthcare apps.

3.3. The IoHE Conceptual Framework in the Healthcare

The significant and most widely used theories for technology acceptance that have been used to study IoT adoption by a great deal of peer-reviewed research are explained using sociology and psychology models associated with six independent variables, as shown in Figure 3, and are as follows: (i) performance expectancy; (ii) effort expectancy; (iii) facilitating condition; (iv) social influence; (v) perceived credibility; and (vi) attitude.

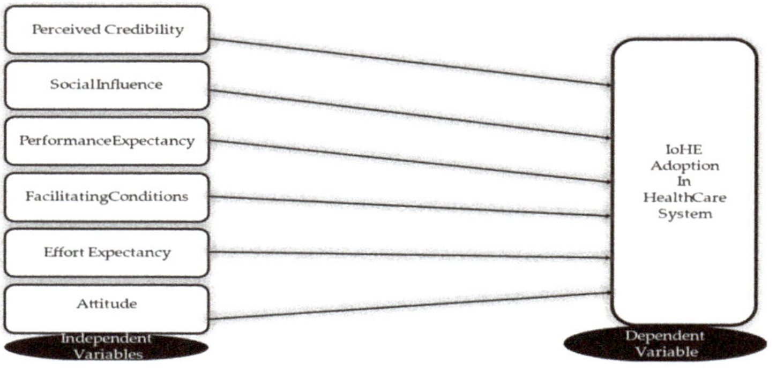

Figure 3. Conceptual framework for the adoption of IoHE in healthcare.

Considering the above, the IoHE conceptual framework was found to be most suitable for establishing relationships and outcomes between independent and dependent variables, which in turn affect the user behavior of the healthcare services.

3.4. The IoHE Pillars in Healthcare

The IoHE pillars comprise data, things (sensors), people, and processes, as shown in Figure 4, that are intelligently connected with billions of sensors to distinguish calibration and appraise their conditions.

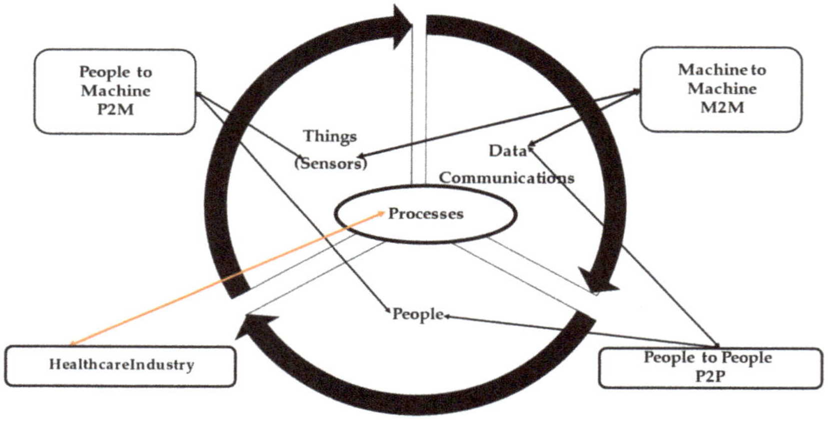

Figure 4. IoHE Pillars in healthcare.

The "Process" is key among "people, data, and things", which interact with each other to deliver value for bestowing immersive patient experience.

3.4.1. Sensors

Medical sensors are the eyes and the ears of the IoHE network. Sensors convert physical signals into electrical signals. As wireless body sensors (WBS) including medical super sensors (MSS) sense data from the patient, the data need to be transmitted to the processing node via the communication network. The functioning IoHE technology depends upon precise measurement by medical sensors. Therefore, it is vitally significant to select appropriate sensors suitable for a healthcare task.

3.4.2. Communication

M2M communication is the key technology behind the IoHE. Values that are sensed by the medical sensors need to be sent for the computing. With the help of 6G Tactile Internet communication, it is easy to establish communication among the various networks using standard IP protocols.

3.4.3. Cognitive Computing (CC)

CC is taking center stage for healthcare, helping to address issues around privacy, and refers to the process of analyzing a problem the way the human brain does. Advances in sensor technology and tiny ML and IoHE devices are integrated to mimic the human brain in solving problems. CC in an IoHE system enables the analysis of hidden patterns that are included in a massive amount of data and improves the ability of a sensor to process healthcare data. The functioning of humans and machines where computers and the human brain truly overlap can be used to improve human decision-making in healthcare applications, bestowing significant value by enhancing the quality of patient care. Single-chip CC is called a microcontroller that functions at the speed of tera- or gigahertz, with low power consumption necessary for the IoHE applications.

3.4.4. Data Analysis

Based on the billions of sensors utilized, the sensors generate massive data. With the mining of a massive amount of data availability and technologies, the world is witnessing the next phase of healthcare. Cloud computing is needed to store the data, but the difficult part is doing an analysis of massive data and identifying useful data. This process is strenuous and developing smart algorithms to gain useful information in real time is a challenging task. The application of MDA and tiny ML can reduce the complexities and the uncertainties associated with the healthcare industry.

3.5. CDI Framework

CDI, which constitutes the integration of CC, tiny ML, and MDA, can be used to palpate a patient's health status, physiologically and psychologically transforming the current healthcare system.

3.5.1. CC, as a Sub-Field of AI

CC applications in healthcare produce superior, best-practice, decision-relevant information for everyone, and computing is analogous to human cognition, rationale, and judgment and has the capacity to deal with symbolic and conceptual information to offer a path to more individualized healthcare decisions.

3.5.2. Tiny ML

Tiny ML is an emerging discipline to revolutionize the healthcare industry that intersects between ML and embedded IoHE devices. The ability to distribute AI resources to memory-constrained devices could have significant advantages for the healthcare sector. Figure 5 shows the AI systems with tiny ML achieving efficiency improvement in computation, engineering manpower, and data efficiency.

This study proposes a novel tiny ML healthcare framework that includes the following levels:

Level 1: Integrate tiny ML techniques into standard analytic tools for one-click access.
Level 2: Select or the tailor predefined predictive models or build a customized model.
Level 3: Optimize predictive models to maximize insight and drive better decision-making.
Level 4: Compare performance over time or against peers to refine improvement interventions.
Level 5: Learn from data to drive future improvements.

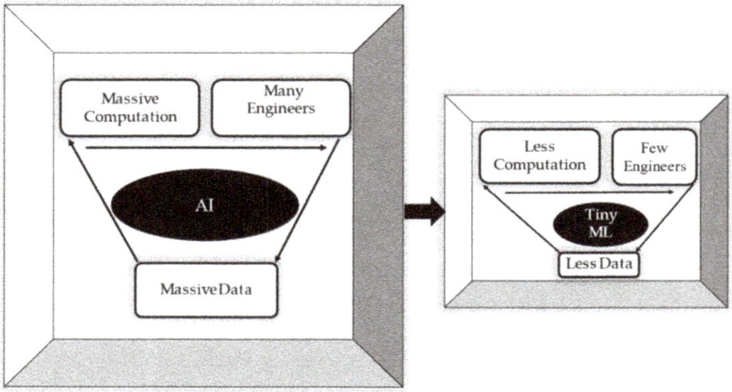

Figure 5. From AI to tiny ML with significant benefits for healthcare systems.

3.5.3. MDA

MDA utilizes huge multiple sets of real time data through tiny ML-based algorithms in this healthcare system to provide a better service for the welfare of society, revolutionizing the healthcare industry.

Tiny ML-based MDA rapidly translates the core areas of a patient's health conditions with comprehensive and accurate reports. There are enormous sources available for MDA to analyze for future endeavors such as electronic health records (EHR), social media platforms, mobile apps, pharmaceutical research, payer records, campaigns, and smart medical devices. Due to the availability of high-quality real time data, the results generated from MDA are more accurate and effective. While medical professionals cannot make sense of billions of data points across millions of patients, tiny ML can improve healthcare with more types of data and more cases to process. Cognitive healthcare has dual benefits: (i) improving patient health and (ii) reducing the costs of sub-optimal treatment plans.

Based on Section 3.5.1–Section 3.5.3 mentioned above, Figure 6 shows the basic elements of the CDI framework that include understanding, verifying, planning, evaluating, attention, and perception to serve as objectives in a specific cognitive task.

3.6. Relevant Theories Supporting the Theoretical Foundation

Since a theoretical framework is defined as the set of existing theories, models, concepts, and relevant definitions that are used in a specific field of study, we structured the theoretical foundation based on the following relevant theories as the proof to support and hold the 6GCIoHE ecosystem for healthcare.

3.6.1. Theory of Mind (ToM)

ToM is a psychological construct that displays the capacity for characteristic thoughts, feelings, and ideals toward others and the ability to apprehend different humans' intellectual states. ToM and intelligence are stimulated through domain-precise tasks, and overall performance is associated with numerous cognitive capacities [32].

3.6.2. Social Cognitive Theory (SCT)

SCT indicates a triadic relationship among the individual, the individual's private behaviors, and outside factors. It is used to apprehend behaviors amongst humans guided through functions and objectives which can be encouraged via their private ideals of self-efficacy and through purpose expectancies from their behaviors within a selected environment. Research affirms the uses and gratifications paradigm and increases it to a theory of the Internet with attendance grounded in SCT [33].

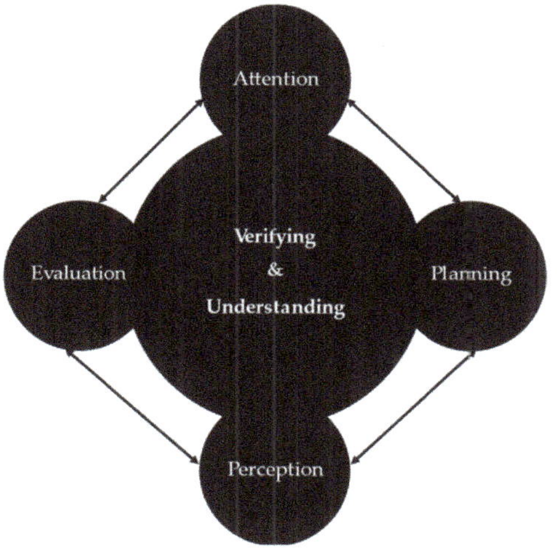

Figure 6. Basic CDI framework.

3.6.3. Theories of Inference and the Mathematics of Probability (TIMP)

TIMP applications are for data statistics evaluation. None of the classical theories of statistics come near discovering new data in an actual scientific problem. Most formal theories, including Bayesian, decision-theoretic, Neyman–Pearson, and others, work with prespecified possibility models. In practice, hypotheses frequently emerge after the data have been examined [34].

3.6.4. The Interface Theory (TIT)

TIT investigates an architecture for IoT applications where so-called "accessors" offer an actor-oriented proxy for devices ("things") and services, and the composition of those interfaces permits the combination of a timed actor model and the pattern, enabling careful assessment of design selections for the 6GCIoHE applications, where "everything" and services interact with the physical world [35].

3.6.5. The Communication Theory (TCT)

TCT studies the process of information, interdisciplinary disciplines of interpersonal communications, psychological paradigm, and philosophical and social dimensions [36].

3.6.6. The Systems Theory (TST)

TST is implemented in the 6GCIoHE systems for evaluation applications. One of the essential mechanisms of systems analysis is systems thinking, enabling the contouring of systems from a broad perspective as opposed to precise activities within the system [37].

3.6.7. The Theory of Transparency (TOT)

The TOT wherein perpetual experience is frequently stated to be transparent and privy to the properties of the objects around an environment, and the fundamentals of transparency require actions and decisions. Since a theoretical framework is defined as the set of existing theories, models, concepts, and relevant definitions that are used in a specific field of study, we structured the theoretical foundation based on the following relevant theories as the proof to support and hold the 6GCIoHE ecosystem for healthcare [38].

3.6.8. Ethical Theory (ET)

The ET offers moral concerns inherent to all the stages of the healthcare sector and drives the ethics of care based primarily on ethical interpersonal relationships and care as a virtue [39].

4. Methodology

As the healthcare industry is under pressure for return on investment and deficiencies in digital technology for healthcare applications, there is an extensive effort from the research community for novel and profitable automation process management as well as semantic management of healthcare platforms. A new era of the IoHE so-called "Cognitive IoHE" was introduced that aims at integrating cognitive technologies into the IoHE-based systems to ensure smart management through enabling cooperation and interaction between the IoHE and humans. Autonomic CC sheds light on unprecedented opportunities for developing smart CIoHE systems and a strong focus on managing complex systems through automating tasks based on specific patterns (monitoring, analysis, plan, execution, and knowledge).

Considering the above, we introduced a collaborative model-driven methodology [40] to map the 6GCIoHE system functions which monitor and perform analysis, planning, and execution of the management process. Within this methodology, a set of cognitive design patterns was suggested to drive the 6GCIoHE system architect to provide flexible, smart IoHE-based applications. These patterns deal with functional and non-functional requirements, ensuring the management of CDI capabilities and scalability issues within a smart healthcare system. The patterns need smart manageability, interoperability, and scalability of the 6GCIoHE system. These patterns are represented to design the structure, to draw the behavior, and to delineate how the management processes should be coordinated to meet the system's functional requirements based on the knowledge pattern. Once the management processes are identified and modeled, the next level is semantic integration management, where mainly information about the system and its environment as well as procedural knowledge (know-how) for decision-making is formalized to be automatically reused by the management processes. However, the ability to manage IoHE data variety, velocity, volume, and system performance in terms of response time and scalability management is a paramount concern. Data quality is crucial, therefore, five traits within data quality—accuracy, completeness, reliability, relevance, and timeliness—are imperative. Figure 7 portrays an overview of the proposed methodology that includes two phases: (i) identification and (ii) formalization.

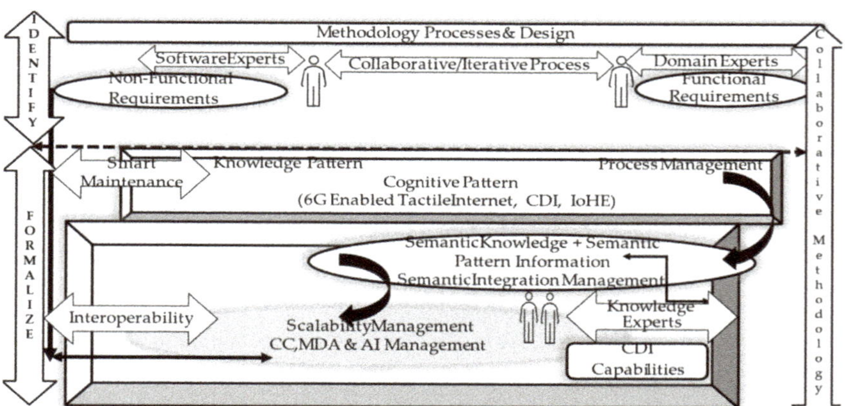

Figure 7. A model driven methodology for the design of 6GCIoHE system.

4.1. Identification

Identification is to retrieve the system functions and to identify the non-functional requirements. It is an iterative process, where the functional requirements are incrementally refined and represented using a use case describing the functions of the system without specifying any implementation details.

4.2. Formalization

Formalization is to focus on formalizing and structuring the identified requirements into concrete models describing the system processes' interactions. Within the formalization phase, sub-levels are introduced incrementally to deal with challenges related to the design of the 6GCIoHE system such as the coordination of management processes and the semantic integration of CDI and IoHE.

5. 6GCIoHE System Paradigm

5.1. The Next Leap Evolution

As shown in Figure 8, the Tactile Internet adds a new dimension to P2M interaction by tactile and haptic sensations, enabling an unforeseeable plurality of new applications, products, and services. Ubiquitous availability, better reliability, and high security with extremely low latency will define the character of the emerging 6G enabled Tactile Internet, introducing numerous new opportunities.

The 6G vision connectivity, as shown in Figure 9, is to brace the development of a global, ubiquitous intelligent mobile society (GUIMS) with the following achievements: (i) solve the limitations of 5G including system coverage and IoE; (ii) achieve up to a hundred times higher data rate, greater system capacity, better spectrum efficiency, and lower latency than 5G to serve the interconnection of everything; (iii) introduce ubiquitous, intelligent, and integrated network with holographic, broader, and deeper coverage, including terrestrial communication, satellite communication, and device-to-device communications;

(iv) serve airspace, land, and sea, realizing a global mobile broadband communication system; (v) work on higher frequency to achieve wider bandwidth, such as better mmWave, terahertz, visible light, etc.; (vi) produce a personalized intelligent network combined with artificial intelligence (AI), especially tiny ML, to offer virtualized personal mobile communication with an endogenous security scheme or function security to offer the capability of self-awareness, real time dynamic analysis, and confidence evaluation, realizing cyberspace security; (vii) merge computation, navigation, and sensing with communications; (viii) adopt a more open architecture with software defined network (SDN), virtualized network functions (VNF) and radio access network (RAN) to realize self-intelligent development and rapid dynamic deployment of network functions; (ix) generate massive data through the IoE combining with novel technologies such as cloud computing, edge computing, tiny machine language (ML), blockchain, etc., realizing group collective intelligence (swarm intelligence); (x) act as a "global wireless power grid"; and (xi) develop a better massive MIMO.

In the emerging 6G and beyond of wireless communication, an increasing number of ultra-scale intelligent factors will result from interference exploitation. To manage this exploitation, challenges for detection algorithms in uplink MIMO systems exist, especially for higher-order quadrature amplitude modulation (QAM) signals.

5.2. The 6G Enabled Tactile Internet End-to-End Architecture

As shown in Figure 10, the architecture can be split into three distinct domains: (i) a master domain; (ii) a controlled domain; and (iii) a network domain.

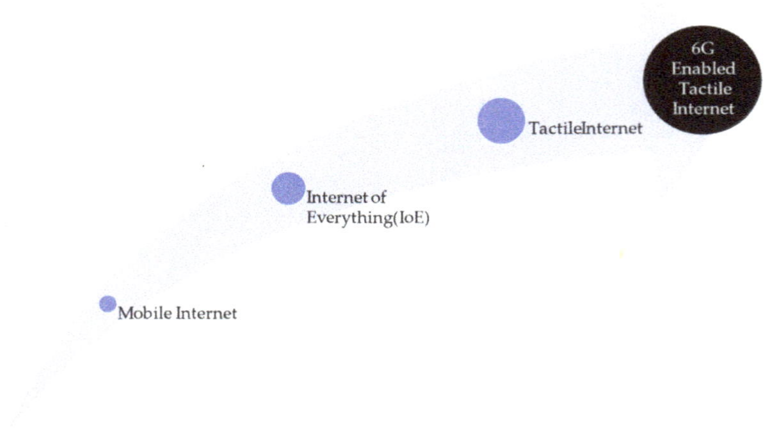

Figure 8. Leap evolution of 6G enabled Tactile Internet.

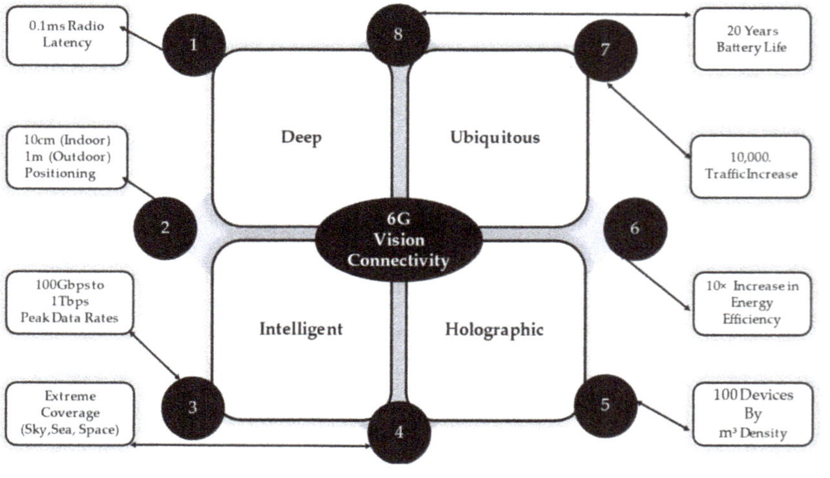

Figure 9. 6G vision connectivity.

5.2.1. The Master Domain

The master domain constitutes a human (operator) and a human–system interface, a haptic device, and has the provisioning for audio and visual feedback which converts the human input to tactile input through tactile coding methods, allowing the user to touch, feel, and manipulate objects in real and virtual environments. This primarily controls the operation of the controlled domain.

5.2.2. The Control Domain

The control domain includes a teleoperator (controlled robot) and interacts with objects in the remote environment. It is directly controlled by the master domain via various command signals. The energy is transferred between the master and the controlled domains via command and feedback signals to achieve the control loop closure.

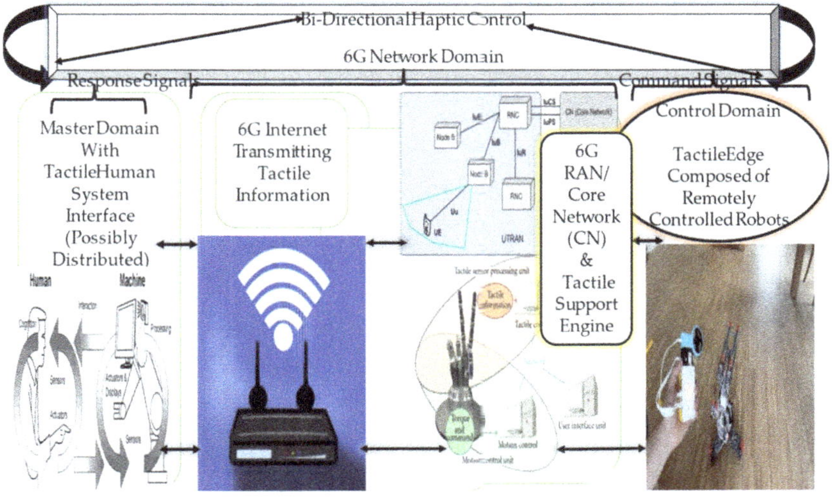

Figure 10. Functional representation of the 6G enabled Tactile Internet architecture.

5.2.3. The Network Domain

The control domain facilitates the medium for full-duplex communication via 6G between the master and the controlled domains to attain the coupling between the humans and the remote environment.

5.2.4. RAN/CN

The 6G-driven communication architecture composed of RAN and CN meets the basic requirements to realize the 6G enabled Tactile Internet. To achieve this objective, the essential functions of the 6G RAN in the Tactile Internet ecosystem are as follows: (i) radio access technologies (RATs) such as mmWave, massive MIMO, and full-duplex; (ii) Tactile QoE/QoS for tactile applications in conjunction with M2M and smart grids applications; (iii) efficient packet delivery through reliable radio protocols and physical layer; and (iv) novel medium access control (MAC) techniques, where the air–interface conflicts are optimally controlled. The key functionalities of the 6G CN associated with the Tactile Internet are as follows: (i) dynamic application of QoS provisioning; (ii) edge–cloud access; and (iii) security.

5.2.5. End-to-End Latency

End-to-end latency of less than 1 ms remains the main challenge related to the system realization of a Tactile Internet. As the human reaction time is of the order of 1 ms, the P2M communication faces no cyber issues; therefore, it is a vital requirement for a reliable Tactile Internet system. The following are the factors that affect the end-to-end delay: (i) propagation delay; (ii) transmission delay; (iii) queuing delay; (iv) coding and decoding process; (v) routing process; and (vi) protocol stack optimizations. The Tactile Internet must handle the tactile data in the same manner as the conventional audio/visual information. Hence, the communication between smart sensors or devices for healthcare requires encoding mechanisms which provide transmission of tactile data over packet-switched networks.

5.3. Revolutionizing Healthcare with CDI

5.3.1. CC Architecture for Cognitive Healthcare

Computing architecture based on massive data/CC system architecture includes network technologies (6G Tactile Internet, IoHE), data analytics, and cloud computing.

The primary applications of the CC system, as shown in Figure 11, include cognitive healthcare, health monitoring, etc. Every layer in CC system architecture faces technological challenges and system needs. The development of the IoHE collects a variety of valuable information to bestow greater understanding and transmission of data, where it can provide an important source of information for the realization of CC. This system technology can process relevant information using intelligent computing techniques such as cloud computing, tiny ML, and data mining to make decisions. Since the emergence of tiny ML, cloud computing resources have provided tremendous benefits for advancing CC. The IoHE and cloud computing can present software- and hardware-based CC, while MDA provides novel ways to explore opportunities for data such as human massive data thinking.

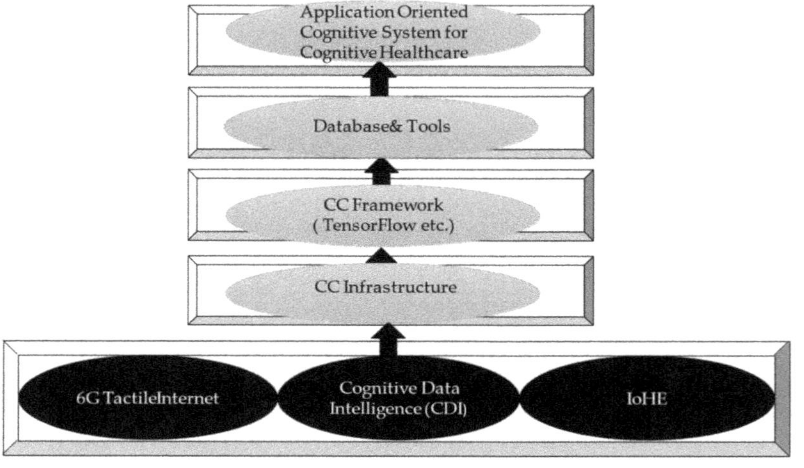

Figure 11. Cognitive system architecture for cognitive healthcare.

5.3.2. Linking CC, MDA, and Tiny ML for Healthcare Value Creation

CC is a key enabler of these human-centered smart systems and is a subdivision of AI, and anything that is cognitive is also AI. CC is also an AI-based system that enables it to interact with humans as a fellow human and interpret the contextual meaning to help the humans in decision-making. The mapping between features of massive data and cognitive computing and the features associated with the link between massive data and CC are shown in Figure 12.

CC can be used to process large volumes of data and comprises the concepts of observation, interpretation, evaluation, and decision (described below) that are mapped to five attributes of massive data: (i) volume; (ii) variety; (iii) veracity; (iv) velocity; and (v) value:

1. Observation is imperative in a CC system in which data aggregation, integration, and examination are done, hence, data volume must be available for observation.
2. Interpretation provides a better understanding and solving of complex problems in the presence of a variety of information sources, hence, variety indicates that data can be sourced in a variety of ways, such as IoT, social media, email, etc.
3. Evaluation is the natural ability of a human being to produce information, therefore, processing a massive amount of data needs evaluation in real time by the CC system. Velocity is a feature of big data, where speed is a vital requirement for production control and processing.

4. Decision feature is to make decisions by a CC system as per the analyzed data. Veracity defines quality prediction, uncertainty, and reliability of data. The value attribute indicates the valuation of massive data prior to it becoming knowledge creation.

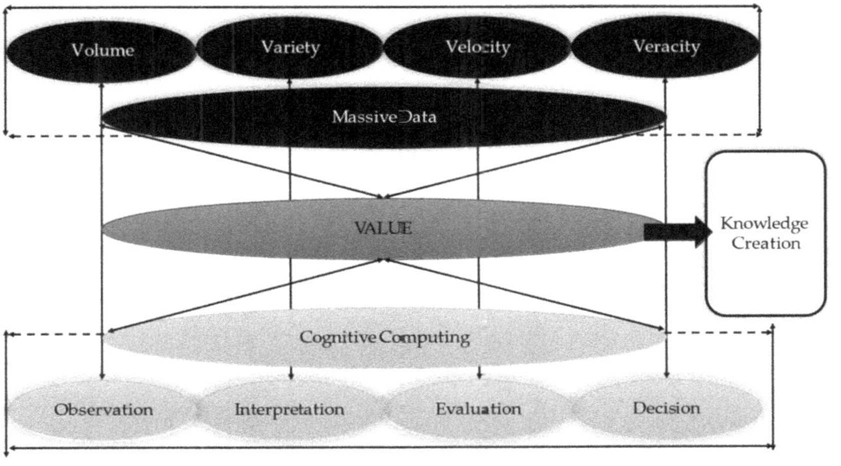

Figure 12. Linking between CC and massive data for knowledge creation.

Quantifying the health condition and the associated risk of an individual is one of the biggest innovations. The next phase of healthcare will bring new inventions into the picture which will not only increase the lifespan of individuals but will also predict diseases through eXR analysis. This revolution has just started, and the benefits of these technologies in the healthcare sector are immense. Technology innovation has been a natural phenomenon for many decades, and advancement of IoT and AI, especially tiny ML algorithms, has driven human-centered smart systems to higher-quality services such as smart healthcare. Analyzing massive data by humans is a lengthy process. IoT devices and connections will grow in the course of time. This exponential growth is leading to an explosion of data, and the enormous amount of data being produced on a continuous basis is termed "massive data". The concept of massive data has brought out many definitions by other researchers in the healthcare domain.

Healthcare organizations seek to analyze raw data, as they want to identify the trends for further profit maximization. Data analysis by humans can be time-consuming. Therefore, use of sophisticated CC can be applied to crunch the massive amount of data. The dire need to address the concerns of data deluge led to the emergence of MDA. Hence, MDA has gained significant importance, as it enables organizations to achieve a sustained competitive advantage.

AI and data analytics are not new concepts. The advancements in MDA and AI have enabled the early detection and prediction of such diseases. Tiny ML emerged as one of the most critical game-changing technologies and propels healthcare organizations to provide better security and a deeper understanding of the relationships and trends in inpatient data. Healthcare organizations are at the helm of ML and must use AI to remain competitive and secure in the IoHE healthcare platforms. When selecting an IoHE platform, organizations must understand which platform functions use tiny ML and IoT data, testing and validation protocols for ML, and how the ML updates. If a healthcare organization uses tiny ML for security, the organization must confirm if the patient data are used to retrain and update the ML's knowledge base to best reflect and improve the ML's trustworthiness.

The application of CDI (CC, MDA, and tiny ML) is going to reduce considerably the complexities and the uncertainties associated with the healthcare industry. With the mining of a massive amount of data availability and technologies, the world is witnessing the next phase of healthcare applications. With the availability of a vast amount of daily data and tiny ML, the algorithms bestow better results with higher accuracy and in less time for the healthcare sector. Thus, the integration of CC, MDA, and AI, especially tiny ML, bestows the co-creation of value by the process of transferring data to information to knowledge to wisdom, as shown in Figure 13, that can help to better understand the complexity of data deluge.

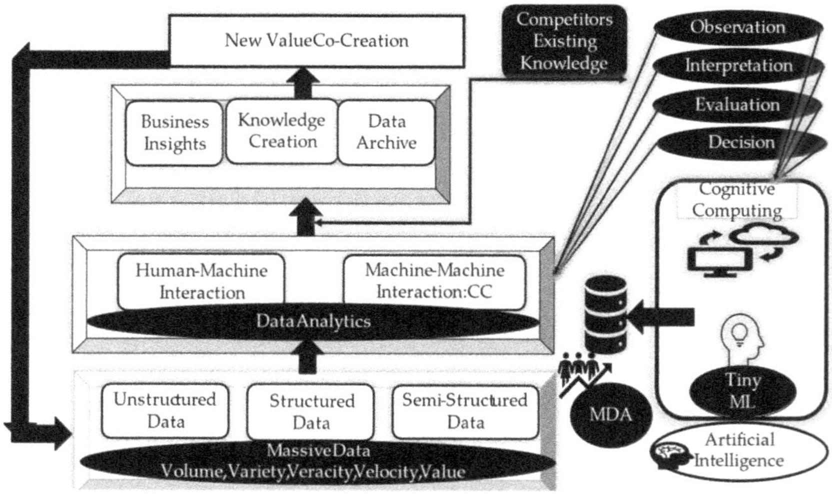

Figure 13. Linking CC, MDA, and AI for healthcare value creation.

6. The 6GCIoHE System Architecture

Research shows that there is no consensus on the 6GCIoHE system architecture. The three-layer architecture defines the main idea of the CIoHE, but it is not sufficient for researching the 6GCIoHE system. Healthcare applications have different requirements; therefore, the five-layer architecture is chosen to directly affect the application performance. Furthermore, when a healthcare project is implemented with cutting-edge technologies (6G Tactile Internet, CDI, IoHE) and broad application areas, the five-layer architecture is considered best. That is why we proposed the five-layer architecture. The five-layer 6GCIoHE architecture proposed is inspired by the layers of processing in the human brain and the ability of human beings to think, feel, remember, make decisions, and react to the physical environment. The proposed 6GCIoHE system architecture addresses the latency issues present in cloud-based solutions. This architecture also supports time-critical healthcare applications that require an emergency response and contains five layers, as shown in Figure 14, namely: (i) the sensing layer (perception layer) comprises the physical elements such as sensors, devices, and machines; (ii) the communication layer (transport layer) includes the communication systems such as 6G Tactile Internet; (iii) the processing layer (middleware layer) consists of communication protocols, cloud computing, and CDI software to analyze data; (iv) the application layer deals with intelligent applications; and (v) the business layer defines the business model.

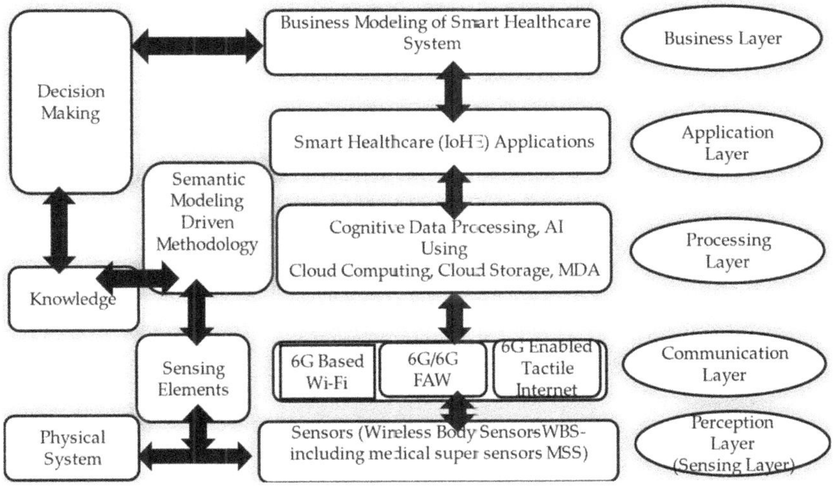

Figure 14. The 6GCIoHE system architecture.

To capture real time data concerning the physical world, the CIoHE phenomenon is fast gaining momentum in different application domains, especially in healthcare. Inspired by the human nervous system and cognitive abilities, a set of autonomic cognitive design patterns alleviate the design complexity of smart IoHE-based systems while taking into consideration CDI and scalability management. These patterns are articulated within a model-driven methodology that we proposed in Section 4 for developing a flexible cognitive monitoring system to manage a patient's health based on heterogeneous sensors.

The usage of the 6GCIoHE system technology in the healthcare domain is still in its nascent stage. As such, many challenges need to be addressed by the research communities and the industry. Some challenges and the existing solutions are discussed in the next section.

6.1. Sensing Layer

The sensing (perception) layer contains different sensors for monitoring the health parameters of patients and transmits the data to application devices. As a WBS, including MSS, senses data from the patient, it needs to be transmitted to the processing node via the communication network. Some of the key technologies useful for transferring data from the MSS to the closest processing node are 6G Tactile Internet or 6G-based fixed wireless access (FWA) or 6G Wi-Fi [1] (p. 15).

6.2. Transport Layer

The technology at the heart of communication for the transport layer in the IoHE architecture would be the 6G Tactile Internet that offers speed, accuracy, real time latency, energy conservation, and reliability among sensors and actuators close to a human body. The 6G Tactile Internet, 6G Wi-Fi, or 6G FAW can help to reduce healthcare costs and improve quality by using different sensors to read a patient's health parameters, thus improving the patient's quality of life.

6.3. Processing Layer

Cloud computing is a computing model that bestows a pool of configurable resources that can be accessed ubiquitously through the 6G enabled Tactile Internet to provide cloud services such as Software-as-a-Service, Platform-as-a-Service, Infrastructure-as-a-Service, and Database-as-a-Service. Cloud computing is one of the enabling technologies for IoHE and offers significant benefits to healthcare organizations such as rapid elasticity, self-healing, and self-configuration. It is employed at the processing layer, also called the middleware layer, in the five-layer 6GCIoHE architecture. In the IoHE, the multitude of sensors generates a massive amount of data that can be stored and analyzed with the help of cloud infrastructure. Medical professionals can monitor a patient's health, whose information is collected through various sensors and is stored in the cloud. Even though the cloud provides many advantages, one of the key challenges is latency. The process of data collection and analysis by the sensors is done through the Internet and needs to reach physicians. This process takes a significant amount of time, which might not be suitable for emergency healthcare services. The ideal solution to reduce time delay is the latency offered by 6G enabled Tactile Internet, or 6G Wi-Fi, or 6G FAW networks.

6.4. Application Layer

In this layer, state-of-the-art applications of the 6GCIoHE system-related technologies in the healthcare domain are presented. Healthcare applications have different requirements; therefore, the five-layer architecture is chosen to directly affect the application performance. Applications with similar themes and scope are therefore grouped into categories as follows: (i) real time monitoring and alerts generation; (ii) telemedicine; (iii) chronic disease detection and prevention; (iv) home and elderly healthcare. Each of these healthcare application categories and the role of the 6GCIoHE system and the associated technologies are as follows: (i) through the IoHE, WBS can be deployed on a human body to measure different parameters, and data can be analyzed for prescribing the necessary medication to the patients in emergencies; and (ii) an e-health system is used for continuous monitoring of ECG, temperature, foot pressure, and heart rate.

6.5. Business Modeling Layer Value Co-Creation through Business Models

The business model concepts underlying the disease have their precedents in technology-based businesses that closely mirror those seen in the ICT industry. The evolution transformed through product innovations that compelled it to reinvent its business model, shifting to provide services and solutions for smart healthcare. As with healthcare today, ICT a generation ago was part of a classically maturing market with disruptive innovation. One must re-imagine business models to find ways to innovate at scale and with speed as a differentiator for improving healthcare and earning superior returns.

7. 6GCIoHE System for Healthcare Applications and Challenges

7.1. Applications

Currently, the amount of healthcare data is doubling every 73 days. CC is defined as a system that learn at scale, allowing it to interact with humans more naturally. CC can understand the natural language and can process massive amounts of data to comprehend and learn from them, helping healthcare providers with care plan enhancements. Cognitive systems bestow advice to individual patients and caregivers by developing deep domain insights and adducing this information to patients in a usable, natural, and timely manner. The ICT infrastructure needs to be malleable enough to harmonize applications with distributed devices and fast-track digital applications with IoHE systems to ensure data protection and powerful security. There are several new IoHE solutions that help doctors and patients communicate without needing in-person visits to get better treatment and save money [33] (p. 3).

The remote symptom monitoring has improved for both patient outcomes and experience versus patients who had routine in-person visits. The symptom monitoring can also allow for real time adjustments to treatment besides simplifying data collection and transmission between a patient and doctor.

The 6GCIoHE system can address the following wide-reaching capabilities used across disciplines, sectors, or treatment efforts and medical conditions: (i) symptom monitoring/tracking; (ii) medicine adherence; (iii) ingestible sensors; (iv) hygiene monitoring;

(v) body scanning; (vi) smart labs; (vii) diabetes; (viii) Parkinsons; (ix) asthma; (x) cognitive/mental health; and (xi) Alzheimer's.

The future of population health management (PHM), as shown in Figure 15, will be trussed to CC, which converts unstructured data into structured data by utilizing massive parallel processing and AI to search the healthcare literature.

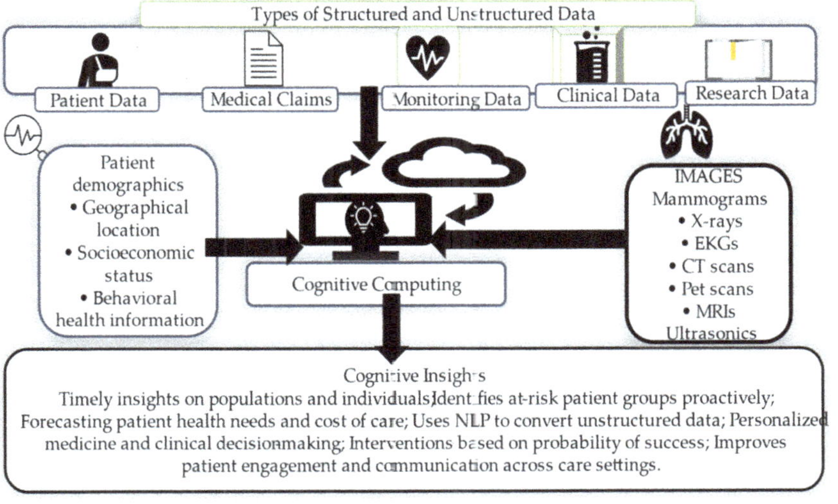

Figure 15. Cognitive computing for population health management.

A new era is emerging, from digital to cognitive, for healthcare applications where clinicians can collaborate with cognitive computing systems to improve healthcare. With ever-increasing healthcare information from connected medical devices, personal fitness trackers, implants, and other sensors that collect a massive amount of real time data, one person in their lifetime is likely to create more than 1 million gigabytes of health-related data.

A smart monitoring system is essential to manage patients' health evolution based on WBS, including examples such as: (i) in the context of managing diabetes, WBS can be used to measure the "Blood Sugar". Thus, the cognitive monitoring management pattern identifies the interactions among the WBS and the physicians for visualization and anomaly detection; (ii) the semantic knowledge pattern manages data heterogeneity and enables the collaboration among the WBS and the physicians; and (iii) the massive data stream detection pattern manages data heterogeneity and velocity.

7.2. Challenges

At present, the healthcare sector is facing numerous challenges, however, emerging cutting edge digital technology, such as the 6GCIoHE system, can assist to stay ahead of the curve and lend the following solutions: (i) healthcare mobile and web applications driven by 6G Tactile Internet can offer powerful virtual consultations; (ii) social networking applications can be used by healthcare professionals across the globe via the 6GCIoHE system; (iii) innovation in the Medicare EHR system can make reimbursements easily; (iv) data management systems can automatically oversee the inventory; (v) cloud plus data analysis capabilities enables efficient management; (vi) block-chain based database solutions need to be implemented; (vii) prescription based online medical stores need to be established; (viii) remote surgery is one of the typical application scenarios for the 6GCIoHE system; and (ix) the 6G Tactile Internet communication system is also a typical application scenario for emotional remedy purposes.

The following principles offer stakeholders the opportunity to address the IoHE security challenges.

(i) Security implementation at the design stage: in smart healthcare, it becomes a reality of an innovative concept that offers the best service to patients. The IoHE takes new challenges in the healthcare field to create excellent support systems for patients and medical professionals. Comprehension of consequences for the failure of a device enables healthcare stakeholders to suggest informed risk-based security decisions. Hence, one should design with system and operational disruption in mind so that the failure does not lead to greater systemic disruption, e.g., WBS is used to sense and capture the data related to patient health/disease via the Internet, and devices display continuous process monitoring; (ii) promotion of security updates and vulnerability management by leveraging cryptographic integrity and authenticity protections to address vulnerabilities via software updates and develop automated mechanisms; (iii) develop proven security practices enabling security by default through unique, impossible to hack usernames and passwords. Failure to execute adequate security measures could be damaging in terms of reputational costs and financial costs; (iv) security measures prioritization must be incorporated based on potential influence. The awareness of a device's intended use and environment helps stakeholders to consider the technical characteristics of the IoHE device and the security measures that may be essential. The mitigation planning and analysis should help prioritize decisions to execute additional security measures. Identify and authenticate the devices connected to the network; (v) propagate transparency across the IoHE spectrum: stakeholders should include vendors and suppliers in the risk assessment process to create transparency, enabling an increase of awareness of potential third-party vulnerabilities and promoting trust. Development of a software bill of materials that can be used as a means of building shared trust among the stakeholders should be realized. Furthermore, it can serve as a valuable tool in the IoHE ecosystem to manage risk and patch any vulnerabilities; and (vi) connectivity should be done carefully regarding nature and the purpose of connections to enable healthcare stakeholder decisions. One should build in controls to allow stakeholders to disable network connections when needed or desired to enable selective connectivity.

8. Toward a Theoretical Framework

8.1. Theory's Focal Phenomena

Theories offer an illustration of one's notion of a subset of real global phenomena. The high-quality theory is thought to enhance the knowledge of the researcher and other scholars' know-how within the theory's domain. It additionally served to enhance practitioners' abilities to perform efficiently and efficaciously within a theoretical framework. A theory makes novel contributions to a discipline in the following ways: (i) prior theories are not covered by a theory's focal phenomena; (ii) a theory is probably taken into consideration as novel because it frames or conceives existing, well-known focal phenomena in new ways; and (iii) a theory's novelty arises due to essential modifications it makes to an existing theory, more precisely, adding or deleting constructs or specifying the boundary of the theory more concisely.

Thus, the theoretical framework proposed in this paper falls under a theory's focal phenomena that were not included in prior theories.

8.2. Security, Privacy, and Familiarity Affecting Trust in the 6GCIoHE System

To date, no study investigated the approaches where the 6GCIoHE system security, privacy, and familiarity can affect trust and, in turn, how trust can affect risk perception and attitudes toward use in healthcare applications. No research considered ways in which risk perception can mediate to bolster the relationship between trust and users' attitudes toward using the 6GCIoHE system technology at the heart of communication for the transport layer in the IoHE architecture. This would be the 6G Tactile Internet that offers speed, accuracy, real time latency, energy conservation, and reliability among sensors and actuators close to a human body. The 6G Tactile Internet, 6G Wi-Fi, or 6G FAW can reduce healthcare costs and improve its quality by using different sensors to read a patient's health parameters, thus improving the patient's quality of life.

Cloud computing is one of the enabling technologies for IoHE that offers significant benefits to healthcare organizations such as rapid elasticity, self-healing, and self-configuration. It is employed at the processing layer, also called the middleware layer, in the five-layer 6GCIoHE architecture. In the IoHE, the multitude of sensors generates a massive amount of data that can be stored and analyzed with the help of a cloud infrastructure. Medical professionals can monitor a patient's health, whose information is collected through various sensors and is stored in the cloud. Even though the cloud provides many advantages, one of the key challenges is latency. The process of data collection and analysis by the sensors is performed through the Internet and should reach physicians. This process takes a significant amount of time, which is not suitable for emergency healthcare services. The ideal solution to reduce the time delay is the latency offered by 6G enabled Tactile Internet, 6G Wi-Fi, or 6G FAW networks. In this layer, state-of-the-art applications of 6GCIoHE system-related technologies in the healthcare domain are presented.

The business model concepts underlying the healthcare ecosystem have their precedents in technology-based businesses that closely mirror those seen in the ICT industry. As with healthcare today, ICT a generation ago was part of a classically maturing market with disruptive innovation. One must re-imagine business models to find ways to innovate at scale and with speed as a differentiator for improving healthcare business models and earning superior returns. Consequently, this study sheds insights on developing a novel framework to measure the causes and the effects of contingency factors and how they can influence users' attitudes toward using the 6GCIoHE systems. Security, privacy, and familiarity factors are discussed below in more detail to pave the way for developing the theoretical framework.

8.2.1. Security

Researchers reveal significant insecurity in the IoHE devices. The IoHE wireless devices pose many security challenges such as intrusion, denial of service, forgery, or heterogeneous network threats. Hackers can use ransom ware to target vulnerabilities in the Windows operating system to prevent healthcare professionals from accessing affected devices.

Security is defined as the protection of software and hardware from misuse, malfunction, unauthorized access, damage, disruption, and misdirection. Security is important to the IoHE applications because of sensitive health data privacy; hence, the following security solutions are imperative to protect data from attacks: (i) access control is a key step in protecting IoHE applications and health data, therefore, well-designed access control and strong access management must be implemented to ensure healthcare data security and privacy. Through access control systems, an organization can restrict and monitor the use of critical data and protect privacy and security. Furthermore, healthcare stakeholders should have training and awareness of information security to provide security of IoHE applications and health data; (ii) the IoHE devices collect data from various environments, hence, physical security is paramount for the IoHE devices. Physical security of IoT health devices and health data involves protection against environmental threats, accidents, physical sabotage, and theft; (iii) IoHE devices should have replacement devices for protecting physical attacks so that the IoHE devices continue collecting and transferring data without interruption; (iv) network security, such as 6G Tactile Internet, is an important issue for IoHE devices and applications. All IoT devices connect to networks and communicate with each other over a network. Firewalls, filtering structures, internet protocol (IP) security, and secure sockets layer/transport layer security should be used to ensure network security.

Ethical hacking can identify security vulnerabilities and risks for an organization. Moreover, most of the devices are designed without security, making them easy goals for safety breaches. By using the IoHE hacking tools, one can secure the system and the infrastructure within the organization. This process helps to create an ideal strategy that effectively and efficiently incorporates tools to identify and resolve security vulnerabilities.

Ways to secure the IoHE devices include: (i) developing an exhaustive map of healthcare assets; (ii) following the best security practices; (iii) effective authentication implementation; (iv) insulating devices that do not have a built-in control; and (v) usage of appropriate tools.

8.2.2. Privacy

Privacy has many definitions in the literature. Privacy is an important topic for information security in the healthcare system in the world. In this study, privacy is defined as freedom from unauthorized intrusion and an important challenge in the IoHE environment due to the availability of sensory devices and the speed and the volume of information flow. Any compromise of privacy leads to problems such as eavesdropping, unauthorized access to or alteration or destruction of information, hacking, identity theft, forgery, and social engineering.

Data privacy is a fundamental issue for IoE health devices and applications because of the ubiquitous character of the IoE environment. IoE health devices connect to each other for data transmission and a strong encryption algorithm that is used for data to be encrypted over a secure network such as 6G Tactile Internet. Healthcare data are collected from the IoE devices via remote access mechanisms which have some challenges regarding privacy and security. Data collected by the sensor are transmitted to the database or the cloud over a network such as 6G Tactile Internet. Since healthcare data include essential significant information, the world's hackers want to capture health data. Therefore, the privacy of data must be protected. Some health organizations are reluctant to adopt the IoE because of fears of privacy compromises, particularly in cases that involve medical data in which maintaining privacy and anonymity of the user is of the utmost importance due to legal and statutory requirements, which in turn affects trust to adopt the IoE in the healthcare domain.

The IoHE application must have a "need-to-know" principle for authorization management. The stakeholders should have enough information about procedures, guidelines, standards, and policies related to data security and privacy. All IoHE devices, IoHE applications, and network components log must be collected with central log management systems. Besides, central log management or security information and event management must have auditing to ensure security. Undesirable events must be reported to a security team quickly to interfere with unwanted events. Central logs management or security information and event management must have strong authentication and authorization to monitor the audit log. The log should be checked continuously.

8.2.3. Familiarity

Familiarity means good knowledge of some fact or knowledge primarily based on preceding interactions. Scholars verified the significance of familiarity and trust in an e-commerce potential and argue that familiarity has an indirect positive influence on the intention to undertake endorsed agents.

8.3. Trust in the IoHE

Researchers use trust in the IoHE as a conventional trust framework to discuss security in cell networks as the important anchor of monitoring device behaviors, identifying devices, connection protocols, and the related procedures to devices. Security measures at the device level could be adopted to enhance security. At the network level, security could be improved by using point-to-point encryption techniques based on cryptography algorithms, message integrity verification strategies, and trusted routing mechanisms. Security measures to prevent data security and privacy are required to be adopted at the cloud level, and suitable training concerning awareness is needed at a human level.

Researchers proffered the following to attain trust in the IoHE: (i) design of a trust framework to be included within the development of any IoHE entity; (ii) advanced dynamic protocol for trust management which enables IoHE systems to deal with misbehaving nodes whose status or conduct might change dynamically; (iii) offered an extensible trust model that was seamlessly integrated into the IoHE ontology and focused on the IoHE-trust modeling, an ontology for fuzzy semantics reusing existing trust models; (iv) designed meta-models for contractors by defining privacy and quality of context conventions independently of those of the users and the creators for the independent management of privacy in the IoHE; and (v) established a proper trust–management control mechanism primarily based on the architecture modeling of the IoHE.

However, the dearth of consideration of trust in relation to the quality of experience (QoE) is also seen as a shortcoming and is one of the significant challenges concerning trust in the IoT establishment of remote IoHE devices.

Devices on the Internet must be trusted to ensure privacy and security. Trust management is important for IoHE devices and applications to provide security and privacy of data. Hackers could connect devices to manipulate data in the IoHE applications. Data collection trust is a serious issue because a massive amount of data is collected from devices and are utilized by IoHE applications, which need make the right decisions about patients to improve the quality of healthcare.

8.4. Risk Perception

"Risk perception" is described as the subjective judgment that people make about characteristics and severity of a risk. Researchers investigated a well-known perceived risk scale to undertake the IoHE applications. Research shows a correlation between user's risk perception related to the IoHE device being able to prevent one from being hacked (one of the risk components). Researchers also found that (i) respondents are not in consensus concerning the perception of risk; (ii) perceived risk influenced consumers' online behavior; (iii) perceived risk is a major obstacle in the IoHE adoption; (iv) analysis of risk perception among users of smart devices linked to the IoT at home and found that risk perception is associated with knowledge of and anxiety regarding the devices; (v) risk perception also is a key factor in determining the IoHE adoption; and (vi) trust is a crucial factor in adopting the IoHE satisfactorily and in regard to expected transaction outcomes.

8.5. Attitude towards Using the 6GCIoHE System

Attitude can be described as an experience or opinion regarding something or someone or a way of behaving caused by something or someone. Scholars discovered the following associated with mindsets in the healthcare domain: (i) doctors embrace notable attitudes towards the IoHE based medical devices, which meant that they are privy to and are prepared to adopt technology and ascribe an exceptional quality to the information transmitted; (ii) maximum customers of the IoHE in healthcare support progressive perspectives concerning valuable features and desire solutions in regions which include inventory or material tracking, identification, and authentication that could make healthcare offerings more effective, convenient, and safe; and (iii) even patients consider favorable perspectives employing the IoHE devices.

8.6. Building Blocks of 6GCIoHE System Theoretical Framework

Based on Sections 2–7 and Sections 8.1–8.5 above, the advancements of the proffered theoretical framework focused on the building blocks and the practices, as shown in Figure 16, are the following:

(i) The design of the 6GCIoHE system paradigm based on the theoretical foundation which includes the relevant theories; (ii) the 6GCIoHE architecture to exploit the challenges and the opportunities in the cognitive healthcare applications; (iii) the collaboration among the healthcare stakeholders within the 6GCIoHE system value chain with the interactive actions between 6G Tactile Internet, CDI, and IoHE applications to achieve optimum performance in the healthcare sector; (iv) the security, the privacy, and the familiarity affecting trust in the IoHE implementation; (iv) the risk perception as an important factor in determining the IoHE adoption; and (v) the risk perception mediating the relationship between trust in the IoHE and users' attitude towards using the IoHE.

The proffered theoretical framework would assist in understanding, evaluating, and accessing the emerging 6GCIoHE system for millions of people for higher overall performance of the healthcare applications.

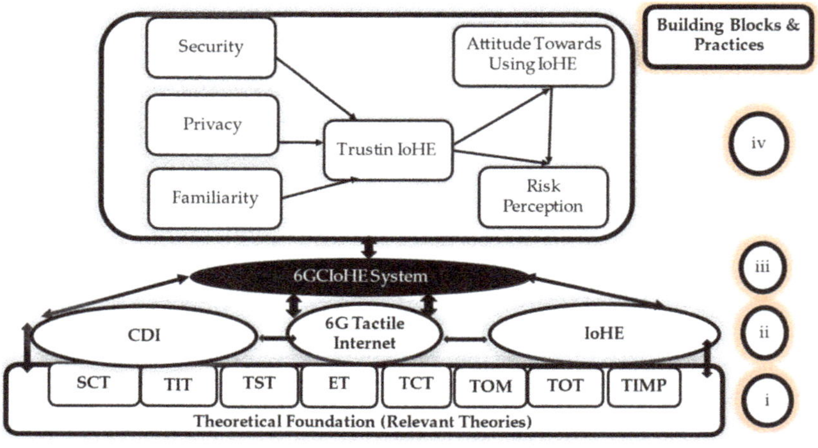

Figure 16. The 6GCIoHE system theoretical framework.

8.7. Theoretical Framework Contributions

An overall contribution of our framework is the establishment of theory for the 6GCIoHE system adoption and implementation, which was previously non-existent in the literature.

The additional contributions of the framework components as related to the healthcare industry and research are as follows: (i) even though the 6GCIoHE system involves M2M communication, the human component cannot be ignored from the model; and (ii) the 6GCIoHE system, a multilayer stack of technologies, contributes new knowledge regarding privacy and security to gain user trust.

9. Consolidated Lessons Learned and Future Research Direction

Currently, the radio channel above 140 GHz and future gigabyte Wi-Fi with IoT is little understood. Hence, 6G wireless and beyond should be pursued with cutting edge work in propagation measurements and modeling above 140 GHz. One important new technology being researched is 6G cell-free massive MIMO, a radical network architecture with the goal of eliminating interference between cells with higher-frequency spectrum bands, including THz bands.

The IoHE network demands massive and smart connectivity, huge bandwidth, and real time latency with an ultra-high data rate for a quality healthcare experience. The emerging 6G communication system is expected to provide intelligent IoHE services ubiquitously at any time to improve the quality of life of human beings.

The proliferation of eHealth, including cognitive healthcare services, challenges the ability to meet their stringent quality of service (QoS) requirements, i.e., continuous connection availability (99.99999% reliability), extremely low latency (sub-ms), and mobility support. The swanky intelligence of emerging 6G networks combined with a wider spectrum will guarantee key performance indicators (KPIs) and gains in spectral efficiency by up to 10 times. The 6G enabled Tactile Internet and its aggregation with multidimensional techniques offers a newly distributed security paradigm in the context of intelligent IoHE applications with less than 1 ms latency for connecting homes to hospitals, healthcare people, medical devices, and hospital infrastructure, etc., and 6G will also eliminate time and space barriers through remote surgery.

Massive data management is a challenge for healthcare researchers for many reasons, such as huge volume, high production velocity, different sources, data quality, resource reliability, etc. CC is in its nascent stages, and the success of using CC technologies for MDA in cognitive healthcare can be interesting for future studies.

The ever-growing diffusion of WBS and IoHE devices is heavily changing the way healthcare is approached worldwide. The 6GCIoHE system architecture needs to be exploited further to make smart healthcare systems capable of supporting real time applications when processing a massive amount of data produced by WBS networks.

The tiny ML concept leads data scientists to develop artificial neural networks and deep learning algorithms through biomimicry. Such technologies in the context of practical clinical research show how tiny ML can act as a tool to support and amplify human cognitive functions for physicians delivering care to increasingly complex patients.

Currently, the IoE connections are disparate and clunky, and connecting devices do not create value like connecting people. A central problem with IoE's current architecture is that users are forced to trust platforms.

The literature on the cognitive healthcare topic lacks a holistic view of the current state of research and application. Future research should consider the philosophical perspectives underlying theoretical accounts to ensure that their techniques are consistent with the cognitive-developmental models.

10. Recommendations

Today's avant-garde technology is evolving in the age of unparalleled academic research. Assuming incantation performs the role of the supernatural, the avant-garde technology should play the "instinctive" position. "Telekinesis" [41] is the potential to perform real-time physical action over a remote object without having a physical connection, specifically Tactile Internet and IoHE. The Tactile Internet will permit remote manipulation of objects, as in remote surgery, while imparting the feeling that the remote object is at the fingertips because of the extremely low latency of the 6G connection to offer that real time excitement. The integration of these technologies can augment the tactile capabilities of the individual toward distances that are beyond the usual human reach. Hence, we recommend the possibility of redesigning the manner to engage within a smart healthcare system for diverse factors of living.

Human-centric mobile communications must be one of the most important applications of 6G enabled Tactile Internet and consequently desires careful attention by the wireless research community. Data rate requirements of one terabit per second, end-to-end reliability with latency much less than one millisecond, and higher security, secrecy, and privacy must be key features of 6G Tactile Internet and must address application types: (i) MTRLLC; (ii) mURLLC; (iii) HCS; and (iv) MPS [1] (p. 15).

Mobile operators globally are accelerating their 5G RAN investments. Open RAN will capture a certain percentage of the total RAN market. Many operators are keen to accelerate the pace of open RAN developments and enable these open interface-based systems to achieve performance and functional parity with "purpose-built" systems. We recommend open RAN to become a fundamental part of the 6G mobile system solutions.

The 6GCIoHE system ought to meet healthcare utility necessities on latency, reliability, connectivity density, and gaining KPIs commonly 50 to 100 times higher than 5G. CDI must gain collating pathogenic and affected person information consisting of affected person history, pleasant practices, and diagnostic gear to investigate huge quantities of information to offer advice primarily based on real time desires describing the four salient characteristics: (i) understand; (ii) reason; (iii) learn; and (iv) engage. The IoHE must re-invent the healthcare industry at three levels: (i) a business process with digital technology that drives to improve products, services, and processes as well as customer and healthcare constituent experiences; (ii) a business model where digital products and the process should drive new ways of doing business for the healthcare industry with transformational changes as digitalization re-invents at the business model level; and (iii) a business moment where digital re-invention is created through the need to compete with unprecedented business velocity and agility, specifically the "business moment" would provide relevant and efficient ethical and security measures. Stronger cooperation is needed between the relevant stakeholders involved in the IoHE solutions for ethical and secure use of the IoHE, which is as follows:

1. User consciousness and trust should be built.
2. Consumers should know their rights to secure medical information.
3. To guard medical data and information of the population, the government must establish appropriate standards in terms of information security and apply them to all medical institutions.
4. From the manufacturers' perspective, a reputable system for IoHE devices should be provided for qualitative and objective benchmarks for trustworthiness.

We also recommend the following for the healthcare industry to achieve safe and secure connectivity: (i) manage security at every level of the 6GCIoHE system; (ii) protect the identity of objects and users; (iii) execute multi-factor authentication; and (iv) protect identities and not gateways.

11. Conclusions

This research satisfactorily performed the specific goals as stated in the introduction Section 1. A novel theoretical framework was advanced based on the perspectives outlined in Sections 2–7 and Sections 8.1–8.5. The key findings consist of proof of the 6GCIoHE system paradigm and associated dialog on the five-layer architecture and its applications, challenges, and beneficial effects on cognitive healthcare. The emerging 6G Tactile Internet is anticipated to be deployed on or after 2030 and shall facilitate the provisioning of tiny ML-as-a-Service to end-users via pervasive intelligence applied to enforce exceedingly efficient network transmission, optimization, control, and management of valuable resources. The use of cognitive systems must be to minimize the deficiencies of the MDA concerns. The research bestowed unique insights within the discipline of CDI, especially on the traits of CC that were mapped with the perspectives of massive data, specifically volume, variety, veracity, velocity, and value for knowledge creation. Inspired by the aid of the human cognitive process, we presented a comprehensive definition for the 6GCIoHE system paradigm security solutions to mitigate potential security threats in healthcare applications.

Author Contributions: Conceptualization, P.K.P.; Data curation, P.K.P.; Formal analysis, P.K.P.; Investigation, P.K.P.; Methodology, P.K.P.; Project administration, P.K.P.; Resources, P.K.P.; Software, P.K.P.; Supervision, F.C.-S.; Validation, P.K.P.; Visualization, P.K.P.; Writing—original draft, P.K.P.; Writing—review & editing, P.K.P. Both authors have read and agreed to the published version of the manuscript.

Funding: This research received no external funding.

Institutional Review Board Statement: Not applicable.

Informed Consent Statement: Not applicable.

Conflicts of Interest: The authors declare no conflict of interest.

Appendix A

Acronym List	
AI	Artificial Intelligence
AR	Augmented Reality
CC	Cognitive Computing
CDI	Cognitive Data Intelligence
CH	Cognitive Healthcare
CIoHE	Cognitive Internet of Healthcare Everything
CN	Core Network
CT	Cognitive Technology
EHR	Electronic Health Records
ET	Ethical Theory
eXR	eXtended Reality
5G	Fifth Generation Mobile System
FWA	Fixed Wireless Access
GUIMS	Global Ubiquitous Intelligent Mobile Society
HCS	Human Centric Services
ICS	Industrial Control System
ICT	Information Communication Technology
IoT	Internet of Things
IoE	Internet of Everything
IoHE	Internet of Healthcare Everything
IoMT	Internet of Medical Things
IS	Information System
KPI	Key Performance Indicators
M2M	Machine to Machine
MDA	Massive Data Analytics
MIMO	Multiple-input multiple-output
ML	Machine Learning
mURLLC	Massive Ultra-Low Latency Communications
MPS	Multi-Purpose Services
MR	Mixed Reality
MSS	Medical Super Sensors
MTRLLC	Mobile Teraband Reliable Low Latency Communications
NFV	Network Function Virtualization
PHM	Population Health Management
P2P	People-to-People
P2M	People-to-Machine
QAM	Quadrature Amplitude Modulation
QoE	Quality of Experience
QoS	Quality of Service
RAN	Radio Access Network
RAT	Radio Access Technologies
SCT	Social Cognitive Theory
SDN	Software Defined Network
6G	Sixth Generation Mobile System
TF	Theoretical Framework
TCT	The Communications Theory
TI	Tactile Internet
TIMP	Theories of Inference and the Mathematics of Probability
Tiny ML	Tiny Machine Learning
TOM	Theory of Mind
TOT	Theory of Transparency
TST	The Systems Theory
URLLC	Ultra-Reliable and Low-Latency Communication
VNF	Virtual Network Function
VR	Virtual Reality
WBS	Wireless Body Sensors

References

1. Padhi, P.K.; Charrua-Santos, F. 6G enabled industrialinternet of everything: Towards a theoretical framework. *Appl. Syst. Innov.* **2021**, *4*, 11. [CrossRef]
2. Padma, P. Unlocking the Power of Tactile Internet via 5G Network, Internet of Things Latest News. 2020. Available online: https://www.analyticsinsight.net/unlocking-the-power-of-tactile-internet-via-5g-network/ (accessed on 19 April 2021).
3. ITU-T Technology Watch Report. The Tactile Internet. 2014; pp. 1–24. Available online: https://www.itu.int/dms_pub/itu-t/oth/23/01/T23010000230001PDFE.pdf (accessed on 19 April 2021).
4. Ahmed, M.N.; Toor, A.S.; O'Neil, K.; Friedland, D. Cognitive computing and the future of health care cognitive computing and the future of healthcare: The cognitive power of IBM watson has the potential to transform global personalized medicine. *IEEE Pulse* **2017**, *8*, 4–9. [CrossRef] [PubMed]
5. Lin, J.; Chen, W.M.; Lin, Y.; Cohn, J.; Gan, C.; Han, S. MCUNet: Tiny deep learning on iot devices. *arXiv* **2020**, *1*, 10319.
6. Chen, M.; Yang, J.; Hao, Y.; Mao, S.; Hwang, K. A 5G cognitive system for healthcare. *Big Data Cogn. Comput.* **2017**, *1*, 2. [CrossRef]
7. Ray, P.P.; Kumar, N. SDN/NFV architectures for edge-cloud oriented IoT: A systematic review. *Comput. Commun.* **2021**, *169*, 129–153. [CrossRef]
8. IBM. Watson Health-Population Health Management Beyond the HER, IBM White Paper. 2017; pp. 1–12. Available online: https://www.ibm.com/downloads/cas/YPRP9BLM (accessed on 22 May 2021).
9. Alsharif, M.H.; Kelechi, A.H.; Albreem, M.A.; Chaudhry, S.A.; Zia, M.S.; Kim, S. Sixth generation (6G) wireless networks: Vision, research activities, challenges and potential solutions. *Symmetry* **2020**, *12*, 676. [CrossRef]
10. Dang, S.; Amin, O.; Shihada, B. From a Human-Centric Perspective: What might 6G Be? 2020. Available online: https://arxiv.org/pdf/1906.00741.pdf (accessed on 17 July 2021).
11. Gui, G.; Liu, M.; Tang, F.; Kato, N.; Adachi, F. 6G: Opening new horizons for integration of comfort, security, and intelligence. *IEEE Wirel. Commun.* **2020**, *27*, 126–132. [CrossRef]
12. Chowdhury, M.Z.; Shahjalal, M.; Ahmed, S.; Jang, Y.M. 6G wireless communication systems: Applications, requirements, technologies, challenges, and research directions. *arXiv* **2019**, arXiv:1909.11315v1.
13. Janbi, N.; Katib, I.; Albeshri, A.; Mehmood, R. Distributed artificial intelligence-as-a-service (DAIaaS) for smarter IoE and 6G environments. *Sensors* **2020**, *20*, 5796. [CrossRef] [PubMed]
14. Zhang, L.; Liang, Y.-C.; Niyato, D. 6G Visions: Mobile ultra-broadband, super internet-of-things, and artificial intelligence. *China Commun.* **2019**, *16*, 1–14. [CrossRef]
15. Ateya, A.A.; Muthanna, A.; Gudkova, I.; Abuarqoub, A.; Vybornova, A.; Koucheryavy, A. Development of intelligent core network for tactile internet and future smart systems. *J. Sens. Actuator Netw.* **2018**, *7*, 1. [CrossRef]
16. Simsek, M.; Aijaz, A.; Dohler, M.; Sachs, J.; Fettweis, G. 5G-enabled tactile internet. *IEEE J. Sel. Areas Commun.* **2016**, *34*, 460–473. [CrossRef]
17. Antonakoglou, K.; Xu, X.; Steinbach, E.; Mahmoodi, T.; Dohler, M. Toward haptic communications over the 5G tactile internet. *IEEE Commun. Surv. Tutor.* **2018**, *20*, 3034–3059. [CrossRef]
18. Dai, H.-N.; Zheng, Z.; Zhang, Y. Blockchain for internet of things: A survey. *IEEE Internet Things J.* **2019**, arXiv:1906.00245v56, 8076–8094. [CrossRef]
19. Atlam, H.F.; Azad, M.A.; Alzahrani, A.G.; Wills, G. A review of blockchain in internet of things and AI. *Big Data Cogn. Comput.* **2020**, *4*, 28. [CrossRef]
20. Cisotto, G.; Casarin, E.; Tomasin, S. Performance requirements of advanced healthcare services over future cellular systems. *arXiv* **2019**, arXiv:1906.01503v1.
21. CISCO-Internet of Everything (IoE): Top 10 Insights from Cisco's IoE Value at Stake Analysis for the Public Sector. Available online: http://www.cisco.com/web/about/ac79/docs/IoE/IoE-VAS_Public-Sector_Top-10-$$\backslash$backslash$Insights.pdf (accessed on 17 July 2021).
22. Siemens Digital Industries Software. 2020. Available online: https://www.plm.automation.siemens.com/global/en/industries (accessed on 17 July 2021).
23. Wan, S.; Gu, Z.; Ni, Q. Cognitive computing, and wireless communications on the edge for healthcare service robots. *Comput. Commun.* **2020**, *149*, 99–106. [CrossRef]
24. Bini, S.A. Artificial intelligence, machine learning, deep learning, and cognitive computing: What do these terms mean and how will they impact health care. *J. Arthroplast.* **2018**, *33*, 2358–2361. [CrossRef] [PubMed]
25. Norden, A.D.; Dankwa-Mullan, I.; Urman, A.; Fernando, S.K.; Rhee, K. Realizing the promise of cognitive computing in cancer care: Ushering in a new. *JCO Clin. Cancer Inform.* **2018**, *2*, 1–6. [CrossRef] [PubMed]
26. Khalil, K.; Asgher, U.; Ayaz, Y.; Ahmad, R.; Ruiz, J.A.; Noriyuki Oka, N.; Ali, S.; Sajid, M. Cognitive computing for human-machine interaction: An IBM watson implementation. In *Advances in Intelligent Systems and Computing*; Springer: Cham, Switzerland, 2021. [CrossRef]
27. Polson, P.G.; Lewis, C.; Rieman, J.; Wharton, C. Cognitive walkthroughs: A method for theory-based evaluation of user interfaces. *Int. J. Man-Mach. Stud.* **1992**, *36*, 741–773. [CrossRef]
28. Nord, J.H.; Koohang, A.; Paliszkiewicz, J. The internet of things: Review and theoretical framework. *Expert Syst. Appl.* **2019**, *133*, 97–108. [CrossRef]

29. Falkenberg, E.D.; Lyytinen, K.; Verrin-Stuart, A.A. Limitations of information systems theory and practice: A case for pluralism. In *Information System Concepts. In IFIP Advances in Information and Communication Technology*; Springer: Boston, MA, USA, 1995; pp. 14–15, ISBN 9780412539500.
30. Aquilani, B.; Piccarozzi, M.; Abbate, T.; Codini, A.P. The role of open innovation and value co-creation in the challenging transition from industry 4.0 to society 5.0: Toward a theoretical framework. *Sustainability* **2020**, *12*, 8943. [CrossRef]
31. Mechlova, E.; Malcik, M. ICT in changes of learning theories. In Proceedings of the IEEE 10th International Conference on Emerging eLearning Technologies and Applications (ICETA), Stará Lesná, Slovakia, 8–9 November 2012; pp. 253–262. [CrossRef]
32. Kelemen, O.; Máttyássy, A.; Kéri, S. Improvement of theory of mind in schizophrenia: A 15-year follow-up study. *Psych* **2019**, *1*, 420–428. [CrossRef]
33. Tsai, C.H. Theory, social cognitive theory, and the technology acceptance model to explore a behavioral model of telehealth systems. *Int. J. Environ. Res. Public Health* **2014**, *11*, 4905–4925. [CrossRef]
34. Frank, S.A. The price equation program: Simple invariances unify population dynamics, thermodynamics, probability, information, and inference. *Entropy* **2018**, *20*, 978. [CrossRef] [PubMed]
35. Lohstroh, M.; Lee, E.A. An Interface theory for the internet of things In *SEFM 2015 Collocated Workshops*; Springer: Cham, Switzerland, 2015; pp. 20–34.
36. Slepian, D.; Wyner, A.S.O. Rice's contributions to Shannon theory. *IEEE Trans. Inf. Theory* **1988**, *34*, 1374. [CrossRef]
37. Dang, H.-S.; Nguyen, T.-M.-T.; Wang, C.-N.; Day, J.-D.; Dang, T.M.H. Grey system theory in the study of medical tourism industry and its economic impact. *Int. J. Environ. Res. Public Health* **2020**, *17*, 961. [CrossRef]
38. Averill, E.W.; Gottlieb, J. *Two Theories of Transparency*; Springer: New York, NY, USA, 2019; pp. 1–21.
39. Paley, J. Virtues of autonomy: The kantian ethics of care. *Nurs. Philos.* **2002**, *3*, 133–143. Available online: https://uwethicsofcare.gws.wisc.edu/wp-content/uploads/2020/03/Paley-J.-2002.pdf (accessed on 29 June 2021). [CrossRef]
40. Mezghani, E.; Expósito, E.; Drira, K. A model-driven methodology for the design of autonomic and cognitive IoT-based systems: Application to healthcare. *IEEE Trans. Emerg. Top. Comput. Intell.* **2017**, *1*, 224–234. [CrossRef]
41. Fiorillo, C.D. Beyond bayes: On the need for a unified and jaynesian definition of probability and information within neuroscience. *Information* **2012**, *3*, 175–203. [CrossRef]

applied system innovation

MDPI

Article

Value of the 3D Product Model Use in Assembly Processes: Process Planning, Design, and Shop Floor Execution

Carl Kirpes, Dave Sly and Guiping Hu *

Industrial and Manufacturing Systems Engineering (IMSE), Iowa State University, Ames, IA 50011, USA; cjkirpes@gmail.com (C.K.); davesly@iastate.edu (D.S.)
* Correspondence: gphu@iastate.edu

Abstract: Organizations can enhance the value of their assembly planning, assembly design, and assembly shop floor execution through the use of the 3D product model. Once a tool targeted at product design, the 3D product model, enabled by current and emerging manufacturing process management technologies, can create additional value for organizations when used in assembly processes. The research survey conducted and described in this paper demonstrates the value organizations have seen in using the 3D product model in the assembly process. The paper also explores the current state of those organizations who have not yet implemented the use of the 3D product model in their assembly processes and the value that they foresee for possible future implementation. The essential findings of this research are the five qualitative areas in which value is derived from using the 3D product model in complex assembly processes and how those value drivers apply across various industries and organization sizes. These results provide a framework for future research to develop quantitative models of the value of the 3D product model use in assembly processes.

Keywords: assembly; process planning systems; concurrent engineering; automotive industry applications; manufacturing industry applications

Citation: Kirpes, C.; Sly, D.; Hu, G. Value of the 3D Product Model Use in Assembly Processes: Process Planning, Design, and Shop Floor Execution. *Appl. Syst. Innov.* **2021**, *4*, 39. https://doi.org/10.3390/asi4020039

Academic Editor: Claudio Zunino

Received: 25 May 2021
Accepted: 15 June 2021
Published: 18 June 2021

Publisher's Note: MDPI stays neutral with regard to jurisdictional claims in published maps and institutional affiliations.

Copyright: © 2021 by the authors. Licensee MDPI, Basel, Switzerland. This article is an open access article distributed under the terms and conditions of the Creative Commons Attribution (CC BY) license (https://creativecommons.org/licenses/by/4.0/).

1. Introduction

The first adoption of 3D product models among the engineering disciplines was in product design. Whereas 2D drawings previously required engineering expertise and an object-oriented mindset to imagine a 3D object based on the shapes depicted, the 3D product model provides a non-engineering trained person with the same opportunity to visualize the product before it is built. Such functionality is of great benefit as engineers work on the design of the product with marketing, sales, suppliers, potential customers, and other functions inside and outside of the organization. However, if the use of the 3D product model stops at the design phase, the benefit of the 3D product model is not fully realized.

To better understand the value of the 3D product model in the assembly process, a literature review was conducted. Journal papers, articles, and conference papers were sourced using the Compendex and Scopus databases. Articles searched for included reference to "3D" and "assembly" but not "printing" (in order to remove articles on 3D printing, which is not the focus of this study). Then, an analysis was run on the number of times these key words were used throughout the article. Noting that some articles are longer than others, the total count was divided by the article page count to get an understanding of the density of the use of the "3D" and "assembly" keywords in each article. Each article was also reviewed for mentions of the value that using 3D product models in the assembly processes could create. Such value might include improving the efficiency and reliability of the assembly process design, reducing process planning problems, and shortening the process planning cycle [1,2]. Then, the same approach was taken in terms of dividing the number of mentions of value creation by the total page count to get a density value for value creation mentions in each paper. The plot of these relative densities of "3D" to "assembly", where "value creation" is the size of each bubble, is shown in Figure 1 [1–34] where each different colored bubble represents a different journal paper, article, or conference paper. Figure 1 demonstrates that while there was a lot of research published on 3D product models, and an equivalent amount published on assembly processes, less literature had been published on the use of 3D product models in assembly. This was especially evident in the lack of data points in the upper right corner of Figure 1 and the fact that of those papers that do have a high density of "3D" and "assembly", there is still a gap in the literature in terms of the value that the use of 3D product models creates in assembly processes. Exploring this gap and better understanding the value that 3D product models create in the assembly process provided the guiding purpose for this research.

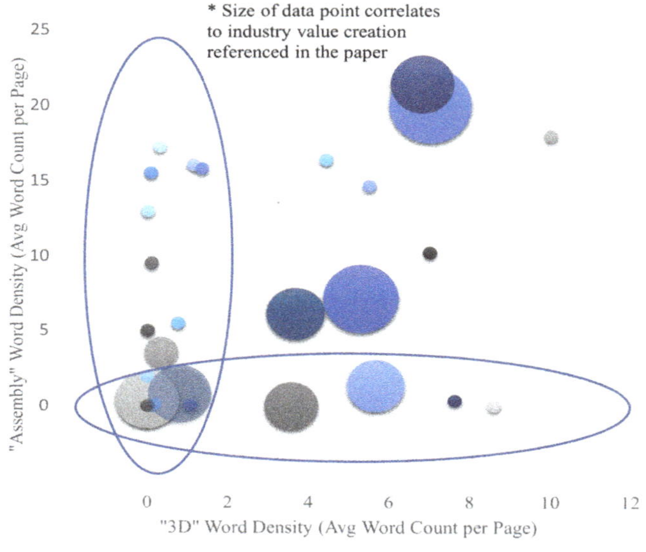

Figure 1. 3D product model in the assembly process: literature review.

The ability to capture the value that 3D product models create in assembly processes has been limited by the availability of software to bring the 3D product model to the assembly processes [3,4]. Enabling factors including innovative education and training along with data-sharing systems and standards have created an environment for technology to advance in such a way that smart manufacturing and the use of the 3D product model in the assembly process becomes feasible [5]. As such feasibility came to fruition, researchers and industry alike have found ways to transform the engineering Bill of Materials into the Manufacturing Bill of Materials [6–8]; develop manufacturing process management software [9]; and extract 3D product model data and transfer it to virtual manufacturing and assembly software [10–15]. Researchers and industry also advanced this virtual assembly knowledge further, developing tools to automate assembly planning for complex products [16,17].

Once technology enabled the use of the 3D product model in the assembly process, research was conducted demonstrating how the 3D product model could be used in the assembly process planning, assembly design, and assembly shop floor execution processes (referred to in individual and in aggregate as "the assembly process" for the sake of brevity in this paper). This can be done by integrating the 3D product model with assembly line balancing via process consumption and leveraging the 3D product model in the link between the Bill of Process and the Bill of Materials including Model/Option dependency [18,35]. A similar approach includes the creation of a digital mock-up to transmit models and attribute data between the Engineering Bill of Materials and the Process (or Manufacturing) Bill of Materials and serve as a unified data source in assembly planning [19]. By implementing such approaches, the use of the 3D product model in assembly processes allows for engineers to check the product's configuration for assemblability [20]. The 3D product model can also be used on the shop floor to enhance shop floor work instructions, improve process planning, and increase efficiency of assembly work [21,22]. Each of these studies indicate ways that companies can utilize the 3D product model developed by design teams in assembly processes. Given that some research has shown that such a feat is possible, then, the question becomes whether industry sees value in using the 3D product model beyond product design, and if so, in what areas industry believes value exists.

The objective of this research is to evaluate the use of 3D product models in assembly planning, assembly design, and assembly shop floor execution. The authors note that the 3D product model can have a large array of meanings from the virtual computer-aided design (CAD) model, to a digital mock-up, to a digital twin, or even a physical 3D model. The 3D product model application focused on in this paper is the value of bringing the virtual CAD model and digital 3D assembly mock-up to the assembly process beyond the engineering design. To illustrate an example, Figure 2 shows an assembly process where a fire truck is being built. Each octagon represents a station on the assembly line; some operations are completed by operators (the blue dots in the figure), and others are completed by robots. The operators have written work instructions which they use to know what parts to assemble and where to assemble those parts as different models and options of the fire truck progress down the assembly line. Figure 3 shows this same assembly line, but the written work instructions are replaced with 3D product models (virtual CAD models) of the part with visual assembly guidance. This is an example application of the 3D product model in the assembly process.

The paper explores the value organizations see in implementing the 3D product model in assembly processes. This paper will go through the methodology and results of a survey that was conducted to understand the value that organizations are seeing by implementing the 3D product model in assembly processes. The paper concludes with a review on the value of 3D product model use in assembly processes based on the survey results and a discussion on future research recommendations.

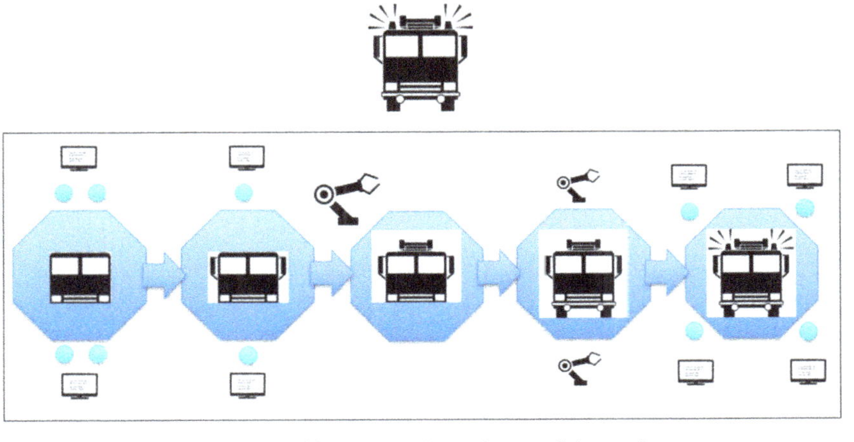

Figure 2. Assembly process using written work instructions.

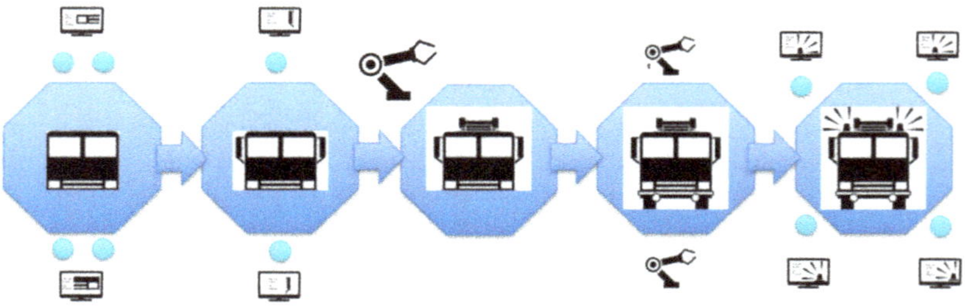

Figure 3. Assembly process deploying the 3D product model to the shop floor.

2. Materials and Methods

In this study, we adopt a survey to reach industry practitioners and understand if and how they have deployed the 3D product model in assembly planning, assembly design, and assembly shop floor execution processes. In addition, the survey aims to capture the value that the respondents have seen, or plan to see, in using the 3D product model in these assembly processes. The following are example survey questions developed accordingly:

(1) Please indicate your level of knowledge about the use of 3D product models in the assembly process, assembly design, or assembly shop floor execution at your organization:
- (a) Not at all knowledgeable
- (b) Somewhat knowledgeable
- (c) Knowledgeable
- (d) Very knowledgeable

(2) Has your company implemented 3D product models into the assembly process planning/design?
- (a) Yes
- (b) No

(3) Has your company implemented 3D product models into the assembly shop floor execution?
- (a) Yes
- (b) No

(4) To what extent does your company use 3D product models in the assembly process?
- (a) Screenshots of 3D model
- (b) 3D model mapped to process activities
- (c) 3D model mapped to parts consumption within process activities
- (d) 3D model fully integrated with Engineering Change Order/Manufacturing Change Order Process
- (e) Other (please describe):

(5) What technology is your company deploying to accomplish the use of 3D product models in the assembly process?
- (a) Proplanner's Assembly Planner
- (b) PTC's MPMLink
- (c) Siemens Teamcenter
- (d) Software developed in-house
- (e) Other (please describe):

(6) What are the limitations of the technology your company is using to communicate the information contained in the 3D product model to the assembly team?

(7) If your company has implemented the 3D product model in the assembly process planning/design, what was your company doing before implementing the use of 3D product models for assembly process planning/design? (Select all that apply)
   (a) Excel
   (b) Simulation
   (c) Paper/Post-It notes
   (d) Physical mock-ups
   (e) Other (please describe):
   (f) Not applicable
(8) If your company has implemented the 3D product model in the assembly shop floor execution, what was your company doing before implementing the use of 3D product models in the shop floor execution?
   (a) Virtual work instructions without the 3D product model
   (b) Hard copy work instruction with 2D drawings
   (c) Hard copy work instructions with pictures
   (d) Work instructions without visuals, words only
   (e) Operator training with no shop floor work instructions
   (f) Other (please describe)
   (g) Not applicable
(9) How did your company move from the former state to implementing 3D product models in the assembly process? (Select all that apply)
   (a) Pilot study
   (b) Wholesale cutover
   (c) Started small and scaled implementation
   (d) Other (please describe):
(10) What is the value your company has seen in implementing 3D product models in the assembly process? (Select all that apply)
   (a) Accuracy of assignment of the right parts, tools, work allocation, and work instructions
   (b) Faster new product/model roll out to production
   (c) Less time updating work instructions
   (d) Quicker operator training
   (e) Other (please describe):
(11) What are the savings/estimated value/return (quantitative and/or qualitative) of implementing the use of 3D product models in the assembly process?
(12) What is the size of the company you work for?
   (a) 1–999 employees
   (b) 1000–9999 employees
   (c) 10,000–49,999 employees
   (d) 50,000+ employees
(13) What industry is your company a part of?
   (a) Aerospace
   (b) Agricultural Equipment
   (c) Automotive (e.g., Light Car and Truck)
   (d) Consumer Appliances
   (e) Heavy Industrial (e.g., Heavy Equipment, Trucks, and Buses)
   (f) Recreational Vehicles (e.g., Motorcycles, RVs, Four Wheelers, and Boats)
   (g) Supply Chain
   (h) Other:
(14) What is your title?
   (a) C-level Executive

(b) Vice President
(c) Director
(d) Manager/Supervisor
(e) Engineer
(f) Other Individual Contributor
(g) Other:

To ensure these survey questions would be applicable to industry, the research team connected with a Manufacturing Process Management (MPM) software company. Discussions with the engineers and customer-facing employees in the organization allowed the research team to refine the questions around what level organizations were deploying 3D product model functionality in the assembly process planning, assembly design, and assembly shop floor execution; what the limitations were of the software used to deploy the 3D product model in the assembly process; and what plans the organization had for further deployment of the 3D product model in the assembly process. Using this framework, the research team came up with three categories that companies would be divided into based on their responses to the set of survey questions. These categories were (1) companies that had implemented the use of the 3D product model in the assembly process; (2) companies that had not yet implemented the use of the 3D product model in the assembly process but planned to; and (3) companies that did not plan to implement the use of the 3D product model in the assembly process. The later survey questions were tailored accordingly to fit within the context of the category an organization was classified into based on their responses to the initial survey questions, while still retaining the following overarching themes: what the company was doing before they implemented the use of the 3D product model in the assembly process, how that functionality was rolled out, what software was used, what the limitations are of that software, and what value the company was seeing or planned to see based on the use of the 3D product model in assembly processes. The research team used Qualtrics to develop and administer the survey. The survey hierarchy is shown in Figure 4.

The research team sent the survey out to individuals who were a specific subset of the population. Potential survey candidates were first screened based on the company that they worked for. Since the focus of the study is specific to companies that have manufacturing production and assembly operations, only individuals who worked for these types of organizations were included as potential survey respondents. As an example, an individual that worked for an automobile manufacturer or an agricultural equipment manufacturer would be considered a potential survey respondent, as organizations in those industries are engaged in assembling complex products and thus could use the 3D product model in the assembly planning, assembly design, or assembly shop floor execution processes. However, an individual who works at a hospital or a food/beverage company would not be considered a potential applicant for the survey. Those organizations do not have assembly as a part of their business processes and thus are not viable survey candidates. Including respondents that do not have assembly operations could skew the survey results, as those organizations would likely answer that they do not plan to implement the use of the 3D product model in the assembly process (simply because they do not have an assembly process through which to implement the use of 3D product models).

Once potential survey respondents were screened based on the industry in which they worked, they were then filtered down based on their role in the organization. For example, an engineer or director-of-operations is likely to have knowledge about the company's plans or current use of 3D product models in the assembly process. However, an individual in accounting or human resources is likely not to have knowledge in that regard. As a result, if an individual worked for a company in an applicable industry, but was not in an applicable role, that individual was contacted in the context of providing an introduction to (or sharing the survey with) someone in their organization who was in an applicable role or was not sent the survey at all.

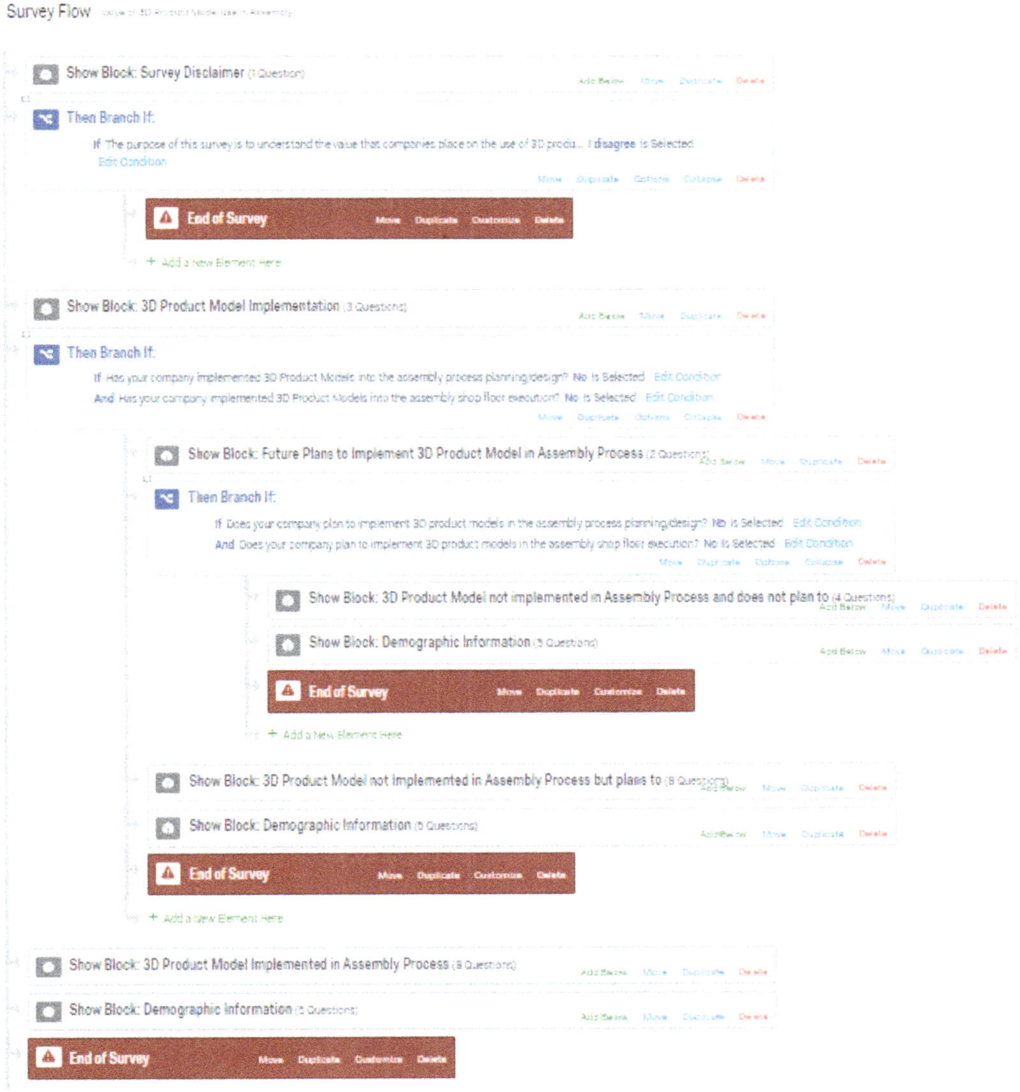

Figure 4. Survey flow hierarchy.

The research team started with a database of over 2700 contacts, which were filtered based on companies in applicable industries and individuals in applicable roles within those industries to 151. Of these 151, the research team found that the contact information was up to date for 85 individuals. These 85 individuals were provided with the link to take the survey on the value of the 3D product model use in the assembly process at their organization. Of the 85 individuals sent the survey, 35 responded, equating to a response rate of 41%. Note that the date between when the survey was administered and the final data were collected was approximately one month. The number of respondents in the first week was 21; in the second week, it was 8, in the third week, it was 6, and in the fourth week, it was 0. The decreasing number of respondents and the fact that by the fourth week there were 0 respondents provided the research team with the indication that the majority of all respondents who were likely to respond had, and therefore, it was a good time to collect and analyze the data. The survey software was also set up to automatically record an individual's response after two weeks of no activity on the survey. The objective of this being that if the survey respondent had answered the majority of the questions but forgot to click submit, as an example, then the survey information would be captured and of use in the analysis. However, this approach also led to some survey responses being logged where, as an example, the survey respondent had started the survey, answered the first couple questions, and then exited the survey. In such a case, not enough data were available for analysis. Of the 35 responses, 7 responses were removed prior to the detailed analysis due to survey responses getting recorded with too few questions answered for meaningful analysis.

Prior to analyzing the survey results, the research team's hypothesis was that most companies would fall under the category of having not yet implemented the 3D product model in the assembly process but planned to do so in the future. This hypothesis was based on the research's team review of the academic literature written on the topic, including the lack of papers that correlate the implementation of the 3D product model in the assembly process to the value such implementation creates for an organization. In addition, the research team conducted several meetings with an MPM software company to gain a further understanding of industry-capable 3D product model software functionality and limitations through the lens of this organization, which helped shape the hypothesis. Given the lack of literature discussing the value of the 3D product model in the assembly process, the research team also hypothesized that the value of using the 3D product model in the assembly process is expressed qualitatively and unknown quantitatively.

3. Results

Of the 28 survey respondents with enough data for analysis, the responses were categorized into three buckets as noted in the Materials and Methods section of this paper. A total of 23 respondents (82% of those analyzed) said that they have implemented the 3D product model in the assembly process. Note that assembly process means assembly planning and design and/or assembly shop floor execution. A total of five respondents said they had not yet implemented the 3D product model in the assembly process; of those, four respondents (14% of those analyzed) said they planned to implement the 3D product model in the assembly process, and one respondent (4% of those analyzed) did not plan to implement the 3D product model in the assembly process. This categorization of the survey responses is shown in Figure 5. It is worth noting that of the 23 respondents that implemented the use of the 3D product model in the assembly process, eight had implemented the use of the 3D product model in assembly planning and design but not shop floor execution, one had implemented the use of the 3D product model in shop floor execution but not assembly planning and design, and 14 had implemented the use of the 3D product model in both.

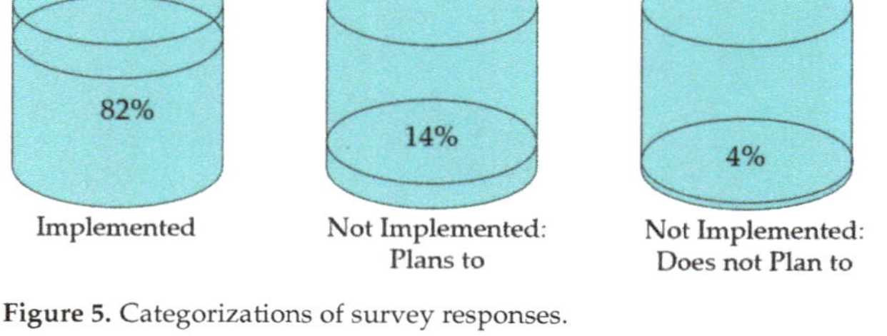

Figure 5. Categorizations of survey responses.

3.1. Company Has Implemented the 3D Product Model in the Assembly Process

As discussed in the Materials and Methods section of this paper, potential survey candidates were screened based on the organization/industry in which they worked and their role within that organization to increase the likelihood that these individuals would have knowledge of the use of the 3D product model and how it was or was not being deployed in the assembly process in their organization. To test the effectiveness of the survey screening approach, survey respondents were asked to indicate their level of knowledge about the use of 3D product models in assembly planning, assembly design, or assembly shop floor execution. The categories and responses were as follows: very knowledgeable (6), knowledgeable (11), somewhat knowledgeable (5), not at all knowledgeable (1). With only one of the 23 respondents indicating that they were not at all knowledgeable about the use of the 3D product model in the assembly process, the survey screening approach appears to have been a success and drives greater confidence in the results described in this paper.

To ensure that results were not biased by any one industry, the researchers checked the distribution of responses by industry type. As shown in Figure 6, there is a relatively even distribution of survey responses across multiple industry types, and all industry types have applicability to an organization in an industry that would be a candidate for the use of 3D product models in the assembly process. The authors note that when viewed as a group, the aerospace, agricultural equipment, and automotive industries comprise a large portion of the survey responses. Based on the authors' experience, and comparison to the relative percentage of these industries to others with complex assembly operations on the 2020 Fortune 500 list (the largest 500 publicly traded companies in America by revenue), these three industries are a large portion of the population of the manufacturing organizations that have complex assembly operations. Of the Fortune 500 companies, only 44 companies are in industries that have complex assembly operations applicable to this research. Figure 7 shows the comparison of the percentage of survey respondents by industry relative to the percentage of companies with complex assembly operations by industry from the Fortune 500 and validates that the sample of industry respondents is representative of the broader population of manufacturing companies with complex assembly operations.

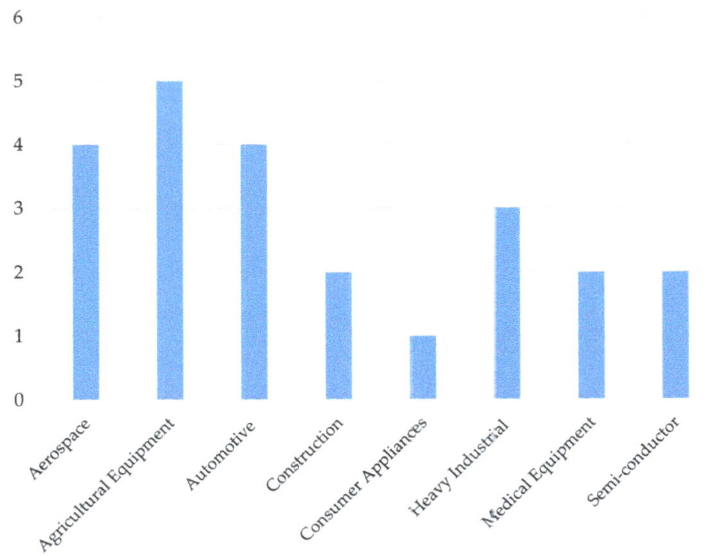

Figure 6. Industry distribution of respondents.

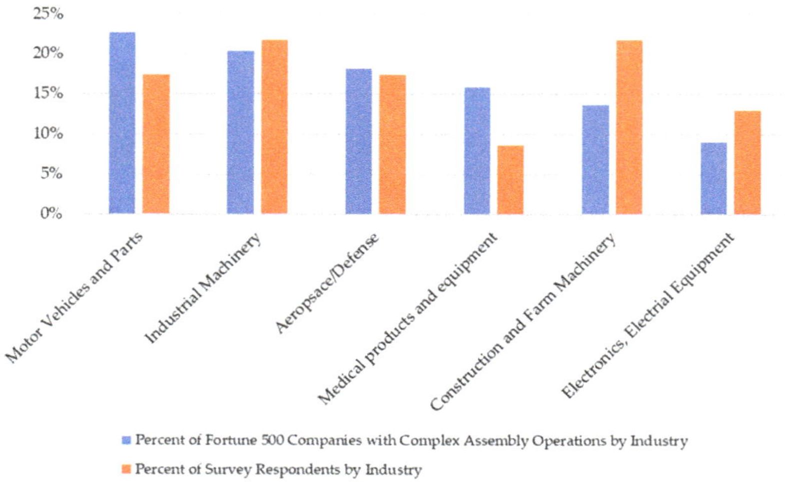

Figure 7. Comparison of survey respondents percentage by industry to Fortune 500 companies with complex assembly operations percentage by industry.

In addition to ensuring that the industries represented by the survey respondents were reflective of the population, the research team also wanted to ensure that the jobs of the survey respondents within those organizations were representative of the population. Due to the nature of organizational hierarchies, there are more engineers than engineering managers/supervisors, more managers/supervisors than directors, more directors than vice presidents, etc. When reviewing the survey respondents, as seen in Figure 8, the categorization of respondents follows this general trend, which demonstrates that the survey responses are not biased by any one job type when compared to the broader population.

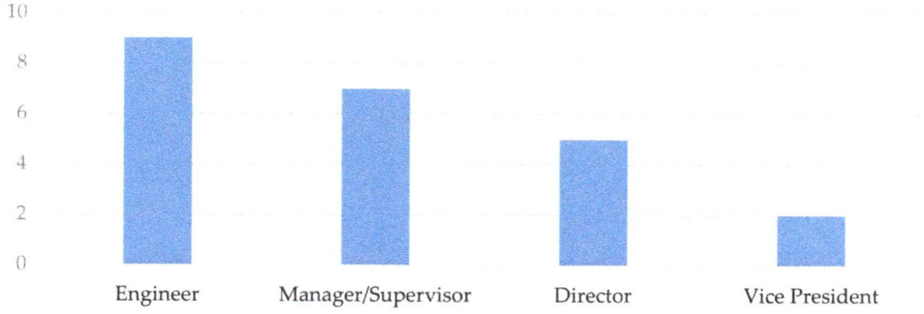

Figure 8. Job title distribution of respondents.

The research team also gathered information on the company size, by employee count, which was represented by the respondents. An interesting finding is that all respondents who worked for a large corporation said that their organization implemented the 3D product model in the assembly process. Figure 9 shows the breakdown of which size company the respondents work for based on employee count.

Figure 9. Company size based on employee count.

Noting that the majority of companies implemented the use of the 3D product model in the assembly process, further segregation into at which levels these companies implemented the use of the 3D product model in the assembly process is warranted. Figure 10 shows the responses in order of the most complex to least complex deployment of the 3D product model implementation in the assembly process.

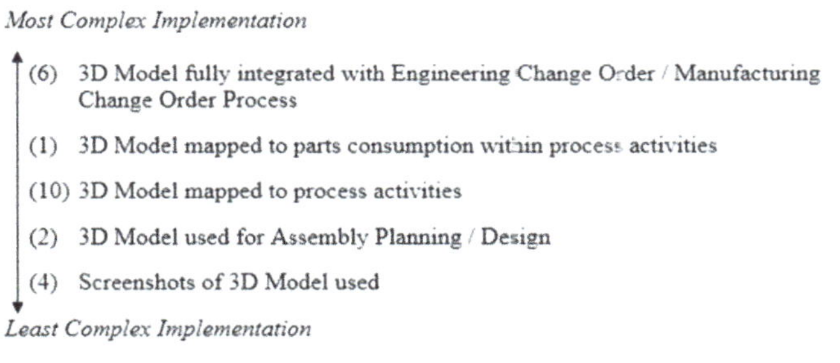

Figure 10. Extent to which 3D product models are used in the assembly process.

The software most commonly used to deploy the 3D product model was software developed in-house (nine total responses) The next two highest response counts for 3D product model software use in the assembly process were PTC's MPMLink (six responses) and Siemens Teamcenter (five responses). Although this demonstrates that technology is available on the marketplace today to use the 3D product model in the assembly process, the respondents did list limitations that these software currently face. In general, the responses show that the ease of use, the ability to handle revision/change control, and the integration with other software/processes were common limitations experienced by the software developed in-house and the software available on the market today.

Prior to deploying the software that allowed these organizations to use the 3D product model in their assembly process, the organizations relied on varying ways to communicate information contained in 3D product models to the process planning and design team. These included physical mock-ups (8), Excel (7), simulation (7), and paper/Post-It notes (4). Note that the sum of these responses (26) is greater than the number of survey respondents (23), as survey respondents were allowed to select more than one way in which this information was communicated prior to implementing the software that enabled the use of the 3D product model in the assembly planning/design. For those respondents who had implemented the 3D product model in the shop floor execution, they were also asked what they were doing to communicate the 3D product model information to the shop floor prior to their current state. All respondents to this inquiry shared that they used work instructions in some form to communicate this information; this included work instructions without visuals, words only (1), hard copy work instructions with pictures (3), hard copy work instructions with 2D drawings (6), and virtual work instructions without the 3D product model (6).

To transition to the 3D product model use in the assembly process, most organizations started small and scaled implementation (15), while some conducted a pilot study first (7). Only one respondent did a wholesale cutover to the 3D product model deployment and use in the assembly process.

Having the context of what companies were doing prior to implementing the 3D product model use in the assembly process and how they transitioned to their current state was important, as it provides insight for those organizations that have not yet implemented the 3D product model in the assembly process direction on how they might do so. However, more importantly is the value that the organizations have seen from implementing the 3D product model in the assembly process, as that is the incentive through which other organizations might consider doing the same. Figure 11 demonstrates that more respondents found value in the accuracy of assignment of the right parts, tools, work allocation, and work instructions relative to the other areas of value when implementing the 3D product model in the assembly process by a factor of more than 50%. This value driver was selected by over 60% of the survey respondents, and at 50% or greater of the total survey respondents at every level in the organization (engineering, manager/supervisor, director, and vice president). In addition, the accuracy of assignment of the right parts, tools, work allocation, and work instructions was selected by more than 67% of the participants across the aerospace, agricultural equipment, automotive, consumer appliances, and heavy industrial industries. These results indicate that the greatest potential for industry and academia to quantify the value of the 3D product model use in the assembly process is within the category of the accuracy of assignment of the right parts, tools, work allocation, and work instructions. Interestingly though, organizations with less than 1000 employees infrequently listed this value driver, meaning that developing models and quantifying the value for the accuracy of assignment of the right parts, tools, work instructions, and work allocation will likely be more beneficial to medium and large organizations. The results indicate that smaller companies would benefit greater by a model that quantitatively calculates the value of the 3D product model in the assembly process as it relates to faster new product/model roll out. The results also show that as the organization size increased, the number of value drivers indicated by the survey respondents increased. This points to larger organizations being able to better scale the deployment of the 3D product model in the assembly process and thus capture more of the value drivers accordingly.

Survey respondents further elaborated on the areas where they were finding value in the use of the 3D product model in the assembly process. Respondents shared comments on value gained such as "substantial", "20:1", and "multi-million-dollar savings in time to market". Respondents shared that implementing the 3D product model in assembly processes "helps ensure no parts are forgotten in the work instructions for large, complex equipment" and created a "more efficient process for manufacturing engineers who develop the assembly process" and "reduced errors in assembly".

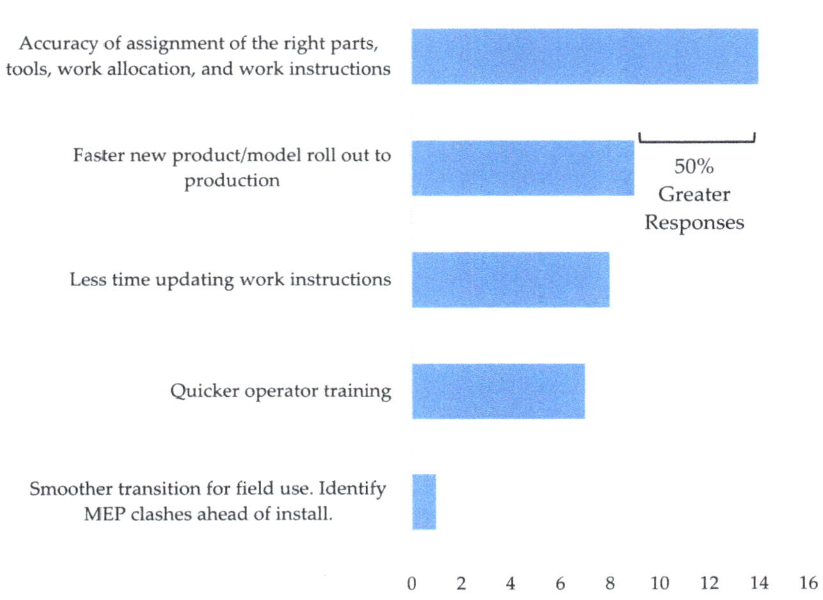

Figure 11. Value of 3D product model in the assembly process by response count.

3.2. Company Has Not Implemented the 3D Product Model in the Assembly Process but Plans to

Although the majority of the survey respondents had implemented the use of the 3D product model in the assembly planning and design or assembly shop floor execution processes, a small number of respondents had not. Of the five survey respondents that had not yet implemented the 3D product model in the assembly process, four said that their organizations planned to. The four respondents self-identified as somewhat knowledgeable (2), knowledge (1), and very knowledge (1) regarding their level of knowledge of 3D product model use in assembly processes. The individuals had varying titles, were from companies that ranged in size, and were from varying industries. As such, similar to the organizations that had already implemented the 3D product model in the assembly process, no one industry, company, or individuals' level within the company seemed to dominate or skew the data.

The plans to implement the 3D product model in the assembly process also varied but were amongst the less complex phases of the implementation of the 3D product model use in assembly process implementations, including using only screenshots of the 3D model and mapping the 3D product model to process activities. This makes sense, as the next logical step for a company that is not using the 3D product model in the assembly process is to move to a phase that allows them to capture value by doing so but does not require the investment/scope of a more complex implementation. The technology that respondents said their organizations planned to use to accomplish this was software that was developed in-house, which is what the majority of the respondents that had implemented the 3D product model in the assembly process had done, and the evaluation of Assembly Planner as the software platform.

Without the 3D product model in the assembly planning and design process, the respondents shared that their organizations are currently using Excel (2), physical mock-ups (1), and 3D model snapshots (1) to communicate the information today to the assembly planning and design team. The information contained in the 3D product model is currently being conveyed to the shop floor execution team through hard copy work instructions (3) and virtual work instructions that do not include the 3D product model (1). When the organizations do move from this current state to the future state of integrating the 3D product model in the assembly process, they expect to see value in areas ranging from less time updating work instructions (1); quicker operator training (1); accuracy of assignment of right parts, tools, work allocation, and work instructions (1); and faster new product/model roll out (1). One organization, who identified themselves as the smallest tier company available on the survey, shared that they believe implementing the 3D product model in the assembly process will save that organization $50,000 per year.

3.3. Company Has Not Implemented the 3D Product Model in the Assembly Process and Does Not Plan to

Only one respondent of the 28 responses analyzed said that they do not plan to implement the 3D product model in the assembly process. While this respondent shared that they are knowledgeable about the 3D product models use in the assembly process, that individual shared that they were unaware of technology solutions to integrate the 3D product model in the assembly process. It is the authors' aim that this paper will demonstrate such options to implement the 3D product model in the assembly process. Through the examples of how those organizations who have implemented the 3D product model in the assembly process transitioned to that state and the value that those companies have seen as a result, now other organizations such as the one noted here that are unaware of solutions to integrate 3D product models into the assembly process can see the value and direction through which they can do so.

4. Conclusions

The use of 3D product models beyond product design is gaining traction in industry. Companies are moving away from 2D drawings and paper-based work instructions to take advantage of the value that 3D product models offer beyond product design in not only the assembly planning and design process but also in shop floor execution. This survey disproved the research team's first hypothesis that the majority of companies would fall under the category of having not yet implemented the 3D product model in the assembly process but planned to do so, as most companies had already implemented the 3D product model in their assembly process.

While companies recognize that there is inherent value in applying the 3D product models beyond just product design, up to this point, much of that value is still described in ways that are very qualitative in nature. The research team failed to disprove the second hypothesis: that the value of using the 3D product model in the assembly process is expressed qualitatively and unknown quantitatively. Thus, future research focuses should include developing a more quantitative approach to the value companies can expect to gain when deploying 3D product models in the assembly planning and design process as well as in assembly shop floor execution.

In one application of the 3D product model in the assembly process, a large agricultural equipment manufacturer transitioned from using Microsoft Excel to reconcile engineering change orders to using the 3D product model in combination with Proplanner's Assembly Planner. In the original process using an Excel sheet, the company spent almost seven hours updating work instructions following an engineering change order. In the new process, the company could complete the same update, with the 3D product model included in the assembly process work instructions, in just over two hours. This resulted in a savings of over four and a half hours per engineering change order and with over 1500 engineering change orders processed each year that is greater than 6500 man-hours saved per year. By integrating the 3D product model into the assembly process, and automating that through Proplanner's Assembly Planner, the organization is able to save hundreds of thousands of dollars as a result.

In another application of the 3D product model in the assembly process, a large automotive manufacturer uses the 3D product model to verify assembly interferences before launching a product model change. The automotive manufacturer has a virtual assembly lab that allows them to complete such studies and check how a 3D product model change will work in the context of how it will integrate with tooling and the existing products already assembled onto the vehicle on the assembly line. The manager of the digital manufacturing group at this organization stated that the application of the 3D product model through the virtual assembly lab provides the company with a key competitive advantage over other automakers, where using the 3D product model in the assembly process saved millions of dollars in a new product launch.

Beyond these examples, the results shown in Figure 11 are substantial because they provide a framework through which industry and academia alike can start to develop a quantitative model for the value of the 3D product model in the assembly process. These five value areas can be used to determine the value of implementing the 3D product model on a given assembly line, which is then summed across assembly lines in a plant and across assembly plants within a company to get the enterprise value of implementing the 3D product model in the assembly process. The paper also demonstrates findings on which value drivers are more applicable to different size organizations and which value drivers are recognized the most in given industries. Then, via the survey results and the five value areas determined, this paper creates the building blocks for quantifying the value of the 3D product model in the assembly process in future research.

Author Contributions: C.K. and D.S. conceived the idea. C.K. formulated the problem, developed the survey, and worked with industry participants to collect and analyze the results. G.H. and D.S. provided guidance throughout the research and G.H. proofread the manuscript. All authors have read and agreed to this version of the manuscript.

Funding: This research received no external funding.

Institutional Review Board Statement: Not applicable.

Informed Consent Statement: Informed consent was obtained from all subjects involved in the study.

Data Availability Statement: The data presented in this study are available as consolidated results in Section 3 of the manuscript. The individual response data are not publicly available due to the right of each survey participant to remain anonymous and only have the data shared as consolidated results.

Acknowledgments: The authors would like to thank Proplanner for their support in developing survey questions that were applicable to industry and critical to this research.

Conflicts of Interest: The authors declare not conflict of interest.

References

1. Sun, J.; Yin, F.; Yang, C. Assembly Process Planning System (APPS) based on 3D product model. In Proceedings of the 2010 International Conference on Digital Manufacturing and Automation, ICDMA 2010, IEEE Computer Society, Changsha, China, 18–20 December 2010.
2. Chen, Z.; Tang, J. Aircraft digital assembly process design technology based on 3D Model. In Proceedings of the 2018 International Conference on Aeronautical, Aerospace and Mechanical Engineering, AAME 2018, Sarawak, Malaysia, 28–30 June 2018. EDP Sciences.
3. Shridhar, J.M.; Ravi, S. Virtual manufacturing: An important aspect of collaborative product commerce. *J. Adv. Manuf. Syst.* **2002**, *1*, 113–119. [CrossRef]
4. Quintana, V.; Rivest, L.; Pellerin, R.; Venne, F.; Kheddouci, F. Will Model-based Definition replace engineering drawings throughout the product lifecycle? A global perspective from aerospace industry. *Comput. Ind.* **2010**, *61*, 497–508. [CrossRef]
5. Mittal, S.; Khan, M.A.; Romero, D.; Wuest, T. Smart manufacturing: Characteristics, technologies and enabling factors. *Proc. Inst. Mech. Eng. Part B: J. Eng. Manuf.* **2017**, *233*, 1342–1361. [CrossRef]
6. Xu, H.C.; Xu, X.F.; He, T. Research on transformation engineering BOM into manufacturing BOM based on BOP. In *Applied Mechanics and Materials*; Trans Tech Publications Ltd.: Stafa-Zurich, Switzerland, 2008; Volume 10, pp. 99–103.
7. Tozawa, Y.; Yotsukura, M. Integration of bills of material towards a communication tool. In Proceedings of the 2009 WRI World Congress on Computer Science and Information Engineering, CSIE 2009, Los Angeles, CA, USA, 31 March–2 April 2009. IEEE Computer Society.
8. Quintana, V.; Rivest, L.; Pellerin, R.; Kheddouci, F. Re-engineering the engineering change management process for a drawing-less environment. *Comput. Ind.* **2012**, *63*, 79–90. [CrossRef]

9. Michel, R. Next on the horizon: The bill of process. *Manuf. Bus. Technol.* **2005**, *23*, 30.
10. Kang, X.; Peng, Q. Integration of CAD models with product assembly planning in a Web-based 3D visualized environment. *Int. J. Interact. Des. Manuf.* **2014**, *8*, 121–131. [CrossRef]
11. Chawla, R.; Banerjee, A. A virtual environment for simulating manufacturing operations in 3D. In Proceedings of the 2001 Winter Simulation Conference, Institute of Electrical and Electronics Engineers Inc., Arlington, VA, USA, 9–12 December 2001.
12. Xiao, H.; Li, Y.; Yu, J.-F.; Cheng, H. Dynamic assembly simplification for virtual assembly process of complex product. *Assem. Autom.* **2014**, *34*, 1–15. [CrossRef]
13. Lang, Y.D.; Yao, Y.X.; Xia, P.J. A survey of virtual assembly technology. In *e-Engineering and Digital Enterprise Technology, e-ENGDET*; Trans Tech Publications Ltd.: Stafa-Zurich, Switzerland, 2007.
14. Zorriassatine, F.; Wykes, C.; Parkin, R.; Gindy, N. A survey of virtual prototyping techniques for mechanical product development. *Proc. Inst. Mech. Eng. Part B: J. Eng. Manuf.* **2003**, *217*, 513–530. [CrossRef]
15. Li, Y.; Dong, L.; Yu, J.F.; Zhang, K.F. A system for 3D digital assembly process planning and simulation in airplane manufacturing enterprises. In Proceedings of the 2nd International Conference on Manufacturing Science and Engineering, ICMSE 2011, Trans Tech Publications, Guilin, China, 9–11 April 2011.
16. Xu, L.D.; Wang, C.; Bi, Z.; Yu, J. AutoAssem: An automated assembly planning system for complex products. *IEEE Trans. Ind. Inform.* **2012**, *8*, 669–678. [CrossRef]
17. Lang, Y.D.; Yao, Y.X.; Xia, P.J.; Li, J.G. Virtual assembly system for large-scale complex products. In *e-Engineering and Digital Enterprise Technology, e-ENGDET.*; Trans Tech Publications Ltd.: Stafa-Zurich, Switzerland, 2008.
18. Sly, D. Integrating 3D product models with assembly line balancing via process consumption. *Procedia Manuf.* **2018**, *17*, 183–189. [CrossRef]
19. He, W.X.; Wan, B.L.; Zhang, B.; Wu, X.H. Research of the technology and application of process digital mock-up for spacecraft assembly based on the MBD. In Proceedings of the 5th International Conference on Computing, Control and Industrial Engineering, CCIE 2014, Trans Tech Publications Ltd., Wuhan, China, 25–26 October 2014.
20. Chao, P.Y.; Chen, T.T. Analysis of assembly through product configuration. *Comput. Ind.* **2001**, *44*, 189–203. [CrossRef]
21. Sun, J.; Yin, F. Design of Assembly Process Information Release System(APIRS) based on 3D product model. In Proceedings of the 3rd International Conference on Information Management, Innovation Management and Industrial Engineering, ICIII 2010, Kunming, China, 26–28 November 2010.
22. Xiao, H.; Duan, Y.; Zhang, Z. Mobile 3D assembly process information construction and transfer to the assembly station of complex products. *Int. J. Comput. Integr. Manuf.* **2018**, *31*, 11–26. [CrossRef]
23. Wright, T.P. Factors affecting cost of airplanes. *USA Air Serv.* **1936**, *21*, 23–25. [CrossRef]
24. Reinhardt, J.; Garrett, J.H., Jr.; Akinci, B. Framework for providing customized data representations for effective and efficient interaction with mobile computing solutions on construction sites. *J. Comput. Civ. Eng.* **2005**, *19*, 109–118. [CrossRef]
25. Morey, B. Chrysler implements flexible manufacturing. *Manuf. Eng.* **2007**, *139*, 63.
26. Ma, Z.; Liu, Z.; Zhang, D. An integrated mobile material management system for construction sites. In Proceedings of the 2013 Architectural Engineering National Conference: Building Solutions for Architectural Engineering, AEI 2013, American Society of Civil Engineers (ASCE), State College, PA, USA, 3–5 April 2013.
27. Angrish, A.; Craver, B.; Xu, X.; Starly, B. A search engine for manufacturers using product manufacturing information (pmi) enhanced 3D model search. In Proceedings of the ASME 2018 13th International Manufacturing Science and Engineering Conference, MSEC 2018, American Society of Mechanical Engineers (ASME), College Station, TX, USA, 18–22 June 2018.
28. Bo, H.; Huang, W.; Liu, X. Bill of manufacturing resource oriented to capacity allocation in steel enterprises. In Proceedings of the 2011 International Conference on Frontiers of Manufacturing Science and Measuring Technology, ICFMM 2011, Chongqing, China, 23–24 June 2011.
29. Chen, Y.; Kamara, J.M. A framework for using mobile computing for information management on construction sites. *Autom. Constr.* **2011**, *20*, 776–788. [CrossRef]
30. Lazzerini, B.; Marcelloni, F. Genetic algorithm for generating optimal assembly plans. *Artif. Intell. Eng.* **2000**, *14*, 319–329. [CrossRef]
31. Miller, D.M.; Park, Y. Simulation and analysis of an automotive assembly operation. In Proceedings of the Southern Automotive Manfacturing Conference and Exposition, SAE International, Nashville, TN, USA, 11–13 August 1998.
32. Silva, W.B.; Rodrigues, M.A.F. A lightweight 3D visualization and navigation system on handheld devices. In Proceedings of the 24th Annual ACM Symposium on Applied Computing, SAC 2009, Honolulu, HI, USA, 8–12 March 2009. Association for Computing Machinery (ACM).
33. Wang, L.X.; Guo, M.Y.; Guo, Y. Exploration of the modeling process of assembly based on MBD and Pro/ENGINEER. In Proceedings of the 3rd International Conference on Advanced Engineering Materials and Technology, AEMT 2013, Trans Tech Publications Ltd., Zhangjiajie, China, 11–12 May 2013.
34. Xu, X.; Wang, L.; Newman, S.T. Computer-aided process planning—A critical review of recent developments and future trends. *Int. J. Comput. Integr. Manuf.* **2011**, *24*, 1–31. [CrossRef]
35. Sly, D.; Kirpes, C. Emerging Frontiers in Industrial and Systems Engineering: Success Through Collaboration. In *Building and Managing the Bill of Process to Streamline the Enterprise–An Emerging Technology-Enabled Systems Approach*; Nembhard, H.B., Cudney, E.A., Coperich, K.M., Eds.; CRC Press: Boca Raton, FL, USA, 2019; pp. 149–165.

Short Biography of Authors

Carl Kirpes is a PhD student in Industrial Engineering at Iowa State University. He holds a Masters in Systems Engineering, a Bachelors in Industrial and Manufacturing Systems Engineering, and a Bachelors in Mechanical Engineering from Iowa State University. His research focuses on the value of emerging technology applications in manufacturing and assembly processes. Carl is a licensed Professional Engineer in the state of Texas and a certified Project Management Professional. He works for Marathon Petroleum Corporation as a Crude Oil Scheduling, Strategy, & Analysis leader and currently serves on the Board of Trustees for the Institute of Industrial and Systems Engineers as the Senior Vice President of Industry.

David Sly, Ph.D. is the Director of Entrepreneurship in the Industrial and Manufacturing Systems Engineering department at Iowa State University (IMSE). He earned a BS, MS, and PhD in IMSE and an MBA in business (emphasis in Marketing) all at Iowa State University (ISU). Dr. Sly is a registered Professional Engineer in the State of Iowa and is also President of Proplanner, which is a cloud-based Manufactory PLM software firm located within the ISU Research Park.

Guiping Hu is an Associate Professor in the department of Industrial and Manufacturing Systems Engineering at Iowa State University. She received her MS and PhD degrees from University of Pittsburgh. Her research interests include operations research and data analytics with applications in supply chain design, manufacturing decision support systems, and energy systems analysis.

applied system innovation

MDPI

Article

Cognitive Manufacturing in Industry 4.0 toward Cognitive Load Reduction: A Conceptual Framework

Adriana Ventura Carvalho [1,*], Amal Chouchene [1,2], Tânia M. Lima [1,3] and Fernando Charrua-Santos [1,3]

1 Department of Electromechanical Engineering, University of Beira Interior, 6200-358 Covilhã, Portugal; a.chouchene@ubi.pt (A.C.); tmlima@ubi.pt (T.M.L.); bigares@ubi.pt (F.C.-S.)

2 LIMTIC—Laboratoire de Recherche en Informatique Modélisation et traitment de l'Information et la Connaissance, Research Team SIIVA, Université de Tunis El Manar, Tunis 1068, Tunisia

3 C-MAST—Center for Mechanical and Aerospace Science and Technologies, University of Beira Interior, 6201-358 Covilhã, Portugal

* Correspondence: adriana.carvalho@ubi.pt

Received: 26 October 2020; Accepted: 26 November 2020; Published: 3 December 2020

Abstract: Cognitive manufacturing utilizes cognitive computing, the industrial Internet of things (IoT), and advanced analytics to upgrade manufacturing processes in manners that were not previously conceivable. It enables associations to improve major business measurements, for example, productivity, product reliability, quality, and safety, while decreasing downtime and lowering costs. Considering all the facts that can prejudice the manufacturing performance in Industry 4.0, the cognitive load has received more attention, since it was previously neglected with respect to manufacturing industries. This paper aims to investigate what causes cognitive load reduction in manufacturing environments, i.e., human–computer interaction technologies that reduce the identified causes and the applications of cognitive manufacturing that use the referred technologies. Thus, a conceptual framework that links cognitive manufacturing to a reduction of the cognitive load was developed.

Keywords: Industry 4.0; cognitive manufacturing; cognitive load; human–computer interaction

1. Introduction

Industry 4.0 (I4.0) is the mainstream name given to the Fourth Industrial Revolution introduced by the digital age. Some experts trust that Industry 4.0 is more evolutionary than revolutionary since computers were the establishment of the last revolution and remain the establishment of Industry 4.0 [1]. Different technological developments resulted in Industry 4.0. This new paradigm faces the challenge of being exceptionally computerized and financially savvy, as well as producing custom products in a large-scale manufacturing condition, and it can change the role of traditional assembly lines by altering how goods are produced and services are offered [2]. The rapid advancement of technologies such as cloud computing, big data, mobile internet [3], augmented reality (AR), virtual reality (VR), and artificial intelligence (AI) has resulted in substantial work in these areas [4]. Furthermore, the combination of these information technologies and the assembly industry has instigated the current fiercely debated issue of intelligent manufacturing (IM) [5].

Cognitive manufacturing (CM) utilizes cognitive computing, the industrial Internet of things (IIoT), and advanced analytics to upgrade manufacturing processes in manners that were not previously conceivable. It enables associations to improve major business measurements, for example, productivity, product reliability, quality, and safety, while decreasing downtime and lowering costs [6].

In 1990, Ref. [7] emphasized how the cognitive imprecision of data and vulnerability in user awareness of the computing environment were significant components of human–computer interaction, which should be imprecision-tolerant take into account the inexact mode of communication. Considering this, before the time of big data, cognitive technologies were unrealistic, as their systems need data to analyze. For most manufacturers, having enough information is no longer an issue. Indeed, most manufacturers deal with more information than they can break down, leading to a use of more seasoned techniques [1]. Cognitive systems are fit for automating routine decisions, and they support Industry 4.0 by creating significant experiences that help human decisionmakers to manage anomalies or other abnormal and complex business decisions. Cognitive systems can comprehend vast quantities of factors that uncover underlying causes of issues or point to more effective courses of action [1].

The enthusiasm for cognitive aspects of human performance has drastically expanded as of late in manufacturing, supplementing the zone of physical ergonomics, and the spotlight on cognitive aspects may offer critical insights to industries. Significantly increased interest has been aimed at cognitive aspects and their impact on human performance and production outcome [8].

Although the term "cognitive manufacturing" has never been directly related to reducing human cognitive load in manufacturing, it is interesting to assess points where the subjects are interlinked. This paper aims to investigate what causes cognitive load reduction in manufacturing environments, i.e., technologies that reduce the identified causes and the applications of cognitive manufacturing that adopt the referred technologies. In this study, an investigation of the subject was performed through the following research question:

RQ1: Is cognitive manufacturing a term that can be extended to a technology that reduces the human cognitive load in manufacturing?

To help deconstruct the main research question, three sub-questions were developed:

- SRQ1: What causes cognitive overload in manufacturing environments?
- SRQ2: Which are the technologies that are able to reduce the cognitive load in manufacturing environments?
- SRQ3: Which are the cognitive manufacturing applications that use technologies that reduce the cognitive load?

The structure of this paper starts, from this point on, with a literature review of cognitive manufacturing, followed by a literature review of the cognitive load in manufacturing, before moving to a description of the materials and methods used to perform this investigation and the subsequent results, concluding with the discussion and conclusions of the work developed.

1.1. Cognitive Manufacturing

Cognitive manufacturing gathers significant data in real time and applies analytics to help attain previously impossible insights into the manufacturing process. It mechanizes reactions dependent on its discoveries and conveys actionable information as continuously updated knowledge to workers [6]. It is powerful since it joins sensor-based data with machine learning and other AI capabilities to discover patterns in data, structured or not, aligning relevant information together in real time [9].

Cognitive technologies can discover importance in these data in manners that, until recently, only the human brain could fathom. This degree of understanding can be viewed as essential in the modern manufacturing era, where competitiveness and cost sensitivities request new degrees of agility, responsiveness, and innovation from manufacturers [6].

These technologies, such as AI, in general, can be characterized as the capacity of machines to comprehend, learn, and resonate so as to emulate the cognitive functions of the human brain [10].

Manufacturers can utilize cognitive technologies to tackle fundamental business challenges, discover new value in their assembled information, improve quality, and upgrade knowledge in their organizations. Cognitive manufacturing empowers organizations to set a focus on quality throughout the life cycle of a product's development [6].

1.2. Cognitive Load in Manufacturing

Traditionally, the focal point of HCI has been on the most effective method to guarantee that the technology serves the users' needs. Throughout the years, HCI has progressed, and the human share of technology is additionally changing, whereby humans have become increasingly attentive and demanding. Therefore, human methodologies face increased difficulties to underlie an increasingly trustworthy and valuable connection between humankind and technology [11].

Considering all the facts that can prejudice the manufacturing performance, the cognitive load has received more attention, since it was previously neglected with respect to manufacturing industries.

When exposed to stimuli, the cognitive system experiences what is commonly referred to as cognitive load [12]. Cognitive load refers to the mental load that performing a specific task imposes on the human's cognitive system [13].

The theory of cognitive load states that effective instructional material encourages learning by coordinating cognitive assets toward exercises that are significant to learning as opposed to preliminary to learning [14]. This theory is concerned with the way in which cognitive assets are engaged and utilized during learning and problem solving [15].

In cognitive load theory, three types of cognitive load are considered [16]:

- Intrinsic—cognitive load related to a topic or task. We can consider this type as the objective difficulty of a task;
- Extraneous—the manner in which the data or tasks are exhibited. How we find data decides the assets we have available to interpret it;
- Germane—the germane load is created by the development of schemas; it helps in learning new skills and other data.

For manufacturing purposes, the intrinsic and extraneous loads are the most relevant [17].

High intrinsic difficulty occurs in manufacturing work because most routine tasks are performed by automated systems, while assigning complex and variable tasks to the human worker. Furthermore, significant extraneous load is experienced by manufacturing workers once the structure and type of work are not helpful for precise execution, in addition to a lack of technological support, often resulting in written instructions and manual data collection. Traditional workflows and tools augment the cognitive load on workers [17].

Human performance is affected by a cognitive load that is excessively high, whereby information concerning the job and the importance of a cognitive human in an assembly domain could possibly have a noteworthy impact on the production result (quality and productivity). Significant causes of quality defects in manufacturing are currently appointed to both product and process errors [8].

There may be similarities among humans and machines; however, human cognition is the result of a human's interactions with the environment, demonstrating that there are cognitive activities that rely upon the cooperation of the human body (musculoskeletal system and peripheral nervous system) with sensory inputs from the environment, as well as the working of the brain, commonly referred to as distributed cognition [18].

2. Materials and Methods

Cognitive manufacturing remains a yet moderately new subject in academic literature, and gaps were identified in relating the topic to a reduction of the cognitive load experienced by workers in the manufacturing field. Along these lines, there is a need to get a handle on what has been investigated and where results are scarce. As this investigation was being developed, the current literature was reviewed, finding that no author previously investigated cognitive manufacturing and cognitive load, highlighting the importance of this study.

A comprehensive investigation and preliminary study were the first steps of this research to comprehend and identify the problem, thereby obtaining deeper knowledge of the research area. This led to the development of the research question. After the preliminary examinations, a literature review was carried out to obtain a more extensive comprehension of the exploration territory, leading to a restraint of the scope and the generation of a theoretical framework. This also allowed a better understanding of the concept of cognitive load. The perceptions of cognitive manufacturing, its applications, and the technologies that reduce the cognitive load were also deepened.

To address SRQ2 and SRQ3, the following keywords were defined as criteria for the inclusion and exclusion of articles: (cognitive manufacturing) AND (cognitive load), as well as ("human–computer interaction") AND ("cognitive load"), which were combined using Boolean operators. The search was conducted on the basis of the title, abstract, and keywords. The search was also limited to studies in English and Portuguese. For this, two recognized databases were selected: Scopus and Web of Science. As explained previously, due to the lack of relevant results, other sources were also consulted (books, white papers, and gray literature). A graph was developed to enhance comprehension of the publications used to support the answers to SRQ2 (Figure 1) and SRQ3 (Figure 2).

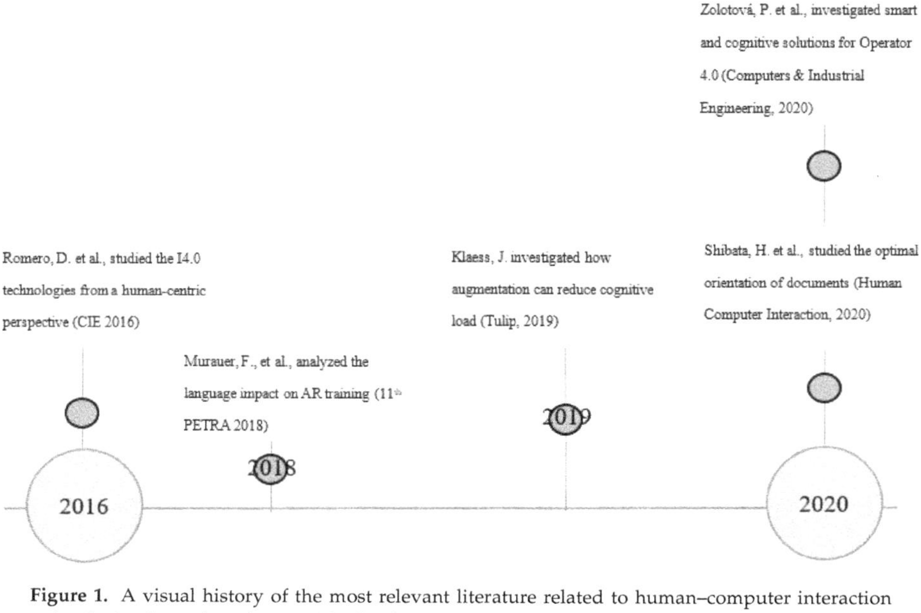

Figure 1. A visual history of the most relevant literature related to human–computer interaction technologies that reduce the cognitive load.

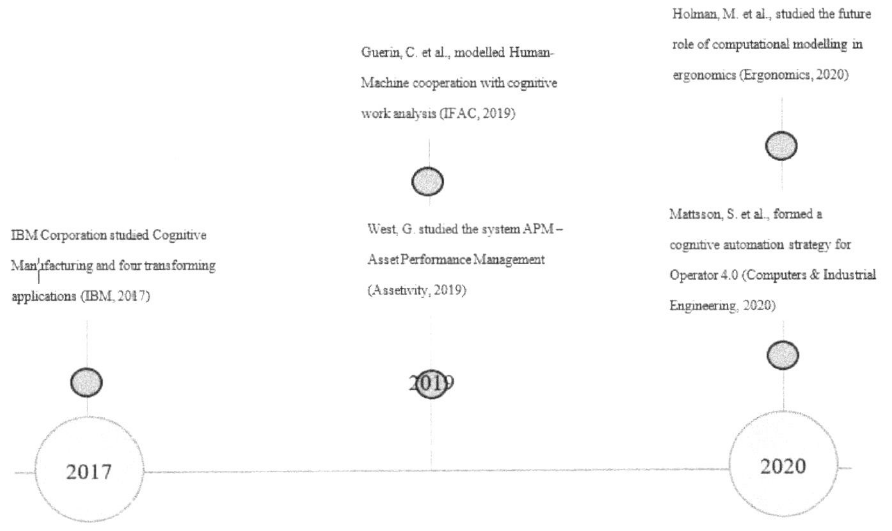

Figure 2. A visual history of the most relevant literature related to cognitive manufacturing applications using technologies that reduce the cognitive load.

To address SRQ1, the available literature was reviewed through the use of the Scopus database, using the following string: TITLE-ABS-KEY ("cognitive overload" OR "cognitive load") AND (LIMIT-TO (SUBJAREA, "COMP") OR LIMIT-TO (SUBJAREA, "ENGI")). Additional records were identified through other sources. The flow chart of the literature search for this first research question can be found in Figure 3.

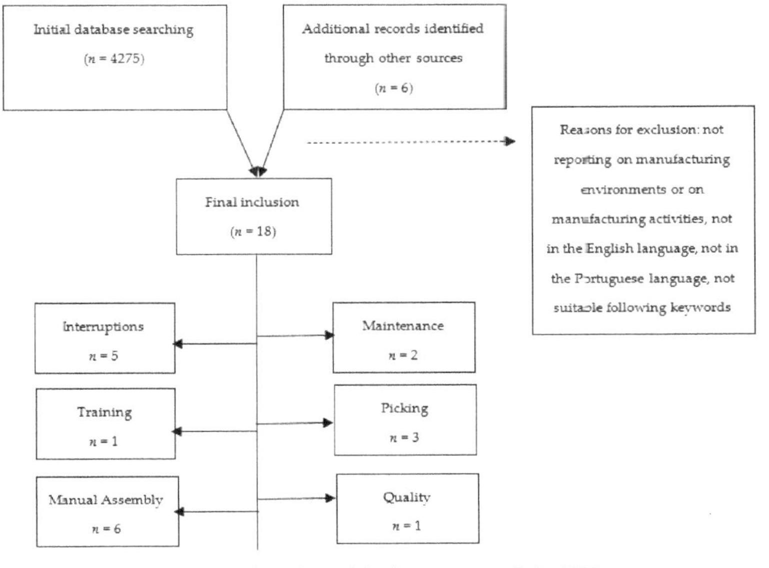

Figure 3. Flow chart of the literature search for RQ1.

A graph was developed to enhance comprehension of the publications used to support the answer to this sub-question (Figure 4).

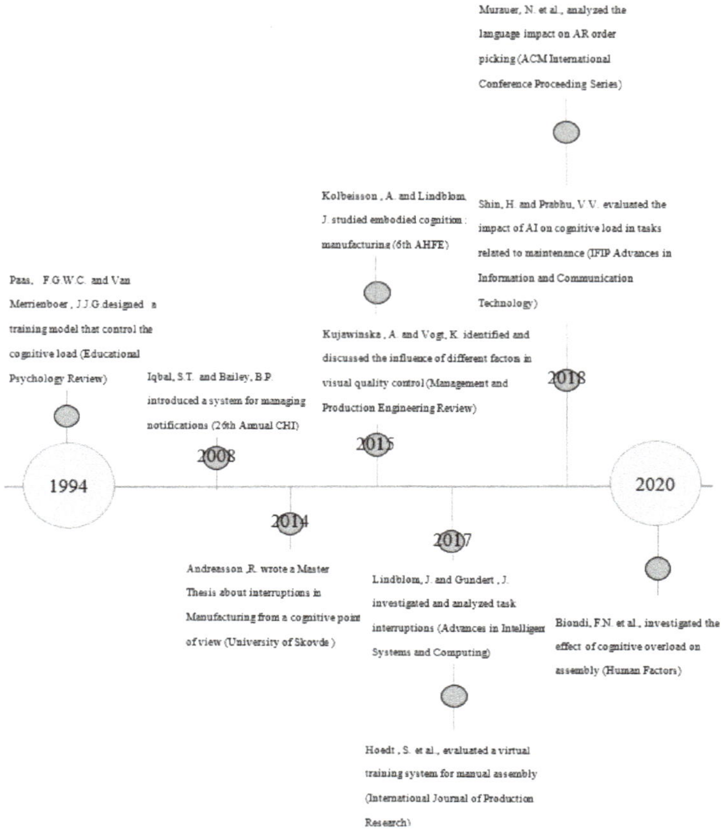

Figure 4. A visual history of the most relevant literature related to cognitive overload in manufacturing environments

Through the analyses of the available literature and according to the conceptual framework developed, a system to discuss and analyze the results was established to answer the research question, as presented in Section 4.

3. Results

In this section, a discussion and analyses of the findings from the sub-questions are presented.

3.1. SRQ1: What Causes Cognitive Overload in Manufacturing Environments?

According to [19], the performance of a job is influenced by its nature, i.e., if the assignment requires the operators to use more or less cognitive power. Table 1 shows the causes of cognitive overload in manufacturing environments, according to the available literature. It is important to consider that manufacturing environments include different tasks that might or might not require excessive use of the human cognitive function. Thus, by identifying the causes of cause cognitive overload, it is possible to access what technological tools can be used to reduce that overload and in what sectors of the manufacturing environment these tools should be implemented.

Table 1. Causes of cognitive overload in manufacturing environments.

Causes of Cognitive Overload	Literature
Interruptions	According to [20–23], interruptions are identified as being an essential driver to cognitive overload, which influences the human's capacity to perform effectively. The authors of [24] stated that interruptions are identified as being an essential driver to cognitive overload, which influences the human's capacity to perform effectively.
Training/instructional situations	The authors of [25] stated that training in some areas commonly speaks to circumstances that are near the breaking point of trainees' capacities, forcing cognitive overload.
Manual assembly	The authors of [9] mentioned that, because of the strategies of manufacturing companies, manual assemblers face a bigger cognitive load than in past times. The authors of [26] demonstrated that the work performed under cognitive overload affects assembly task completion times. The authors of [27] studied a reduction in cognitive load in complex assembly systems. The authors of [28] mentioned the information management strategies in manual assembly. The authors of [29] evaluated the guidelines for assembly instructions. The authors of [8] developed a method for cognitive load assessment.
Maintenance activities	The authors of [30] developed an AI tool to test if, among other aspects, the cognitive load of the maintenance workers using an AI-based system would be lower. The authors of [31] studied a cognitive perspective and methodology for reverse engineering tools.
Order picking	The authors of [32] stated that order picking is a demanding task at the cognitive level. The authors of [33] demonstrated the significant effect of for pinking tasks on human capacities and error rate by recording the human cognitive load, while the authors of [34] studied the order picking process using a projector helmet.
Visual inspection/quality inspection	The authors of [35] based their work on the knowledge that visual control does not ensure a completely correct assessment, due to constrained human reliability that is influenced by several elements which impact the capacity of a human to properly evaluate the quality of the procedure and/or product.

3.2. SRQ2: What Are the Human–Computer Interaction Technologies That Reduce the Cognitive Load in Manufacturing Environments?

Creating a working framework by empowering workers to boost their mental and physical assets might be the secret to decreasing the cognitive load. For this, manufacturers ought to consider equipping their lines with tools that allow workers to concentrate on the job that needs to be done. Innovative technologies can limit the impacts of stress and time pressure while bringing the numerous factors that the worker cannot control under management [17]. The cognitive work of smart factories can be supported by the technologies of Industry 4.0 [36].

As the transfer of cognitive burden continues rising, society will undergo a cognitive revolution, characterized by technology's capability to augment the cognitive potential of humans [30]. Industry 4.0 is rich in new technologies; however, for the purpose of this research, it was important to separate the technologies related to human–computer interactions and a reduction in cognitive load, in order to keep track of the actual technologies that concerned this investigation.

Different types of stimulus material are perceived differently in terms of cognitive processes. For example, remembrance of information from complex messages is often superior when the material is read rather than heard; however, simple material is better retained when it is heard [37]. Furthermore, traditional HCI focuses on preventing usability problems and, according to [38], it should also generate remarkable quality experiences and add to a better quality of life. The ideal approach to decrease cognitive load is through augmentative technologies, which does not necessarily refer only to augmented reality when it comes to manufacturing [32]. The following technologies help workers on the job, and they can be integrated into the environment to improve worker capacity [17]:

- Digital work instructions guide workers through complex procedures, progressing with them, introducing them to the data they need when they need it, decreasing stress, and removing common sources of error [17]. As summed up in [39], reading on high-quality computer displays can be done as fast as reading on paper. A smarter operator [40] interacts with an AI personal assistant, thereby reducing the probability of mistakes happening [41]. This interaction can happen through virtual or augmented reality (virtual operator or augmented operator [40]). Some relevant Industry 4.0 technologies are in-view instructions using head-mounted displays, tablet instructions, projection-based in situ instructions, and step-by-step instructions that guide the worker through the whole process [42].
- Digital training applications help streamline the learning procedure by exhibiting data to the learner through focused, interactive modules. These applications can be designed explicitly for the assignment being referred to, so that workers can be instructed on the exact task that they will perform [17]. Industry 4.0 technologies related to training might include virtual, augmented, and smarter operators [40], whereby workers can be trained using, for example, e-learning [43], virtual reality [44], and augmented reality [45].
- Real-time analytics dashboards can help lessen the attention and energy given to pre-analysis of data by indicating expert data on the performance of humans and machines, thereby simplifying how data are gathered and introduced [17]. This can only be performed due to the use of Industry 4.0 technology such as machine learning, turning "big data" into "smart data" [46], and using AI incorporated into human–machine interfaces to support decision-making [47].
- Augmented reality (AR) reduces human errors and lightens the memory use of the operator, safely replicating the environment [47]. With AR, there is no compelling reason to change focus between the digital and physical worlds and no compelling reason to withdraw from a task to chase applicable data about what to do straight away or how to do it [30]. Augmented operators [40] have their working environment enriched by digital information, which reduces human error and improves decision-making by displaying feedback in real time [41].
- Inline quality checks allow addressing some quality issues that are extremely small and barely detectable by eye, as well as others that are the consequence of worker fatigue. Regardless of the reason, numerous quality issues are accepted due to failures in identifying them. All manufacturers have some convention for checking quality inline; however, if the workers have the correct tools, they will be able to catch more nonconformances, prompting fewer rework hours [32]. Some examples of Industry 4.0 technologies for quality checks are automated solutions [48] and machine vision systems [49,50].

It is important to state that not only operators benefit from technologies that reduce the cognitive load; these tools are also important to anybody doing physical or intellectual work on the shop floor [32].

3.3. *SRQ3: Which Are the Cognitive Manufacturing Applications That Use Technologies That Reduce the Cognitive Load?*

Cognitive manufacturing completely uses the data from hardware, systems, and procedures to infer significant knowledge over the whole value chain through various procedures (design, manufacture, and support activities). Cognitive manufacturing is based on the establishments of IoT and employing analytics joined with cognitive technologies [51].

From the SRQ2 above, we were able to identify the human–computer technologies that reduce the cognitive load. Below, we analyze these cognitive manufacturing applications to fulfill the purpose of this investigation and link cognitive manufacturing to a reduction in cognitive load.

According to the available literature, the following cognitive manufacturing applications were identified:

- Asset performance management (APM) frequently catches information and data that are connected with asset condition, to provide a comprehensive perspective of the performance of the asset. The data are then used for reliability analytics and asset health visualization, to help the improvement and tweaking of different asset models [52]. Companies can use cognitive APM to sense, diagnose, and communicate performance issues to lessen unwanted downtime. The application can envision a potential failure and then investigate data from important user manuals or technician logs to comprehend how a previous similar issue was resolved, using this information to prescribe explicit activities or answers to fix the issue [6].
- Process and quality improvement are represented throughout the manufacturing process. The numerous attributes that impact product quality can be monitored and understood by utilizing cognitive manufacturing tools. Potential quality issues can be recognized earlier by using analytics, algorithms, automated visual inspections, and machine learning instead of customary methods [6]. The symbiosis between the operator and the cyber-physical system (CPS) allows new margins for controlling and improving picking activities [36]. New strategies were suggested where operator cognition is enhanced by moving between assembly modes [53].
- Resource optimization in cognitive manufacturing can help in guaranteeing laborer safety and health. Equipment with sensors that identify immediately hazardous circumstances ensure laborer safety and improve operations in energy resource optimization. Moreover, the use of IoT, data analytics, and machine learning allows evaluating the factors that contribute to energy consumption and in improving floor planning and scheduling. This can be done to optimize the configuration of a production line to balance the workload between stations, as well as use labor more efficiently, increase the rate of production, and optimize available plant capacity [6]. To perform this, ergonomic aspects must be considered, since they are changing as the world advances to Industry 4.0; thus, the discipline must adapt to this new paradigm and its new methods [54].
- Supply chain optimization in cognitive manufacturing gathers different data from structured and unstructured data sources so as to limit supply chain costs, disruptions, and risks. Alerts that describe the threat and present the information in a proper manner to help in decision making, as well as search for alternative suppliers and recommend solutions, represent some of the solutions that a cognitive manufacturing tool can offer [6]. Most companies poorly integrate technology into their supply chain, whereas an optimized supply chain will develop new value propositions and allow meeting new business needs [55].

4. Discussion

In order to flourish in the Industry 4.0 era, manufacturers should look toward the potential of legacy, real-time, and unstructured data to settle on everyday choices that equalize quality and throughput. Having in mind that the average manufacturing site runs hundreds of software applications, it is an extremely significant challenge to make that information accessible and actionable.

The human cognitive load is currently being pushed to the limits. The fact that humanity needs cognitive manufacturing shows that the human brain has boundaries that can be suppressed by the use of machine learning, AI, and other technologies. Reducing the cognitive load by merging the digital and physical worlds allows the gap between the two worlds to diminish.

From the literature review in this paper, we can conclude that both terms can be related with respect to cognitive manufacturing technologies developed to reduce the human workload, whereby they are equally developed to perform activities on which the human worker would spend a vast amount of time, resources, and money, consequently reducing the cognitive load associated with the task. Some activities might not even be possible to execute if not through the use of technology, although there is a lack of literature stating that cognitive manufacturing might lessen the cognitive load of workers.

This investigation was dedicated to cognitive load reduction in Industry 4.0 and cognitive manufacturing issues and it can be qualified as a work in progress. We intend to keep searching for new efficient methods to demonstrate the link between cognitive manufacturing and reduced cognitive load, working with industrial manufacturers through case studies. The next steps of this work will concern the construction and evaluation of data analyses in order to characterize the usefulness of the suggested conceptual framework.

5. Conclusions

To explore the problem found during the literature review, the conceptual framework in Figure 5 was developed. The framework demonstrates the relationships among the factors affecting cognitive overload causing manufacturing errors and how that overload can be lessened using cognitive technologies applied in cognitive manufacturing applications.

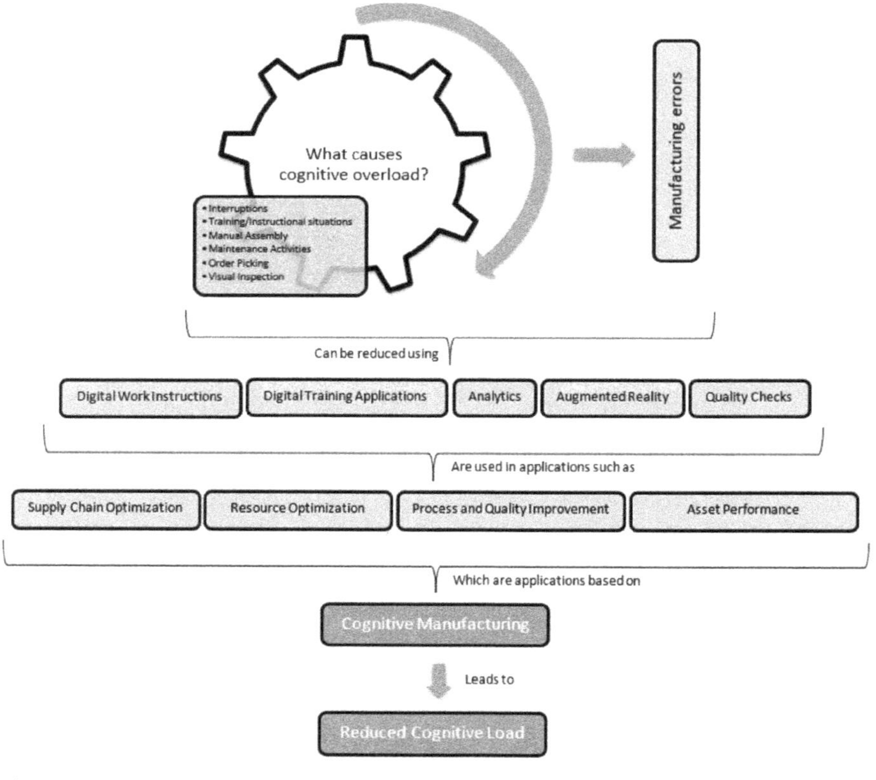

Figure 5. The conceptual framework for cognitive load minimization using cognitive manufacturing applications.

Further research will explore how Industry 4.0 can benefit from a lower burden on human cognitive load, using different tools such as Augmented Reality (AR), allowing investment in cognitive manufacturing to be extended. It will also be interesting to analyze if an investment in cognitive manufacturing tools, aimed solely at reducing the cognitive load of the working force, translates into a good investment for a company.

However, it is safe to conclude that Industry 4.0 technologies play a very important part in the way that cognitive aspects of the operator are processed.

Although this investigation evaluated the relationship between cognitive manufacturing and reduced cognitive load, another aspect which must be taken into consideration is the financial investment in these technologies.

To develop this conceptual framework, the relationship between human–computer interaction technologies and cognitive manufacturing applications that reduce the cognitive load was established on the basis of Industry 4.0 technologies. This investigation represents the starting point for further research on the subject of the relationship between the use of cognitive manufacturing applications and a reduced cognitive load, which was not previously identified in the literature.

Author Contributions: Conceptualization, A.V.C. and T.M.L.; methodology, A.V.C. and A.C.; investigation, A.V.C. and A.C.; writing—original draft preparation, A.C.; writing—review and editing, T.M.L. and F.C.-S.; supervision, T.M.L. and F.C.-S.; project administration, F.C.-S. All authors have read and agree to the published version of the manuscript.

Funding: This research was funded by the project 026653|POCI-01-0247-FEDER-026653—INDTECH—New technologies for smart manufacturing, co-financed by the Portugal 2020 Program (PT 2020), Compete 2020 Program, and the European Union through the European Regional Development Fund (ERDF). The authors wish to thank the relevant bodies for the opportunity and financial support that permitted carrying out this project: Fundação para a Ciência e Tecnologia (FCT) and C-MAST (Center for Mechanical and Aerospace Science and Technologies), under project UIDB/00151/2020.

Conflicts of Interest: The authors declare no conflict of interest.

References

1. Stephen DeAngelis, Industry 4.0: The Emergence of Cognitive Manufacturing. March 2019. Available online: https://www.enterrasolutions.com/blog/industry-4-0-the-emergence-of-cognitive-manufacturing/ (accessed on 1 September 2019).
2. Santos, B.P.; Charrua-Santos, F.; Lima, T.M. *Industry 4.0: An Overview. Lecture Notes in Engineering and Computer Science*; Newswood Limited: London, UK, 2018; Volume 2236, pp. 415–420, ISBN 978-988140489-3.
3. Chen, J.; He, K.; Yuan, Q.; Chen, M.; Du, R.; Xiang, Y. Blind filtering at third parties: An efficient privacy-preserving framework for location-based services. *IEEE Trans. Mob. Comput.* **2018**, *17*, 2524–2535. [CrossRef]
4. Ge, X.; Pan, L.; Li, Q.; Mao, G.; Tu, S. Multipath cooperative communications networks for augmented and virtual reality transmission. *IEEE Trans. Multimed.* **2017**, *19*, 2345–2358. [CrossRef]
5. Hu, L.; Miao, Y.; Wu, G.; Hassan, M.M.; Humar, I. iRobot-Factory: An intelligent robot factory based on cognitive manufacturing and edge computing. *Future Gener. Comput. Syst.* **2019**, *90*, 569–577. [CrossRef]
6. IBM Corporation. *Cognitive Manufacturing: An Overview and Four Applications that are Transforming Manufacturing Today*; IBM Corporation: Armonk, NY, USA, 2017.
7. Karwowski, W.; Kosiba, E.; Benabdallah, S.; Salvendy, G. A framework for development of fuzzy GOMS model for human-computer interaction. *Int. J. Hum. Comput. Interact.* **1990**, *2*, 287–305. [CrossRef]
8. Thorvald, P.; Lindblom, J.; Andreasson, R. On the development of a method for cognitive load assessment in manufacturing. *Robot. Comput. Integr. Manuf.* **2019**, *59*, 252–266. [CrossRef]
9. Hoedt, S.; Claeys, A.; van Landeghem, H.; Cottyn, J. The evaluation of an elementary virtual training system for manual assembly. *Int. J. Prod. Res.* **2017**, *55*, 7496–7508. [CrossRef]
10. Chouchene, A.; Carvalho, A.; Charrua-Santos, F.; Lima, T.M.; Osório, G.J.; Barhoumi, W. Artificial Intelligence for Product Quality Inspection toward Smart Industries: Quality Control of Vehicle Non-Conformities. In Proceedings of the 9th International Conference on Industrial Technology and Management, University of Oxford, Oxford, UK, 11–13 February 2020.
11. Stephanidis, C.; Salvendy, G.; Antona, M.; Chen, J.Y.; Dong, J.; Duffy, V.G.; Fang, X.; Fu, L.P.; Mori, H.; Guo, Y.; et al. Seven HCI Grand Challenges. *Int. J. Hum. Comput. Interact.* **2019**, *35*, 1229–1269. [CrossRef]

12. Bannert, M. Managing cognitive load—Recent trends in cognitive load theory. *Learn. Instr.* **2002**, *12*, 139–146. [CrossRef]
13. Lindblom, J.; Thorvald, P. Towards a framework for reducing cognitive load in manufacturing personnel. *Adv. Cogn. Eng. Neuroergon.* **2014**, *11*, 233–244. Available online: Ibm.com/downloads/cas/VDNKMWM6 (accessed on 1 September 2019).
14. Chandler, P.; Sweller, J. Cognitive Load Theory and the Format of Instruction. *Cogn. Instr.* **1991**, *8*, 293–332. [CrossRef]
15. Sweller, J. Cognitive load during problem solving: Effects on learning. *Cogn. Sci.* **1988**, *12*, 257–285. [CrossRef]
16. Paas, F.; Renkl, A.; Sweller, J. Cognitive Load Theory: Instructional Implications of the Interaction between Information Structures and Cognitive Architecture. *Instr. Sci.* **2004**, *32*, 1–8. [CrossRef]
17. John, K. How Augmentation Can Reduce Cognitive Load and Improve Decision Making in Manufacturing. July 2019. Available online: https://tulip.co/blog/augmentation/cognitive-load/ (accessed on 1 September 2019).
18. Hutchins, E. *Cognition in the Wild*; MIT Press: Cambridge, MA, USA, 1995. Available online: https://mitpress.mit.edu/books/cognition-wild (accessed on 1 December 2020).
19. Lim, Y.M.; Ayesh, A.; Stacey, M. Continuous Stress Monitoring under Varied Demands Using Unobtrusive Devices. *Int. J. Hum. Comput. Interact.* **2020**, *36*, 326–340. [CrossRef]
20. Andreasson, R. *Interruptions in Manufacturing from A Distributed Cognition Perspective*; University of Skövde: Skövde, Sweden, 2014.
21. Kolbeinsson, A.; Lindblom, J. Mind the Body: How Embodied Cognition Matters in Manufacturing. *Procedia Manuf.* **2015**, *3*, 5184–5191. [CrossRef]
22. Iqbal, S.T.; Bailey, B.P. Effects of intelligent notification management on users and their tasks. In Proceedings of the Twenty-Sixth Annual CHI Conference on Human Factors in Computing Systems—CHI '08, Florence, Italy, 5–10 April 2008; p. 93. [CrossRef]
23. Lindblom, J.; Gündert, J. Managing mediated interruptions in manufacturing: Selected strategies used for coping with cognitive load. In *Advances in Neuroergonomics and Cognitive Engineering*; Springer: Cham, Switzerland, 2017; Volume 488.
24. Lee, B.C.; Duffy, V.G. The effects of task interruption on human performance: A study of the systematic classification of human behavior and interruption frequency. *Hum. Factors Ergon. Manuf.* **2015**, *25*, 137–152. [CrossRef]
25. Paas, F.G.W.C.; van Merriënboer, J.J.G. Instructional control of cognitive load in the training of complex cognitive tasks. *Educ. Psychol. Rev.* **1994**, *6*, 351–371. [CrossRef]
26. Biondi, F.N.; Cacanindin, A.; Douglas, C.; Cort, J. Overloaded and at Work: Investigating the Effect of Cognitive Workload on Assembly Task Performance. *Hum. Factors* **2020**. [CrossRef]
27. Bläsing, D.; Hinrichsen, S.; Bornewasser, M. Reduction of Cognitive Load in Complex Assembly Systems. In *International Conference on Human Interaction and Emerging Technologies*; Springer: Cham, Switzerland, 2020; Volume 1152, p. 500.
28. Hinrichsen, S.; Adrian, B.; Bornewasser, M. Information Management Strategies in Manual Assembly. In *International Conference on Human Interaction and Emerging Technologies*; Springer: Cham, Switzerland, 2020; Volume 1152, p. 525.
29. Mattsson, S.; Fast-Berglund, Å.; Li, D. Evaluation of Guidelines for Assembly Instructions. *IFAC-PapersOnLine* **2016**, *49*, 209–214. [CrossRef]
30. Shin, H.; Prabhu, V.V. Evaluating Impact of AI on Cognitive Load of Technicians during Diagnosis Tasks in Maintenance. In *Advances in Production Management Systems. Smart Manufacturing for Industry 4.0*; Moon, I., Lee, G.M., Park, J., Kiritsis, D., von Cieminski, G., Eds.; Springer: Cham, Switzerland, 2018; Volume 536, pp. 27–34.
31. Zayour, I.; Lethbridge, T.C. Adoption of reverse engineering tools: A cognitive perspective and methodology. In Proceedings of the 9th International Workshop on Program Comprehension 2001, Toronto, ON, Canada, 12–13 May 2001; pp. 245–255. [CrossRef]
32. Murauer, N.; Müller, F.; Günther, S.; Schön, D.; Pflanz, N.; Funk, M. An Analysis of Language Impact on Augmented Reality Order Picking Training. In Proceedings of the 11th PErvasive Technologies Related to Assistive Environments Conference on—PETRA '18, Corfu, Greece, 26–29 June 2018; pp. 351–357. [CrossRef]

33. Baechler, A.; Baechler, L.; Autenrieth, S.; Kurtz, P.; Hoerz, T.; Heidenreich, T.; Kruell, G. A comparative study of an assistance system for manual order picking—Called pick-by-projection—With the guiding systems pick-by-paper, pick-by-light and pick-by-display. In Proceedings of the 2016 49th Hawaii International Conference on System Sciences 2016, Koloa, HI, USA, 5–8 January 2016; pp. 523–531. [CrossRef]
34. Funk, M.; Mayer, S.; Nistor, M.; Schmidt, A. Mobile in-situ pick-by-vision: Order picking support using a projector helmet. In Proceedings of the 9th ACM International Conference on PErvasive Technologies Related to Assistive Environments, Corfu Island, Greece, 29 June–1 July 2016. [CrossRef]
35. Kujawińska, A.; Vogt, K. Human Factors in Visual Quality Control. *Manag. Prod. Eng. Rev.* **2015**, *6*, 25–31. [CrossRef]
36. Guerin, C.; Rauffet, P.; Chauvin, C.; Martin, E. Toward production operator 4.0: Modelling Human-Machine Cooperation in Industry 4.0 with Cognitive Work Analysis. *IFAC-PapersOnLine* **2019**, *52*, 73–78. [CrossRef]
37. Allen, R.B. Cognitive factors in human interaction with computers. *Behav. Inf. Technol.* **1982**, *1*, 257–278. [CrossRef]
38. Hassenzahl, M.; Tractinsky, N. User experience—A research agenda. *Behav. Inf. Technol.* **2006**, *25*, 91–97. [CrossRef]
39. Shibata, H.; Omura, K.; Qvarfordt, P. Optimal Orientation of Text Documents for Reading and Writing. *Hum. Comput. Interact.* **2020**, *35*, 70–102. [CrossRef]
40. Romero, D.; Stahre, J.; Wuest, T.; Noran, O.; Bernus, P.; Fast-Berglund, Å.; Gorecky, D. Towards an Operator 4.0 Typology: A Human-Centric Perspective on the Fourth Industrial Revolution Technologies. Available online: https://www.researchgate.net/profile/David_Romero2/publication/309609288_Towards_an_Operator_40_Typology_A_Human-Centric_Perspective_on_the_Fourth_Industrial_Revolution_Technologies/links/58e435e7a6fdccc85bdf36fa/Towards-an-Operator-40-Typology-A-Human-Centric-Perspective-on-the-Fourth-Industrial-Revolution-Technologies.pdf (accessed on 1 December 2020).
41. Madonna, M.; Monica, L.; Anastasi, S.; di Nardo, M. Evolution of Cognitive Demand in the Human–Machine Interaction Integrated with Industry 4.0 Technologies. *Wit Trans. Built Environ.* **2019**, *189*, 13–19. [CrossRef]
42. Pimminger, S.; Neumayr, T.; Panholzer, L.; Augstein, M.; Kurschl, W. Reflections on Work Instructions of Assembly Tasks. In Proceedings of the 2020 IEEE International Conference on Human-Machine Systems (ICHMS), Rome, Italy, 7–9 September 2020; pp. 1–4. [CrossRef]
43. Clarizia, F.; de Santo, M.; Lombardi, M.; Santaniello, D. E-Learning and Industry 4.0: A Chatbot for Training Employees. In Proceedings of the Fifth International Congress on Information and Communication Technology, Advances in Intelligent Systems and Computing, London, UK, 20–21 February 2020; Springer: Singapore, 2020; Volume 1184.
44. Kwegyir-Afful, E.; Kantola, J. Simulation-Based Safety Training for Plant Maintenance in Virtual Reality. In *Advances in Simulation and Digital Human Modeling*; Cassenti, D.N., Scataglini, S., Rajulu, S.L., Wright, J.L., Eds.; Springer: Cham, Switzerland, 2021; Volume 1206, pp. 167–173.
45. Pilati, F.; Faccio, M.; Gamberi, M.; Regattieri, A. Learning manual assembly through real-time motion capture for operator training with augmented reality. *Procedia Manuf.* **2020**, *45*, 189–195. [CrossRef]
46. Luntovskyy, L.; Globa, L.; Shubyn, B. From Big Data to Smart Data: The Most Effective Approaches for Data Analytics. In *Advances in Information and Communication Technology and Systems*; Springer: Berlin/Heidelberg, Germany, 2020; Volume 152.
47. Zolotová, I.; Papcun, P.; Kajáti, E.; Miškuf, M.; Mocnej, J. Smart and cognitive solutions for Operator 4.0: Laboratory H-CPPS case studies. *Comput. Ind. Eng.* **2020**, *139*, 105471. [CrossRef]
48. Dacal-Nieto, A.; Fernandez-Gonzalez, C.; Alonso-Ramos, V.; Antequera-Garcia, G.; Ríos, C. eQUALS: Automated Quality Check System for Paint Shop. In *Advanced Manufacturing and Automation IX*; Wang, Y., Martinsen, K., Yu, T., Wang, K., Eds.; Springer: Singapore, 2020; Volume 634, pp. 402–409.
49. Frustaci, F.; Perri, S.; Cocorullo, G.; Corsonello, P. An embedded machine vision system for an in-line quality check of assembly processes. *Procedia Manuf.* **2020**, *42*, 211–218. [CrossRef]
50. Cicconi, P.; Raffaeli, R. An Industry 4.0 Framework for the Quality Inspection in Gearboxes Production. *CADandA* **2019**, *17*, 813–824. [CrossRef]
51. Kanti, S. How Cogntiive Technologies Are Redefining the Future of Manufacturing? 21 January 2019. Available online: https://www.analyticsinsight.net/how-cognitive-technologies-are-redefining-the-future-of-manufacturing/ (accessed on 21 February 2020).

52. Gary West, Asset Performance Management (APM)—What Is an Asset Performance Management System? 30 April 2019. Available online: www.assetivity.com.au/article/reliability-improvement/asset-performance-management-what-is-an-asset-performance-management-system.html (accessed on 20 February 2020).
53. Mattsson, S.; Fast-Berglund, Å.; Li, D.; Thorvald, P. Forming a cognitive automation strategy for Operator 4.0 in complex assembly. *Comput. Ind. Eng.* **2020**, *139*, 105360. [CrossRef]
54. Holman, M.; Walker, G.; Lansdown, T.; Hulme, A. Radical systems thinking and the future role of computational modelling in ergonomics: An exploration of agent-based modelling. *Ergonomics* **2020**, *63*, 1057–1074. [CrossRef]
55. GCC International Conference on Industrial Engineering and Operations Management, Gulf Cooperation Council, and Industrial Engineering & Operations Management Society. In Proceedings of the International Conference on Industrial Engineering and Operations Management, Riyadh, Saudi Arabia, 26–28 November 2019.

Publisher's Note: MDPI stays neutral with regard to jurisdictional claims in published maps and institutional affiliations.

© 2020 by the authors. Licensee MDPI, Basel, Switzerland. This article is an open access article distributed under the terms and conditions of the Creative Commons Attribution (CC BY) license (http://creativecommons.org/licenses/by/4.0/).

applied system innovation

MDPI

Article

Innovation in Company Labor Productivity Management: Data Science Methods Application

Ekaterina V. Orlova

Department of Economics and Management, Ufa State Aviation Technical University, 450000 Ufa, Russia; ekorl@mail.ru

Abstract: The article considers the challenge of labor productivity growth in a company using objective data about economic, demographic and social factors and subjective information about an employees' health quality. We propose the technology for labor productivity management based on the phased data processing and modeling of quantitative and qualitative data relations, which intended to provide decision making when planning trajectories for labor productivity growth. The technology is supposed to use statistical analysis and machine learning, to support management decision on planning health-saving strategies directed to increase labor productivity. It is proved that to solve the problem of employees' clustering and design their homogeneous groups, it is properly to use the *k*-means method, which is more relevant and reliable compared to the clustering method based on Kohonen neural networks. We also test different methods for employees' classification and predicting of a new employee labor productivity profile and demonstrate that over problem with a lot of qualitative variables, such as gender, education, health self-estimation the support vector machines method has higher accuracy.

Keywords: data science; statistical data processing; predictive analytics; machine learning; classification; clustering; labor productivity; health management; health-saving strategies; electric power industry

Citation: Orlova, E.V. Innovation in Company Labor Productivity Management: Data Science Methods Application. *Appl. Syst. Innov.* **2021**, *4*, 68. https://doi.org/10.3390/asi4030068

Academic Editor: Andrzej Białas

Received: 14 August 2021
Accepted: 13 September 2021
Published: 17 September 2021

Publisher's Note: MDPI stays neutral with regard to jurisdictional claims in published maps and institutional affiliations.

Copyright: © 2021 by the author. Licensee MDPI, Basel, Switzerland. This article is an open access article distributed under the terms and conditions of the Creative Commons Attribution (CC BY) license (https://creativecommons.org/licenses/by/4.0/).

1. Introduction

Rapid modernization and technological innovations inevitably actualize the problem of human capital development. The human capital quality including health quality significant contributes to the labor productivity growth. The main idea of the study is to argue the importance of investment in maintaining the personnel health in order to provide the labor productivity growth of a company, as well as to design a new technology and models for such decisions implementation.

Labor productivity growth and priority economies' sectors modernization are necessary conditions for economic development and important elements of national security. Labor productivity is a key indicator of a countries' economic development and determinant of standard of living. Russian national project "Labor productivity increasing and employment support" [1], supposes target indicator as increasing labor productivity in companies. At the same time, decisions such problems as labor productivity improvement, companies' performance and competitiveness growth, industries and regions development depend on the preservation of human health and its safety.

As an economic characteristic for labor and production efficiency labor productivity shows value of labor costs required to produce a product unit. Traditionally factors, facilitated to labor productivity improvement [2–5], are combined into the following groups:

- Material and technical factors, in that number technology innovations;
- High-performance workspace setting up;
- Development, specialization and concentration of production, using lean production techniques;

- Improvement of production structure and output volumes;
- Advanced training;
- Social and economic factors determined wages and working conditions.

As is shown from the above human potential factor such as employee health quality is not considered as productive and target [6–8]. For a thorough analysis of the substantive foundations of this problem, there are no appropriate methods for quantitative assessment the impact the personnel health on labor productivity, which is to be a basis for decision support and strategies creation in health management in order to increase labor productivity and to provide the growth of operational efficiency of a company.

We determine the human health in accordance with recommendations of the World Health Organization (WHO) [9] as "a state of complete physical, mental and social well-being, and not just the absence of diseases and physical defects." Health state, social skills and knowledge reflect human potential. The growth of the human potential quality directly affects the performance indicators of companies, organizations, institutions and economies in general.

The state and quality of human health as a factor of workforce productivity has not been sufficiently studied in scientific works. Probably, this problem will come to the fore after other mechanisms and sources of labor efficiency growth have been exhausted: production modernization and digitalization, organization improvement and others. Under fourth industrial revolution, the workforce quality moves from insignificant to the most important factors in a labor productivity management.

The issues of personnel health protection in companies' human resource management are presented in [10–12], in which the existing labor protection quality management system is analyzed and mechanisms for its improvement are proposed. Health factors are also taken into account when building quality management systems [13]. Health as a part of human capital, and health preservation as an element of the corporate social responsibility system is considered in [14–17]. Assessment of health care system effectiveness at the macroeconomic level, analysis of the effectiveness of health preservation investments and their impact on economic growth are given in [18–29].

A qualitative analysis based on expert assessments of the impact of the personnel health level on labor productivity is carried out in [30,31], also is regularly presented in the WHO reports and OECD reports [8,9,32]. Many research are devoted to the approaches design for the labor productivity growth as a company competitiveness factor, supposed methods for searching, training and promoting talent [33–36]. A set of leading indicators characterizing the labor conditions and value-motivational environment are presented in [37–39]. The economic returns to health and the impact of health on employment and wages are discussed in the works [40,41]. An assessment of the influence of the employee's age on a labor demand, taking into account the decrease in health potential with the age of a person, is given in the works [42].

In a number of publications, the enterprises is considered with the notion of organizations as complex social systems. Complex interdependent relationship between organizational effectiveness, employees' health and quality culture are explored in [43]. According to the systematic view of work organizations, employee health is closely tied to organizational effectiveness. Building a healthy organizational culture is critical to promoting organizational effectiveness and employee health [44–46]. Considering the interests of employers to improve the employee's health quality in order to company efficiency growth, several studies are examining the impact of measures n future labor market outcomes of employees [47].

Indeed, it is profitable for businesses to invest in employee health for several reasons. First, workers with health problems have a higher probability of receiving payments from the employer for sick leave. This is important not only for the enterprise, but also for the public administration system, since it is of interest for policy makers whether such employer policies reduce dependency rates and if so, whether the effects are sufficiently large to justify their active promotion using public funding. Secondly, reduced turnover due to improved worker-firm matches not only reduces turnover costs for firms. Finally, health-improving measures adopted by firms may increase the labor market attachment of elderly workers, thus alleviating the negative effects of the demographic change in terms of both shortages of skilled workers. In addition to these obvious financial benefits for enterprises to improve the quality of employee health, it should be noted that a comprehensive approach to assessing the quality of employee health and its impact on a labor productivity growth has not been presented to date, and based on the information received, the development of differentiated measures to improve the quality of employee health has not been presented for their homogeneous groups. After all, the grouping of employees into qualitatively homogeneous groups makes it possible to reduce the costs of health-preserving measures, speed up the process of organizing such measures and ensure the fastest growth in labor productivity and efficiency of the enterprise as a whole.

Analysis of existing approaches, methods and models of personnel health and labor productivity management, a number of significant shortcomings of the presented approaches have been identified, limiting the scope of their application: there are no methods for quantitatively assessing the impact of the level and state of health on labor productivity and further recommendations for forming a complex of management decisions, aimed at increasing a labor resources efficiency, taking into account the quality of these resources. The problem associated with the study of heterogeneous factors of labor productivity, including factors of personnel health, with engineering the models to reveal type and nature of the relationships between these factors, with determination homogeneous employees having similar labor productivity profiles in order to manage it, is quite relevant and meaningful over digital transformation of an economy. This necessitates the development of a new approach, technology and supporting models that reflect the essential properties of the socio-economic system of enterprises—the high dynamics of ongoing processes, the uncertainty of the internal and external environment.

In this work, we put forward a hypothesis that for a more complete and comprehensive description of labor productivity as a management object, in addition to economic factors, it is necessary to take into account social, demographic and factors characterizing the personnel health. The study is aimed at justifying the feasibility of financial investment in maintaining a personnel health in order to ensure the labor productivity growth of enterprises as well as design of the technology and models of such investment.

The objective of this study is to develop a technology for labor productivity management of a company, taking into account heterogeneous information about economic, demographic, social factors, as well as information about the quality factors of personnel health, and providing decision-making support in planning labor productivity growth trajectories. To achieve the formulated goal, we solve following problems:

- Identification and substantiation of a set of factors, including factors about health quality that determine labor productivity;
- Design homogeneous employees' groups using all indicated above factors;
- Development a set of management decisions to improve personnel health quality which is differ for each homogeneous group and to contribute to labor productivity growth;
- Assessment of economic efficiency of supposed management decisions to preserve personnel health.

We also note that issues related to occupational health and safety, reduction of occupational injury risks are not considered in this study. Factors reflected employees' health potential and provided their opportunity to perform professional duties, as well as the employers' desire to influence this potential to maintain, to increase and to contribute to the labor productivity growth are considered in this research.

The methodology of the research includes human capital theories, methods of system analysis and modeling of social and economic processes, methods statistical modeling, methods of cluster analysis, methods of decisions making under complexity. The research materials are statistical data and operational reports about large company of the electric power industry, as well as data about employees' survey conducted in this company.

2. Methodology of Research

2.1. Conceptual Approach for Labor Productivity Management

We propose the technology based on step-by-step data processing and modeling. It reflects demographic, social factors and factors about the personnel health quality. The conceptual diagram of this approach is shown in Figure 1 and is aimed to labor productivity management. The technology is implemented by stages.

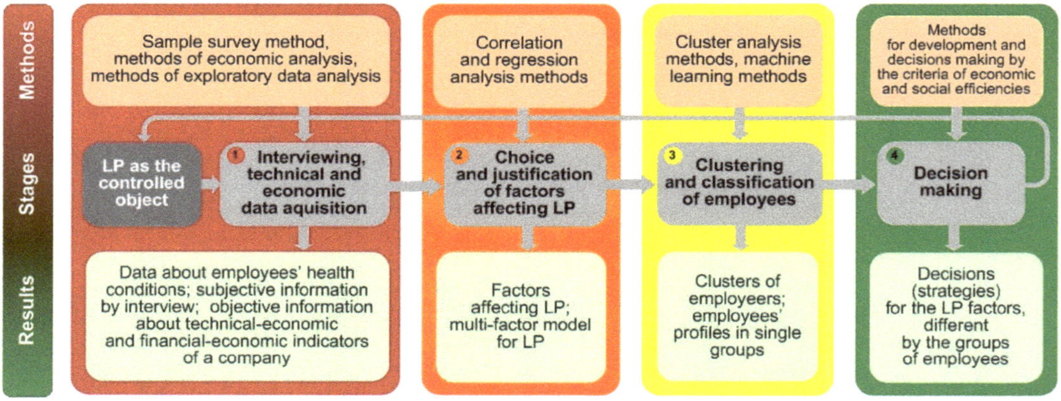

Figure 1. Conceptual scheme of technology for labor productivity (LP) management.

Stage 1. Qualitative analysis of the personnel health status, technical and economic data acquisition. A continuous examination of the company's employees is carried out using medical examinations and questionnaires. The result of this stage is objective data about health state of employees as well as subjective information about the health state and its quality.

Stage 2. Selection and substantiation of factors affecting the personnel individual productivity. An exploratory data analysis is carried out, an assessment of the influence of factors reflecting the social and demographic personnel characteristics as well as factors determining the influence of the health state and its quality on the labor productivity is conducted on the basis of correlation and regression analysis methods.

Stage 3. Employers' clustering (grouping) into homogeneous groups and composition typical personnel profiles for each cluster. The personnel with similar demographic, social factors and health quality are united into homogeneous groups. The result of this stage is the personnel typical profiles by the clusters.

Stage 4. Development of management decisions aimed at improving the state and quality of personnel health for each homogeneous employers' cluster.

2.2. Materials and Methods

The hypothesis and the developed technology are verified on empirical data from a large electric power company in Russia. The experiment was conducted in 2020, in which more than 700 employers from the technical, planning and financial departments took part. To collect data, a survey of all employees was carried out in accordance with the questions, see Appendix A (Table A1). A qualitative analysis of the health status of an employee is carried out using a developed questionnaire consisting of 30 questions, reflecting the self-assessment of the personnel of their health. The processing of the results is carried out on the basis five indicators characterizing different health conditions of the personnel, see Appendix A (Table A2). The assessment procedure of the survey results is assessed by four experts.

In addition to objective data on employees was collected using the company's data bases about employees' education, marital status and number of children. The calculation of labor productivity was implemented in accordance with the algorithm described in Section 3.1. Data analysis and modeling is made in Statistica 10.0 software.

Statistical machine learning methods as a part of data science methods differ from classical statistical methods in that they data driven and do not seek to describe that data with a linear or other general function. Machine learning tends to put a lot of emphasis on developing efficient algorithms that scale to large amounts of data in order to optimize a predictive model. Below is a brief description of the machine learning methods used in the work.

2.2.1. Decision Trees

Decision tree models are a classification model and powerful predictive modeling tool in data science. The decision tree model is based on recursive partitioning—multiple division of data into sections and subsections in order to create homogeneous classes in each summary subsection. A tree model is a set of "if-then-else" implication rules. Trees have the ability to discover hidden patterns corresponding to complex data interactions. A model can be expressed in terms of relationships between predictors that is well interprete. The recursive partitioning algorithm for building a decision tree is rather intuitive. The data is divided multiple times using predictor values, which decompose the data into relatively homogeneous segments. There are various top–down decision trees inducers such as ID3 [48], C4.5 [49], CART [50]. Some consist of two conceptual phases: growing and pruning (C4.5 and CART). Other inducers perform only the growing phase. Detailed algorithms for implementing decision trees can be found in [51].

2.2.2. Support Vector Machines

Support Vector Machines (SVM) is a set of supervised learning methods used for classification and regression analysis. The main idea of the method is to construct a hyperplane that separates the sampled objects in an optimal way. The algorithm works under the assumption that the greater the distance (gap) between the dividing hyperplane and the objects of the shared classes, the smaller the average error of the classifier [52]. The advantages of the method are as follows:

- SVM method is effective in large spaces;
- It is effective if the number of measurements exceeds the number of samples;
- It uses a subset of the training set in the decision function (called support vectors), so it is also memory efficient;
- Versatility: different kernel functions can be specified for the decision function. Common kernels are provided, but one can also specify his own kernels. The disadvantages of the method include:
- If the number of features is much larger than the number of samples, overfit should be avoided when choosing kernel features, and the term regularization is critical;
- SVMs do not provide direct estimates of probabilities, which can be calculated using an expensive five-fold cross-validation.

2.2.3. *K*-Means Method

The *K*-means method is one of the commonly used clustering methods [53]. The algorithm splits the set of elements of the vector space into a predetermined k clusters. It divides the data into k clusters by minimizing the sum of the squared distances of each object to the mean of its assigned cluster. The main idea is that at each iteration the center of mass for each cluster obtained in the previous step is recalculated, then the vectors are divided into clusters again in accordance with which the new centers is closer according to the chosen metric. The algorithm ends when at some iteration there is no change in the intra-cluster distance in a finite number of iterations, since the number of possible partitions of a finite set is finite, and at each step the total standard deviation decreases, so looping is impossible.

2.2.4. Self-Organizing Maps

Self-organizing maps (SOM) are one of the varieties of neural network algorithms [54]. The main difference between this technology and neural networks trained by the back-propagation algorithm is that the teaching method is unsupervised, that is, the training result depends only on the structure of the input data. The algorithm for the functioning of SOM is one of the options for clustering multidimensional vectors. An example of such algorithms is the *k*-means algorithm. An important difference of the SOM algorithm is that in it all neurons (nodes, centers of classes) are ordered into some structure (usually a two-dimensional grid). During training, not only the winning neuron is modified, but also its neighbors, but to a lesser extent. SOM can be considered as one of the methods for projecting multidimensional space into space with a lower dimension. When using this algorithm, vectors that are similar in the original space turn out to be nearby on the resulting map.

3. Experimental Results

3.1. Exploratory Data Analysis and Visualization

We use qualitative and quantitative indicators about social characteristics of personnel (education, marital status, children), anthropometric (gender, age) and characteristics of self-assessment of the state and quality of personnel health (current health problems; chronic diseases; health self-assessment; proper nutrition; bad habits), Table 1.

Table 1. Indicators, variables and range of variable values.

Indicator	Variable	Variable Value or Binary
Age	age	20–65
Gender	gender	female (0), male (1)
Marital status	mar	Not married (0), married (1)
Education	edu	Specialized secondary (0), higher (1)
Children	child	0, 1, 2, ...
Chronic diseases	chron_dis	no (0), yes (1)
Feeling unwell	bad_healh	Infrequently (0), often (1)
Self-reported health as weak and unsatisfactory	health_self-ass	no (0), yes (1)
Inadequate nutrition	nutr	no (0), yes (1)
Bad habits	bad_hab	no (0), yes (1)
Individual labor productivity	labor_pert	136–420 rub (1 ruble is equivalent to 0.01 euro) per hour

The transition of the qualitative values into quantitative ones is carried out on the basis of binary coding (0 and 1), while the quantitative value of a feature increases with its qualitative characteristics intensifies. Individual labor productivity is calculated using the methodology that is used to assess the labor productivity of companies in the basic non-resource sectors of the economy [55]. Within this methodology labor productivity reflects the measure of value added per employee of the company. We receive all data by the special employees' survey with using the developed questionnaires.

Whiskers diagrams visualize the expected impact of qualitative features on labor productivity (LP), Figure 2.

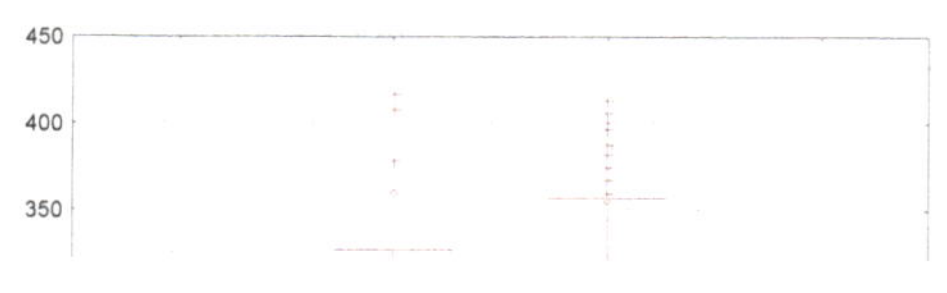

Diagrams show that different levels of the factors gender, child, health_self-ass and bad_hab determine the difference in the labor productivity indicator—labor_perf. Thus, men have higher average labor productivity, about 280 rubles per hour than women—225 rubles per hour. Those personnel who do not have chronic diseases do not often feel unwell and do not have bad habits; their average level of labor productivity is higher than others. Workers with one or two children have more stable level of labor productivity, close to the average, which, in turn, is higher than workers without children.

3D surface plots and contour plots for visualization of labor productivity (labor_perf) depending on the quantitative variables (age) and (child) are shown in Figure 3. From these visualizations, it's obvious that younger workers under the age of 35 with children have higher levels of labor productivity than older workers (over 55 years old) with more than three children, as well as middle-aged workers (from 35 to 55 years old) without children.

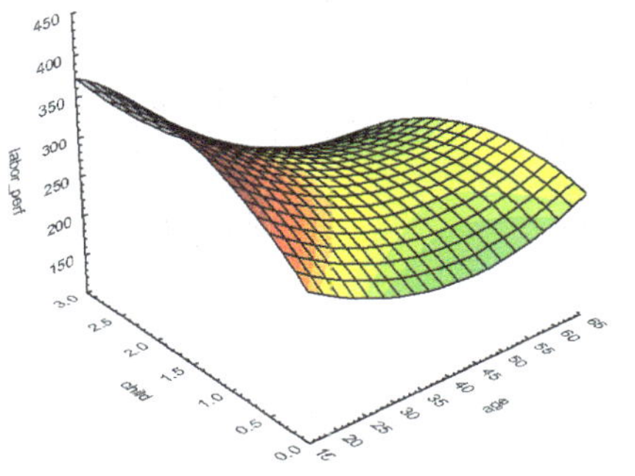

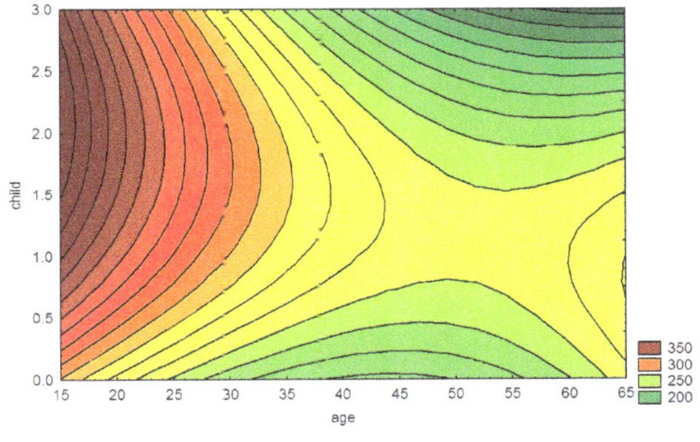

Figure 3. 3D surface plots (**a**) and contour plots (**b**) for the variable "labor_perf" depending on variables "age" and "child".

Descriptive statistics for the variable "labor productivity" characterize the significant variation, non-coincidence of indexes—mean, mode, median and variation demonstrate data heterogeneity, Table 2.

Table 2. Descriptive statistics for the LP variable.

Index	Value	Index	Value
Mean	262.7	Minimum value	123
Mode	208	Maximum value	416
Median	231.5	Standard deviation	84.2
		Variation	0.36

Correlation analysis identifies paired relationships between studied factors and labor productivity. We also note that surveyed indicators were measured on different scales. Labor productivity, age and number of children have a continuous metric scale, education level has an ordinal (rank) scale and other indicators have on a nominal, categorical scale. Therefore, the analysis of these factors' interrelationships in order to identify their significant influence on the modeled indicator—labor productivity—is to carried out using different metrics. So, to measure the relationship between labor productivity, age and the number of children, we use the Pearson correlation coefficient, to assess the effect of education level on labor productivity we calculate Spearman's rank correlation coefficient, and to assess the influence of categorical variables on labor productivity we apply multivariate analysis of variance.

Using multivariate analysis of variance (ANOVA) we test the hypothesis about the equality of the mean values for the corresponding levels of factors pairs (labor productivity and the presence of chronic diseases, labor productivity and gender, etc. for all seven dichotomous variables). We tested hypothesis of equality of mean values and this hypothesis is rejected for three factors—"gender", the presence of chronic diseases ("chron_dis") and the presence of bad habits ("bad_hab"). For the other five factors, this hypothesis is accepted, since the calculated values of the Fisher's F-test are less than the tabular value at a significance level of 0.05. Thus, the factors "gender", "chron_dis" and "bad_hab" are significantly associated with labor productivity "labor_pert", while the other four factors do not have a statistically significant effect on it.

Estimation of the Pearson coefficient for pairs of age—labor productivity and children—labor productivity has shown that age has a statistically significant inverse effect of its average strength on labor productivity (Pearson's correlation coefficient is about 0.32), and the children does not have any significant impact on labor productivity (the calculated value of Student's t-test is less than the tabular value at a significance level of 0.05). Calculation of Spearman's correlation coefficient to assess the influence of education on labor productivity showed that these factors are not correlated, which contradicts the generally accepted thesis that the quality of an employee's education contribute to the growth of his labor efficiency. To exclude false correlations, we construct a matrix of partial correlations, Table 3.

Table 3. Partial correlations of factors (significant parameters are marked in red).

Variable	Age	Gender	Mar	Edu	Child	Chron_ dis	Bad_ healh	Health_ self-ass	Nutr	Bad_ hab	Labor_ pert
age	1.00	−0.23	0.06	−0.22	0.30	0.35	−0.01	0.18	0.17	0.20	−0.32
gender	−0.23	1.00	0.20	0.07	0.03	0.04	0.18	−0.03	−0.10	0.01	0.34
mar	0.06	0.20	1.00	0.07	−0.00	−0.02	0.18	0.09	−0.04	0.07	0.03
edu	−0.22	0.07	0.07	1.00	−0.08	−0.28	−0.00	−0.05	0.00	−0.15	0.02
child	0.30	0.03	−0.00	−0.08	1.00	0.13	−0.14	0.20	0.13	0.26	−0.02
chron_dis	0.35	0.04	−0.02	−0.28	0.13	1.00	0.26	0.26	0.25	0.77	−0.27
bad_healh	−0.01	0.18	0.18	−0.00	−0.14	0.26	1.00	0.48	0.21	0.31	0.02
health_self–ass	0.18	−0.03	0.09	−0.05	0.20	0.26	0.48	1.00	0.54	0.41	−0.31
nutr	0.17	−0.10	−0.04	0.00	0.13	0.25	0.21	0.54	1.00	0.18	−0.20
bad_hab	0.20	0.01	0.07	−0.15	0.26	0.77	0.31	0.41	0.18	1.00	−0.25
labor_pert	−0.32	0.34	0.03	0.02	−0.02	−0.27	0.02	−0.31	−0.20	−0.25	1.00

The table shows that variables as "age", "gender", "chron_dis" and "bad_hab" affect labor productivity (significant correlations are marked in red). In addition, close partial correlations are observed between factors "chron_dis" and "bad_hab", "bad_hab" and "health_self-ass", "bad_hab" and "bad_healh", "nutr" and "health_self-ass" also between a pair of factors "chron_dis" and "age". This explaine an existence of poor nutrition, bad habits cause chronic diseases and poor health.

At the next stage we test several different specifications regression models. The modeling takes into account different predictors scales as well as their multiple correlations. Modeling results for the best model are given in Table 4. The regression model includes a set of predictors, in that number multiplicatively related predictors that have linear correlations with the modeled variable—labor productivity.

The model gives an idea of the quantitative influence of the selected predictors on labor productivity. The combined effects of factors especially increase their negative influence. Thus, such factors combination as the frequency of feeling unwell and the presence of bad habits reduces labor productivity as much as possible (on average by 477 rub/h). The presence of chronic diseases and the self-assessment of one's health are not strong enough reduce labor productivity by an average of 449 rubles per hour, and in married men this effect increases by another 162 rubles per hour. Furthermore, in sample of married men with chronic diseases there is a decrease in labor productivity by an average of 584 rubles per hour. We also mention that there are some shortages of the described regression model. Firstly, some of its parameters are statistically insignificant, and secondly, many qualitative predictors negatively affect the results of model interpretation, as well as interpretation complexity.

Table 4. Regression results for labor productivity modeling (only significant regressors are included).

Predictor (Regressor)	Regression Coefficient	Regression Coefficient Error	T-Statistics	*p*-Value
age	−5.232	2.2219	−2.35490	0.028854
mar*chron_dis	−584.479	120.7443	−4.84064	0.000099
gender*bad_healh	254.329	76.3480	3.33118	0.003329
mar*bad_healh	111.577	49.7048	2.24479	0.036250
edu*bad_healh	121.695	55.9648	2.17450	0.041843
chron_dis*bad_healh	560.109	103.1680	5.42909	0.000026
gender*health_ass	−229.364	78.0956	−2.93696	0.008152
edu*health_ass	−108.392	43.4358	−2.49546	0.021441
chron_dis*health_ass	−488.851	105.3426	−4.64059	0.000158
mar*bad_hab	516.338	108.207[illegible]	4.77176	0.000116
chron_dis*bad_hab	578.543	141.1459	4.09890	0.000558
bad_healh*bad_hab	−476.916	95.9670	−4.96958	0.000074
health_ass*bad_hab	346.146	66.1466	5.23302	0.000040
gender*edu*bad_healh	79.240	34.3925	2.30400	0.032080
gender*mar*health_ass	−162.445	62.3059	−2.60721	0.016862
gender*edu*health_ass	−84.053	34.6773	−2.42387	0.024960

3.2. Clustering and Classification of Companies' Employeers

In order to smooth out the identified data inhomogeneities, as well as to order complex factors interactions we design the technique for dividing personnel into homogeneous groups. This provides a detailed data study and identify patterns in the obtained homogeneous groups. For these we examine various methods.

Clustering has two stages—qualitative analysis using hierarchical methods and analysis using the *k*-means method [56–58]. Analysis of various partitions of the original sample by the method of hierarchical classification showed that it is possible to form from three to six clusters. For the more reasonable grouping of objects (personnel), we use *k*-means clustering method with quantitative criteria to assess the clusterization quality. Table 5 shows the clustering results.

Table 5. Clusters' centers (*K*-means method).

Variable	Average Value of the Variable in the Cluster			
	1	2	3	4
age	34.7	46.3	35.6	33.1
gender	1	0	1	0
mar	1	1	1	0
edu	1	1	0	1
child	1	2	2	0
chron_dis	0	1	1	0
bad_healh	0	1	1	1
health_self-ass	0	1	0	1
nutr	0	1	0	1
bad_hab	0	1	1	0
labor_pert	301.6	221.4	290.8	227.8
cluster size	191	229	184	98

The most numerous cluster (229 employees) is the second, including for about a third of all respondents. These are mostly married women with higher education, above average

age with two or more children. Employees in this cluster have significant health problems, some of them have chronic diseases, bad habits and at the same time, have the lowest labor productivity in the sample.

The first cluster includes mainly married men with higher education, having one child, taking care of their health, has good nutrition and no health problems. The average age of workers in this group is 34.7 years; their average individual productivity is quite high, equal to 301.6 rub. per hour. Employers in this cluster can be designated as "healthy" and do not require special decisions about their health by the company. The third cluster includes 26% of all respondents consist of mainly men with secondary education and two children, they rate their health not very high, they have irregular meals and often have seasonal diseases along with chronic diseases. The fourth cluster is about unmarried women with higher education, who do not have children and bad habits, and they characterize their health as quite satisfactory.

To select the most effective methods for the employers' clustering and then classification by the health quality in order to increase labor productivity, we compare the effectiveness of different statistical and machine learning methods and then choose one with the maximum quality value [59–64]. For this we solve two subproblems:

- Clustering an employers to form homogeneous groups with similar labor productivity profiles. For this we test two clustering methods—*k*-means method and neural network method based on self-organizing Kohonen maps (SOM);
- Classification an employers to identify the performance profile of the new employee, or to determine the cluster to which he is more likely to belong. At this stage we use the following methods: decision trees and support vector machines.

Results of the first subproblem decision. In the clustering problem the number of clusters is not known in advance and the personnel sample is rather heterogeneous. Therefore, to obtain adequate division into clusters, we use another machine learning method—Automated Neural Networks (ANN) based on SOM. Then we compare the effectiveness of two methods for obtaining personnel clusters—*k*-means method and ANN method.

To obtain the number of clusters we use the Self-organizing Kohonen map—a neural network with unsupervised learning. We divide the initial data set into three subsets—the first is a training sample in the amount of 70% of the total sample, which is used to train the neural network and adjust its weights. The second subsample is a test one, is about 15% of the total sample, it is used to check the training and retraining. The third subsample, the validation sample, is used to assess the neural network accuracy on a "new data". First, the topological dimension of the network is set to 25 neurons (the matrix has a dimension of five rows and five columns). As simulation result, it was shown that four clusters can be clearly distinguished, since it was precisely four neurons that described most of the initial data. The clustering results obtained on the basis of the ANN method is shown in Table 6.

Table 6. Clusters' centers (ANN-method).

Variable	Average Value of the Variable in the Cluster			
	1	2	3	4
age	34.2	34.1	41.0	41.7
gender	1	0	1	1
mar	1	1	1	0
edu	1	1	1	1
child	1	1	2	2
chron_dis	0	0	1	0
bad_healh	0	1	1	1
health_self-ass	0	1	1	0
nutr	0	1	1	0
bad_hab	0	1	1	1
labor_pert	294.12	287.35	224.46	265.26
cluster size	170	142	241	149

The clustering results are similar to the result of cluster analysis based on *k*-means. For a more reasonable choice one of the clustering methods and obtaining homogeneous personnel groups we conduct an analysis of used methods robustness. For this, an additional quantitative variable was introduced into the data set, reflecting the number of days spent by an employee on sick. Further, taking into account this new factor, new clustering is formed. Clustering based on *k*-means showed higher stability of groups in terms of their identical composition before and after the introduction of a new factor than the ANN clustering. Therefore, for clustering it is most expedient to use the *k*-means method.

Results of the second subproblem decision. Two classification models are compared. The first model uses a machine learning algorithm based on growing decision trees (boosted trees), and the second model uses the Support Vector Machines (SVM). The classification quality is estimated by the number of correct predictions—the cluster to which the employee belongs in the test sample. Thus, in the model based on boosted trees 96.4% of correct predictions were obtained, and in the SVM method only 89% of correct predictions were obtained. It is shown that the boosted trees method under many categorical predictors show higher accuracy when predict the class (labor productivity profile) of a new employee. The SVM method is most suitable for forecasting when there are many quantitative predictors.

4. Discussion of Results: Management Decisions to Improve Employers' Health Quality

On the next stage of the proposed technology, we examine different regression models of the investigated predictors (regressors) on labor productivity for each constructed cluster. The regression models quality (with a different composition of regressors) corresponding to each cluster is tested and the best ones were selected in terms of quality criteria—maximum determination coefficient (R^2) and minimum moving average approximation error (MAPE). Designed models give a clear idea of the different factors impact on the modeled indicator (labor productivity) and allow making predictions about the factors changes impact on labor productivity, Table 7.

Table 7. Regression results by employees' clusters (significant parameters are marked in red).

Predictor (Regressor)	Regression Coefficient for Models, Different by Employees' Clusters			
	1	2	3	4
age	2.2938	−3.4964	2.4323	−0.3619
child	−3.8365	−5.6638	−6.5767	−76.8798
gender	−3.8608	37.5001	−0.7216	−40.0442
mar	16.3362	−24.5735	38.9882	−22.7630
edu	−9.5457	83.0892	−0.3464	29.0374
chron_dis	−23.6031	38.1560	−0.7990	−5.0536
bad_healh	15.9159	105.3204	−20.1630	−40.0442
health_ass	−30.3469	139.8367	23.9315	−40.0442
nutr	−13.0980	136.2081	−12.1418	26.6200
bad_hab	−23.6031	−80.6174	−3.4200	−46.6946
Regression intercept	139.0025	249.9718	109.6893	460.2615
R^2	0.84	0.73	0.71	0.76
MAPE	10.3	16.6	12.1	11.6
F-test	4.1	4.1	4.6	4.3

Some intercepts in the models are significant for the significance level of 0.05. It means that there are some other predictors for labor productivity modeling such as qualifications level, work experience, possession of the required competencies, production automation and digitalization level etc. The designed models for the analysis and forecasting of employees' labor productivity by their special cluster are totally statistically significant and permit the prediction of labor productivity with a high accuracy, since minimum moving average approximation errors are from 10.3% to 16.6% for different models. These models are used to predict labor productivity under various scenario options for influencing factors about health quality.

For obtaining beneficial effects for business, the idea of a healthy lifestyle and a management system aimed at preserving the employees' health has a number of significant functions. First, it is a motivational function. An investment into employees' health, which is accompanied by certain measures and actions, is an element of the corporate social responsibility system. The existence of the social package for employees is an indicator of a companies' status and its reputation. These increase employees' loyalty to the company and are strategically important to their motivations. Secondly, a company development strategy is an element of its corporate culture. Healthy lifestyle ideas adopted by employers become part of the informational internal corporate environment and have a great team-building effect.

Organization of a set of management decisions to employees' health preserve is based on the following activity types: socio-psychological, financial-economic and material-technical. Activities related to socio-psychological factors provide employees' values orientations, motivation for a healthy lifestyle, training and creation of a favorable moral and psychological climate and an atmosphere of team cooperation, development and implementation of a recreational activities system. Pedagogical tools are used here. Measures focused on financial and economic conditions include ensuring the required costs for material incentives to health-preserving activities. Measures that ensure the material and technical conditions for health preservation include the development of the material and technical base, physical culture and sports events, organization of a psychological relief room.

Employees' homogeneous groups received at the previous stage of the technology, have meaningfully different labor productivity profiles and health status. Therefore, we apply differ management decisions (strategies) for control labor productivity by preserving health quality. Such strategies are developed for four homogeneous clusters characterizing the health status of workers as excellent, good, satisfactory and bad. For each employee's group we propose its own strategy—current control, monitoring and prevention, healthy lifestyle prevention and strong involvement, Table 8.

Table 8. Management decisions (strategies) for improving employee's health quality by their groups.

Cluster	Employee Profile, Reflecting Health Status	Health Improvement Strategy	Management Decisions
1	Excellent	Current control	Regular conversations, trainings
4	Good	Monitoring andprevention	Trainings and monitoring the employees' health
2	Satisfactory	Healthy lifestyle prevention	Regular monitoring and implementation of measures to improve health status
3	Bad	Strong involvement	Regular implementation of activities—promotion of a healthy lifestyle; mass sports; regular spa treatment; wellness programs

We conduct an efficiency assessment for development and implementation of listed strategies. The economic effect for implementation of management decisions is ensured by: reducing the number of technological violations committed by personnel; reduction of sick leave payments, reduction of sick leave; reduction of days of works' incapacity; increasing the quality of work performed, Table 9.

Table 9. Costs assessment before and after management decisions implementation.

Cost Item	Cost Value before Management Decisions Implementation, Rub.	Cost Value after Management Decisions Implementation, Rub.	
		Optimistic Option	Realistic Option
Cost for sick leave	366,125	0	124,593
Cost for the replacement of absent employee	2,295,625	0	1,023,500
Tolal cost	2,661,750	0	1,248,093

For example, the strategy "Strong involvement", developed for the employees in the third cluster, which is about one third of the total sample, includes the following activities: promotion of a healthy lifestyle; mass sports; expanding the list of working specialties provided with regular spa treatment; development of health-improving programs. Total investments for this strategy implementation are 260,000 rubles. The economic effect sources of the proposed program are:

- Reducing a cost of sick leave. Since the employees in the third cluster make 34 percent of all employees, we accept two options an optimistic one, which results in a complete reduction in the incidence of sickness among employees in this cluster, and in the second realistic option, a reduction in the incidence and costs of sick leave payments is ensured by 20%;
- Reducing a cost of replacing an employee who has left for sick leave. Similarly, to assess the effect, an optimistic option is considered, in which these costs are completely eliminated, and a realistic option, in which the cost reduction will be 20%.

The company costs associated with the absence of an employee due to illness will decrease by more than 1,413,000 rub., and the net savings, taking into account the costs of implementing the programs, will amount to more than 1,153,000 rub. Labor productivity for the company, calculated separately for the employees of this cluster, after management decisions implementation, will amount to 6,509,000 rubles per person, which is 2% higher than the initial value.

We also note that as a result of management decisions implementation to preserve an employees' health, there are also social effects associated with motivation increasing and higher satisfied employees with their health quality, which in turn is contribute to labor efficiency growth.

This suggested technology can be used by other companies by different ownership forms and size. The direction of technology improvement could include the factor of "production type"– labor-intensive or capital-intensive, since it affects the elasticity of resource replacement, which is important in the context of economy digital transformation.

5. Conclusions

We have tested and accepted as a correct hypothesis about the need to take into account human health factors as a productive resource of the economy when managing this resource productivity. It is shown that really factors describing the state and quality of human health affect the quality and labor efficiency.

This work is aimed at providing decision making support for manage of planned projects to increase labor productivity and a company performance growth with using data science methodology. To form a computer programs package that decides this problem, it is necessary to determine a set of algorithms and methods used. The search for effective algorithms and methods is the content of this work. Based on the results obtained, system analysts and top managers acquire appropriate toolkits to integrate it into the enterprise management information system.

It is shown that for a more comprehensive description of labor productivity as a controlled object in addition to economic factors, it is necessary to process social, demographic factors and factors about health status of personnel. Using methods of correlation and regression analysis it was found that significant determinants of labor productivity are indicators of personnel health state and health quality.

In order to make decisions about labor productivity management an integrated approach is worked out that systematically describes the patterns of marked factors influencing labor efficiency and sets up adequate management decisions. Under suggested approach many different factors are selected and substantiated on the basis of their correlation on labor productivity. Further the personnel clustering and designing of their homogeneous groups having similar values of demographic, social factors and health quality characteristics are carried out. Then typical personnel profiles are designed that is the basis for development different strategies and management decisions applicated for employees in each cluster separately.

It is proved that to solve the problem of clustering and forming personnel homogeneous groups, it is most expedient to use the *k*-means method, which is more reliable than the clustering based on Kohonen's neural networks. Over problem of classification and prediction a profile of a new employees, featured by many qualitative variables, such as gender, education and a qualitative self-assessment of health, the method based on the boosted trees algorithm is demonstrated higher efficiency.

The developed technology for labor productivity management has been tested and implemented at one of the enterprises of the electric power industry. We proposed management decisions and the following results have been achieved: the enterprise costs related to the incidence of sickness of workers decreased by more than 70%, labor productivity increased by 2%, which ensured an increase in the company's revenue of about 8%.

Funding: This research received no external funding.

Institutional Review Board Statement: Not applicable.

Informed Consent Statement: Not applicable.

Data Availability Statement: Not applicable.

Conflicts of Interest: The author declares no conflict of interest.

Appendix A

Table A1. Health Self-Assessment Questionnaire.

Question Number	Question	Answer	
		Yes	No
1	Does headache bother you?		
2	Do you easily wake up from any noise?		
3	Are you worried about pain in the heart?		
4	Do you think that your eyesight has deteriorated?		
5	Do you think that your hearing has deteriorated?		
6	Do you try to drink only boiled water?		
7	Do the younger ones give way to you in public transport?		
8	Do joint pain bother you?		
9	Does the change in the weather affect your well-being?		
10	Do you have periods when you lose sleep because of anxiety?		
11	Are you worried about constipation?		
12	Are you worried about pain in the liver (in the right hypochondrium)?		
13	Do you have dizziness?		
14	Has it become more difficult for you to concentrate now than in past years?		
15	Are you worried about the weakening of memory, forgetfulness?		
16	Do you feel a burning sensation, tingling sensation, "creeping creeps" in various parts of your body?		
17	Do you have noise in your ears?		
18	Do you keep one of the following medicines at home: validol, nitroglycerin, heart drops?		
19	Do you have swelling in your legs?		
20	Did you have to give up some of the dishes?		
21	Do you have shortness of breath when walking fast?		
22	Are you worried about lower back pain?		
23	Have you ever used any mineral water for medicinal purposes?		
24	Is it possible to say that you have become whiny?		
25	How often do you drink alcoholic beverages?		
26	Do you think that you have become less efficient than before?		
27	Are the periods when you feel joyful, excited, happy disappeared in your life?		
28	How do you assess your state of health (good, satisfactory, bad or very bad)?		
29	Do you often get colds and flu?		
30	Do you smoke?		

Table A2. Correspondence Between Questionnaire Questions and Summary Indicators to be Further Processed in the Technology.

Indicator	Characteristic	Question Number in the Questionnaire
Chronic diseases	Diseases that can be controlled but not completely cured	3,8,11,16,19,21,22,29
Feeling unwell	An employee health state, which does not allow them to fully carry out his labor activity	1,9,10,13,14,17,27
Self-reported health as weak and unsatisfactory	An employee's overall assessment of his health	2,4,5,7,15,18,24,26,28
Bad habits	Habits that negatively affect employee health	12,25,30
Inadequate nutrition	Food that eliminates harmful or useless substances in the diet	6,20,23

References

1. Passport of the National Project (Program). Labor Productivity and Employment Support. 2018. Available online: https://www.government.ru/info/35567/ (accessed on 10 August 2021).
2. Arthur, J. Effects of human resource systems on manufacturing performance and turnover. *Acad. Manag. J.* **1994**, *37*, 670–687.
3. Becker, B.; Gerhart, B. Human resources and 2005 datta, guthrie, and wright 143 organizational performance: Progress and prospects. *Acad. Manag. J.* **1996**, *39*, 779–801.
4. Delery, J.E.; Shaw, J.D. The strategic management of people in work organizations: Review, synthesis and extension. In *Research in Personnel and Human Resource Management*; Rowland, K.M., Ferris, G.R., Eds.; JAI Press: Greenwich, UK, 2001; pp. 165–197.
5. Datta, D.; Guthrie, J. Human resource management and labor productivity: Does industry matter? *Acad. Manag. J.* **2005**, *48*, 135–145. [CrossRef]
6. Goel, V.; Agrawal, R.; Sharma, V. Factors affecting labour productivity: An integrative synthesis and productivity modelling. *Glob. Bus. Econ. Rev.* **2017**, *19*, 299. [CrossRef]
7. The World Bank. Change in the Nature of Labor. World Development Report. 2019. Available online: https://www.vsemirnyjbank.org/en/publication/wdr2019#a (accessed on 10 August 2021).
8. Technical Notes Calculating the Human Development Indices—Graphical Presentation. 2019. Available online: http://hdr.undp.org/sites/default/files/hdr2019_technical_notes.pdf (accessed on 10 August 2021).
9. World Economic Forum. The Global Human Capital Report 2017: Preparing People for the Future of Work. 2017. Available online: http://www3.weforum.org/docs/WEF_Global_Human_Capital_Report_2017.pdf (accessed on 10 August 2021).
10. Global Health Workforce Alliance; Technical Working Group (TWG). Improving Health Worker Productivity and Performance in the Context of Universal Health Coverage: The Roles of Standards, Quality Improvement and Regulation—Final Draft. 2014. Available online: https://www.who.int/ (accessed on 10 August 2021).
11. Galkina, E.E.; Kabanov, A.S.; Khanetsky, A.S. Some aspects of evaluating the effectiveness of the labor protection and safety management system at the enterprises of the aerospace complex. *Qual. Life* **2018**, *17*, 55–60.
12. Sudarkina, H.V.; Glod, O.D. Labor protection management: The system, methods and functions of labor protection management at the enterprise. *Innov. Sci.* **2016**, *6-1*, 245–248.
13. Armstrong, M. *Strategic Human Resource Management*; Kogan Page: London, UK, 2008.
14. Anders, F.; Johan, L.; Yvonne, L.; Bjarne, B. Quality 2030: Quality management for the future. *Total Qual. Manag. Bus. Excell.* **2020**, *17*, 1. [CrossRef]
15. Aldana, S.G.; Pronk, N.P. Health promotion programs, modifiable health risks, and employee absenteeism. *J. Occup. Environ. Med.* **2001**, *43*, 6–46. [CrossRef]
16. Ullah, S.; Malik, M.N.; Hassan, M. Impact of health on labour productivity: Empirical evidence from pakistan. *Eur. Online J. Nat. Soc. Sci.* **2019**, *8*, 117–139.
17. Wei, F.; Xia, Y.; Kong, Y. Public health expenditure and labour productivity: A tentative interpretation based on the science of brain cognition. *Neuro Quantol.* **2019**, *16*, 319–333. [CrossRef]
18. Calvo-Sotomayor, I.; Laka, J.P.; Aguado, R. Workforce ageing and labour productivity in Europe. *Sustainability* **2019**, *11*, 5851. [CrossRef]
19. Chizhova, M.A.; Khayrullina, L.I. Systemic actions in the management of labor protection: Management of safety and health protection of workers. *Bull. Technol. Univ.* **2016**, *19*, 163–167.
20. Bloom, D.E. Governing Global Health. *Financ. Dev.* **2007**, *44*, 31–35.
21. Sheiner, L.; Malinovskaya, A.; Hutchins Center on Fiscal and Monetary Policy at Brookings. Measuring Productivity in Healthcare: An Analysis of the Literature. 2016. Available online: https://www.brookings.edu/wp-content/uploads/2016/08/hp-lit-review_final.pdf (accessed on 10 August 2021).
22. Barro, R.J.; Lee, J.W. International data on educational attainment: Updates and implications. *Oxf. Econ. Pap.* **2001**, *53*, 541–563. [CrossRef]
23. Baumol, W. Productivity growth, convergence, and welfare: What the long-run data show. *Am. Econ. Rev.* **1986**, *76*, 1072–1085.

24. Bhargava, A.; Jamison, D.; Lau, L.; Murray, C. Modeling the effects of health on economic growth. *J. Health Econ.* **2001**, *20*, 423–440. [CrossRef]
25. Boles, M.; Pelletier, B.; Lynch, W. The relationship between health risks and work productivity. *J. Occup. Environ. Med.* **2004**, *46*, 37–45. [CrossRef]
26. Bloom, D.E.; Canning, D.; Sevilla, J. The effect of health on economic growth: A production function approach. *World Dev.* **2004**, *32*, 1–13. [CrossRef]
27. Bloom, D.E.; Canning, D.; Harvard School of Public Health. Health and Economic Growth: Reconciling the Micro and Macro Evidence February. Available online: https://www.anderson.ucla.edu/faculty_pages/romain.wacziarg/demogworkshop/Bloom%20and%20Canning.pdf (accessed on 1 September 2021).
28. Chansarn, S. Labor productivity growth, education, health and technological progress: A cross-country analysis. *Econ. Anal. Policy* **2010**, *40*, 249–261. [CrossRef]
29. Cole, M.; Neumayer, E. The impact of poor health on total factor productivity. *J. Dev. Stud.* **2006**, *42*, 918–938. [CrossRef]
30. Umoru, D.; Yaqub, J.O. Labour productivity and health capital in Nigeria: The empirical evidence. *World* **2008**, *1987*, 1995–2007.
31. Vlegel, M.; Haagsma, J.A.; Munter, L.; Jongh, M.; Polinder, S. Health care and productivity costs of non-fatal traffic injuries: A comparison of road user types. *Int. J. Environ. Res. Public Health* **2020**, *17*, 2217. [CrossRef]
32. The World Health Organization (WHO). The Labor Market for Human Resources for Health in Low- and Middle-Income Countries. 2012. Available online: http://www.who.int/hrh/resources/Observer11_WEB.pdf (accessed on 10 August 2021).
33. Autor, D.; Dorn, D.; Katz, L.F.; Patterson, C.; Reenen, J.V. The fall of the labor share and the rise of superstar firms. *Q. J. Econ.* **2020**, *135*, 645–709. [CrossRef]
34. Job Quality, Health and Productivity. An Evidence-Based Framework for Analysis OECD Social, Employment and Migration. 2018. Available online: https://www.oecd-ilibrary.org/ (accessed on 10 August 2021).
35. Meyer, D.; Liker, J. *Talented Employees. Education and Training of People in the Spirit of Tao Toyota*; Alpina Publisher: Moscow, Russia, 2007.
36. Syarnubi, A.; Sembiring, R.A.; Siswaya, T.; Zuraida, R. Performance, productivity, and safety & health among employee of oil & gas company in qatar. *IOP Conf. Ser. Earth Environ. Sci.* **2018**, *195*, 012075.
37. Orlova, E.V. Methods and models of data analysis and machine learning in the problem of labor productivity management. *Softw. Eng.* **2020**, *4*, 219–229. [CrossRef]
38. Orlova, E.V. Labor productivity management using health factors: Technique and models. *Manager* **2020**, *6*, 57–69. [CrossRef]
39. Mauboussin, M.J. The success equation: Untangling skill and luck in business, sports, and investing. *CFA Inst. Conf. Proc. Q.* **2013**, *30*, 44–51. [CrossRef]
40. Jackle, R. Health and Wages. In *Panel Data Estimates Considering Selection and Endogeneity*; Ifo Institute for Economic Research: Munich, Germany, 2007.
41. Stern, S. Semiparametric estimates of the supply and demand effects of disability on labor force participation. *J. Econom.* **1996**, *71*, 364–406. [CrossRef]
42. Lyashok, V.; Roshchin, S. Young and old workers in the Russian labor market: Are they competitors? *J. New Econ. Assoc.* **2017**, *33*, 117–140. [CrossRef]
43. Sadia, R. The relationship between employee health, quality culture and organizational effectiveness: Findings from the literature. *Int. J. Des. Nat. Ecodyn.* **2016**, *11*, 1–9. [CrossRef]
44. Jiang, J.Y.; Liu, C.W. High performance work systems and organizational effectiveness: The mediating role of social capital. *Hum. Resour. Manag. Rev.* **2015**, *25*, 126–137. [CrossRef]
45. Matthews, J.R. Assessing organizational effectiveness: The role of performance measures. *Libr. Q.* **2011**, *81*, 83–110. [CrossRef]
46. Koys, J.D. The effects of employee satisfaction, organizational citizenship behavior, and turnover on organizational effectiveness: A unit-level, longitudinal study. *Pers. Psychol.* **2001**, *54*, 101–114. [CrossRef]
47. Lechner, M.; Huber, M.; Wunsch, C. Workplace health promotion and labour market performance of employees. *J. Health Econ.* **2015**, *43*, 170–189.
48. Quinlan, J.R. Induction of decision trees. *Mach. Learn.* **1986**, *1*, 81–106. [CrossRef]
49. Quinlan, J.R. *C4.5: Programs for Machine Learning*; Morgan Kaufmann: Los Altos, CA, USA, 1993.
50. Breiman, L.; Friedman, J.; Olshen, R.; Stone, C. *Classification and Regression Trees*; Wadsworth Int. Group: Belmont, CA, USA, 1984.
51. Rokach, L.; Maimon, O. *Decision Trees*; Springer: Boston, MA, USA, 2005. [CrossRef]
52. Awad, M.; Khanna, R. *Support Vector Machines for Classification*; Efficient Learning Machines: Berkeley, CA, USA, 2015. [CrossRef]
53. Adam, C.; Ng, A.Y. *Learning Feature Representations with K-Means*; Stanford University: Stanford, CA, USA, 2012; p. 20.
54. Kohonen, T. *Self-Organizing Maps*, 3rd ed.; Springer: New York, NY, USA, 2001; p. 501.
55. Ministry of Economic Development. Order of the Ministry of Economic Development of Russia. On Approval of the Methodology for Calculating Labor Productivity Indicators of an Enterprise, Industry, Subject of the Russian Federation and the Methodology for Calculating Individual Indicators of the National Project "Labor Productivity and Employment Support". 2019. Available online: http://docs.cntd.ru/document/552128236 (accessed on 10 August 2021).
56. Orlova, E.V. Model for operational optimal control of financial recourses distribution in a company. *Comput. Res. Model.* **2019**, *2*, 343–358. [CrossRef]

57. Orlova, E.V. Decision-making techniques for credit resource management using machine learning and optimization. *Information* **2020**, *11*, 144. [CrossRef]
58. Orlova, E.V. Technology for control an efficiency in production and economic system. In Proceedings of the 30th International Business Information Management Association Conference (IBIMA), Madrid, Spain, 8–9 November 2017; pp. 811–818.
59. Orlova, E.V. Modeling and coordinated control for the production and economic system. In Proceedings of the Mathematical Modeling Session at the International Conference Information Technology and Nanotechnology (MM-ITNT 2017), Samara, Russia, 25–27 April 2017; Volume 1904, pp. 1–6.
60. Shalev-Shwartz Shai, S.; Ben-David, S. *Understanding Machine Learning: From Theory to Algorithms*; Cambridge University Press: Cambridge, MA, USA, 2014; p. 499.
61. Orlova, E.V. Synergetic approach for the coordinated control in production and economic system. In Proceedings of the 30th International Business Information Management Association Conference (IBIMA), Madrid, Spain, 8–9 November 2017; pp. 704–712.
62. Orlova, E.V. The AI model for identification the impact of irrational factors on the investor's risk propensity. In Proceedings of the 30th International Business Information Management Association Conference (IBIMA), Madrid, Spain, 8–9 November 2017; pp. 713–721.
63. Orlova, E.V. Control over chaotic price dynamics in a price competition model. *Autom. Remote Control* **2017**, *78*, 16–28. [CrossRef]
64. Orlova, E.V. Methodology and models for individuals' creditworthiness management using digital footprint data and machine learning methods. *Mathematics* **2021**, *9*, 1820. [CrossRef]

applied system innovation

MDPI

Article

Acting Instead of Reacting—Ensuring Employee Retention during Successful Introduction of i4.0

Steffen C. Eickemeyer [1], Jan Busch [2], Chia-Te Liu [1] and Sonia Lippke [3,*]

1. Department of Business Studies, Jacobs University, 28759 Bremen, Germany; s.eickemeyer@jacobs-university.de (S.C.E.); chandler20708@gmail.com (C.-T.L.)
2. Mechanical Engineering at Miele & Cie. KG, 31275 Lehrte, Germany; jan.busch87@gmx.de
3. Department of Psychology and Methods, Jacobs University, 28759 Bremen, Germany

* Correspondence: s.lippke@jacobs-university.de; Tel.: +49-421-200-4730

Abstract: The increasing implementation of digital technologies has various positive impacts on companies. However, many companies often rush into such an implementation of technological trends without sufficient preparation and pay insufficient attention to the human factors involved in digitization. This phenomenon can be exacerbated when these technologies become highly dependent, as during the COVID-19 pandemic. This study aims to better understand challenges and to propose solutions for a successful implementation of digitized technology. A literature review is combined with survey results and specific consulting strategies. Data from the first wave of the COVID-19 pandemic in Germany were collected by means of an online survey, with a representative sample of the German population. However, we did not reveal any correlation between home office and suffering, mental health, and physical health (indicators of digitization usage to cope with COVID-19 pandemic), but rather that younger workers are more prone to using digitized technology. Based on previous findings that older individuals tend to have negative attitudes toward digital transformation, appropriate countermeasures are needed to help them become more tech-savvy. Accordingly, a software tool is proposed. The tool can help the management team to manage digitization efficiently. Employee well-being can be increased as companies are made aware of necessary measures such as training for individuals and groups at an early stage.

Keywords: change management; COVID-19; decision-support model; digitization; employee motivation; employee satisfaction; human resources; Industry 4.0; software tool

Citation: Eickemeyer, S.C.; Busch, J.; Liu, C.-T.; Lippke, S. Acting Instead of Reacting—Ensuring Employee Retention during Successful Introduction of i4.0. *Appl. Syst. Innov.* **2021**, *4*, 97. https://doi.org/10.3390/asi4040097

Academic Editor: Shahryar Sorooshian

Received: 31 October 2021
Accepted: 23 November 2021
Published: 29 November 2021

Publisher's Note: MDPI stays neutral with regard to jurisdictional claims in published maps and institutional affiliations.

Copyright: © 2021 by the authors. Licensee MDPI, Basel, Switzerland. This article is an open access article distributed under the terms and conditions of the Creative Commons Attribution (CC BY) license (https://creativecommons.org/licenses/by/4.0/).

1. Introduction

In the period of the COVID-19 pandemic, the dependency on digital technologies became heavily apparent, particularly due to the physical distancing policies imposed [1,2], with effects on operations and production management. Overall, digitization in general is progressing steadily and COVID-19 illustrates this clearly. Even traditional entrepreneurs, who have not created options for mobile/home office or similar before COVID-19, have largely changed their views in recent months. However, the importance of digitization in operations and production management was evident long before. Especially due to the high cost pressure (to increase the degree of value creation in business processes), the increased complexity in products and processes, and the demand for higher efficiency in satisfying customer requirements, the use of Industry 4.0 (i4.0) technologies is becoming increasingly important. As i4.0 technologies are amongst others created to deal with relatively complex processes, for example, creating small and highly customized batch sizes, it could enable companies to gain very profitable and competitive advantages. While i4.0 offers success, companies are faced with some challenges when implementing it [3].

One of the many challenges that companies face is the emergence of human symptoms induced by a new and innovative work environment. Under the impact of i4.0, working individuals must confront the autonomy of actions and excessive demands, as well as the adaptation to new responsibilities and roles inside the new processes [4] (Gorecky et al., 2014). This can result in mental stress [5–7], demotivation, and conflicts with superiors, colleagues, and one's team [8,9]. The actual causes of the symptoms are not perceived by companies in depth [10,11]. According to a systematic literature study by Kadir, Broberg, and Conceição, reviewing over 80 studies, there is a lack of attention to human factors in the context of i4.0 [12]. In general, little systematic scientific research is available on the influence of digitization on employees' adverse experiences [12]. In addition, many companies have poor knowledge of i4.0 technologies, which might still increase companies' inappropriate dealings with employees' uncertainties. For instance, a survey study about i4.0 in the manufacturing sector of New Zealand in 2018 shows that 33% of the manufacturers have a "poor" or "very poor" knowledge of i4.0 [13]. Uncertainties and unawareness certainly do not assist employees in the integration of change management. As a result, the impacts of human symptoms and the causes of the symptoms are the key drivers of this fold of the challenge. Accordingly, Hilb expressed his views on digitization and artificial intelligence (AI): "While the business appetite for AI is clear and the advances in technology are certain, it will ultimately be the social dialogue that will be critical. In this respect, companies will have to prove that they are aware of their responsibility in dealing with AI in order to win the trust of society" [14] (p. 867). Based on this, Hilb gives the following management perspective: "Today's boards of directors can play a central role in this process if they are willing and able to take the driver's seat" [14] (p. 867).

The aim of this paper is therefore (a) to review how it is possible to recognize in advance that digitization could overburden employees and therefore might lead to lower productivity. In addition, we aim to (b) investigate how these negative developments can be counteracted, especially in times of crisis, i.e., when the COVID-19 pandemic urged many organizations and individuals to shift to mobile work. Therefore, this paper will (a) review the literature and (b) present empirical data on individuals' perceptions of mobile work and psychological reactions to it, and based on this, (c) explain a self-developed decision-support model with a software tool for top management.

The article is structured in 6 sections: the following, Section 2, presents a literature review to classify the article thematically. Section 3 describes the empirical evidence with the materials and methods used. Based on this, the results of the survey are presented in Section 4. The findings and the relationships described are discussed in Section 5 and used to develop a decision-support model for dealing with employees during digital transformation. Finally, Section 6 contains the conclusions, which provide both a summary of the article and an outlook on subsequent research work.

2. Literature Review

The following is a summary of the areas of literature that are essential to this paper. First, an understanding of digitization (Section 2.1), which is central to this article, is created. In the context of this, the central buzzword, Industry 4.0, which is the driver for many industry initiatives, is brought into focus. Section 2.2 provides a general explanation of human behavior during change processes. These two sections help to provide a basic understanding, so that Section 2.3 can then focus specifically on effects on employees during the implementation of Industry 4.0. These effects on employees and the optimal way to deal with them are the main topic of this article.

2.1. Industry 4.0

The term "Industry 4.0" (originally "Industrie 4.0") was first coined publicly by the German government in the year 2011 [15]. The term described neither existing developments nor current research. Together with representatives from business, politics, and academia, the government initiated the concept of i4.0 to enhance the competitiveness and innovation of Germany's domestic manufacturing industry [16]. The concept of i4.0 as the fourth industrial revolution, was, in fact, predicted ex-ante rather than recognized ex-post [17]. This is unique because the other three great industrial revolutions were always named, described, and analyzed in the aftermath. "Even though Industry 4.0 is one of the most frequently discussed topics these days, I could not explain to my son what it really means", said Audi's production site manager [16]. As a consequence, many studies still claim that there is no generally accepted definition for i4.0 [16,18,19].

In order to ensure a common understanding, we will share short definitions (according to [20]) regarding the core drivers of the four industrial revolutions. The first revolution (1784) was based on water energy and steam engines to power machines. The second revolution (1870) used electrical energy to enable mass production by continuously driven conveyor systems/belts. The third revolution (1969) used programmable logic controllers (PLC) and IT systems for the automation of complete value-added processes. The fourth revolution (today) uses these four main levers: Internet of Things (IoT), Industrial Internet of Things (IIoT), Cloud-based Manufacturing (CbM), as well as Smart Manufacturing (SM) [20,21]. While i4.0 does not revolve around a single innovation that enables the shift, the four main levers comprise various numbers and categories of technologies, such as Big Data analytics, Artificial Intelligence (AI), Cyber-Physical Systems (CPS), and decentralized systems [1,19]. The i4.0 paradigm promotes the connection of sensors, devices, and enterprise assets, both to each other and to the internet [22].

The main difference between i4.0 and Computer-Integrated Manufacturing (CIM) is the concern of the human role in the production environment [20]. In i4.0, human workers play an important role in carrying out production, while CIM considers production without workers [20,23]. This underlines the central role of human resources in value-creation processes in the era of i4.0. Accordingly, these human performance factors should be considered appropriately in change management.

2.2. Human Behavior during Change Processes

In the 2015 research report, "Skills for Digital Transformation", more than 84% of the interviewed individuals shared a strong consensus on business change management skills being of major importance for companies' transformations [24]. The introduction of i4.0 tools usually requires changes in the work processes of employees. Increasing complexity in processes due to i4.0 technologies could have a negative impact on employees (e.g., blurred work–non-work boundaries, higher workload, new responsibilities, fear of job loss, etc.). Accordingly, Cinquini commented on i4.0, AI, and Big Data: "Artificial Intelligence (AI) and Big Data might indicate how decision-making would be uncontextualized, and people and local setting may become irrelevant. Thus, the questions about the consequences for human actors and their roles in pervasive digitalization and how can we develop fair and valid performance management instruments for making managers and employees accountable are important." [25] (p. 847). Respectively, companies need to cope with these concerns to maintain high employee productivity, and with that actually make use of the potentials of i4.0 by means of employees' participation in dynamic operational performance [26].

There are different models for human behavior during change processes. The following discussion utilizes the Kübler-Ross "Change Curve" [27]. This curve (Figure 1) is based on a model originally developed in the 1960s by Elisabeth Kübler-Ross, and is fully in line with the principle to have employees participate in continuous improvement to ensure operational performance [26]. The seven steps of the change curve describe the usual reactions of employees to changes in their work processes.

However, this knowledge of the exact objectives of change measures in the field of digitization is an absolute prerequisite for good integration of employees in the change process. Only if the goal is clear and the possible effects can be modeled, can the path to the goal also be described, which will make it possible to integrate and retain employees (vs. demotivation, termination, or similar). In order to achieve this, the goals and tools for the introduction of i4.0 methods as well as their effects on the employees must be considered in more detail.

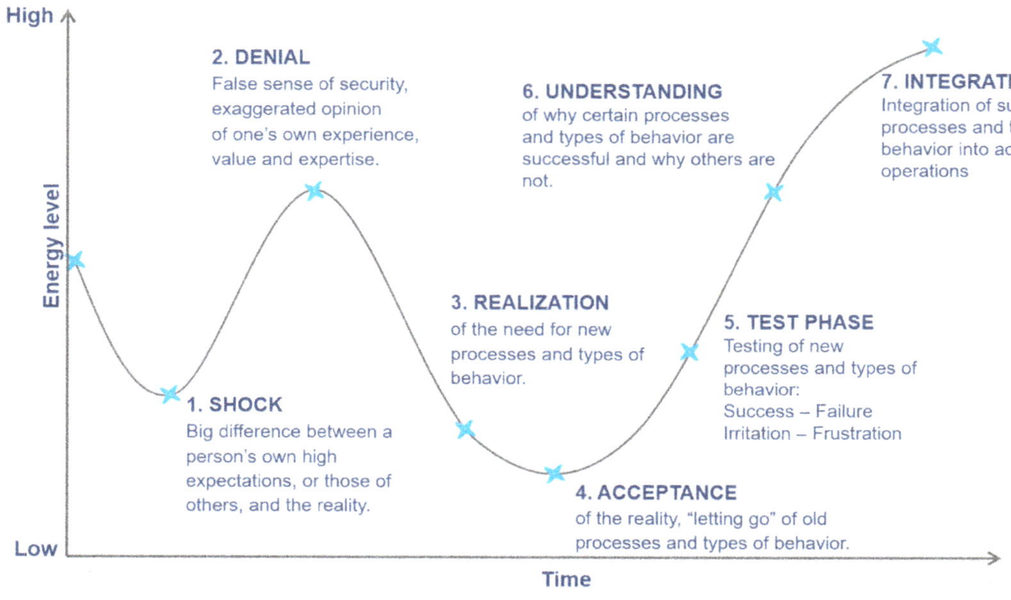

Figure 1. Decision-support model for dealing with employees during digital transformation (based on [27]).

2.3. Effects on Employees during Implementation of Industry 4.0

As described at the beginning, i4.0 is not defined by mutual agreement, nor do complete real value chains exist that fully comply with the usual i4.0 standards. Nevertheless, it is important to understand what experiences of the impact of the implementation of i4.0 on employees have already been described and analyzed. Some studies established maturity models or company-readiness guidelines for i4.0. The process model in small and medium enterprises (SMEs), as a guiding framework to implement i4.0 proposed by Ganzarain and Errasti suggests, is a three-stage maturity model: "Envision", "Enable", "Enact". In summary, it is dedicated to, firstly, defining the understanding of i4.0 in the company's capacity and customers' needs, and secondly, forming a roadmap that enables strategy formulation, which then transforms into the company's actual business model [28].

Another approach by Pessl et al. contains a six-step roadmap as a holistic approach to implementing i4.0. Focusing on the human, it suggests the following procedure: start-workshop, analyze i4.0 maturity, define the target state, define and evaluate measures, prepare and realize decisions, and define projects [29]. Most studies focus on the impact and readiness aspects of the company at the organizational level. In terms of technological transformation, words such as "digitization anxiety" and "technophobia" show up consistently during the research of relevant topics. Researchers have defined the symptoms in broader terms as computer anxiety and technophobia [30,31], especially for older employees.

As rather little systematic research can be found so far on the impact of i4.0 enabling employees to work from home and options for preventing its negative impact, this study aims to better understand how the transition to a remote work environment in Germany due to COVID-19 affected employee experience: Are there differences between employees working from home and employees working on-site regarding age, gender, and children, as well as their perceptions of feeling negatively affected by the COVID-19 pandemic and their well-being/health in general?

3. Materials and Methods

To test, a secondary data analysis was performed with the study described in detail elsewhere [32]. However, the aspects relating to working from home were not analyzed before. The survey, by means of online questionnaire, was launched in June 2020, to collect data from a sample of individuals aged 18 years and above with a sample size of N = 1050, by Bilendi GmbH on behalf of Weleda. A representative sample of inhabitants in Germany was recruited. The questionnaire-based survey was conducted between 8 June and 15 June in the year 2020. Using email, the company contacted individuals in their database by means of stratified sampling. In the inviting email, the purpose of the study was explained. When individuals clicked on the link to the study, they received further participant information which included a confidentiality statement. After reading this, study participants were asked to give informed consent by filling in a form. Only those who provided informed consent were allowed to continue to the questionnaire. The reason for this methodology and advantages of its choice were that the research questions could be investigated at high quality and there was easy access to the sample even in times of restrictions due to the COVID-19 pandemic. The main aim was to be able to generalize the findings to the larger population and to overcome bias in typical respondent behaviors.

Different subject areas were collected, such as socio-demographic data, work situation, perceived stress in times of lockdown, and perceived changes in the "new" everyday life. For the wording of the analyzed questions, see Tables 1 and 2 The way of designing the questionnaire, in order to generate the necessary data to achieve the research objectives, was determined by a theory-based and co-creative process: the last author performed a scoping review of the literature and, in exchange with potential future study participants, items and response formats were piloted and partially adapted. The validation of the questionnaire was performed by means of face validity and pilot analyses due to limited options to cross-validate the measures with other items.

Overall, 60% of the sample reported being employed (625 individuals; not employed: 40%), and only these were contained in the following analyses. Of these, 302 (48.3%) were working from home during the COVID-19 pandemic and 323 (51.7%) were working on-site as usual. Characteristics of these two groups are outlined in Table 1 and were tested for significant differences with frequency analyses (determining Chi^2) and mean differences (MANOVA, reported below).

Table 1. Characteristics of individuals working from home (home office, mobile work, etc.) and those working on-site (office, supermarket, hospital, etc.) regarding socio-demographic variables.

Sample	Worked from Home	Worked on-Site	Total	Bivariate Test Statistic
Gender	-	-	-	-
Female	137 (45.4%)	160 (49.5%)	297 (47.5%)	-
Male	165 (54.6%)	163 (50.5%)	328 (52.5%)	$Chi^2(1) = 1.09; p = 0.30$
Age (years)	M = 38.53 (SD = 13.01)	M = 44.92 (SD = 12.75)	M = 41.61 (SD = 13.26)	$F(1) = 23.27; p < 0.01$
18–29 years	82 (27.2%)	38 (11.8%)	120 (19.2%)	-
30–39 years	80 (26.5%)	60 (18.6%)	140 (22.4%)	-
40–49 years	56 (18.5%)	75 (23.2%)	131 (21.0%)	-
50–59 years	53 (17.5%)	104 (32.2%)	157 (25.1%)	-
60–69 years	16 (5.3%)	37(11.5%)	53 (8.5%)	-
70+ years	15 (5.0%)	9 (2.8%)	24 (3.8%)	$Chi^2(5) = 47.48; p < 0.01$

Table 2. Characteristics of individuals working from home (home office, mobile work, etc.) and those working on-site (office, supermarket, hospital, etc.) regarding socio-demographic variables.

Sample	Worked from Home	Worked on-Site	Total	Bivariate Test Statistic
Number of individuals in the household	M = 2.58 (SD = 1.65)	M = 2.40 (SD = 1.20)	M = 2.49 (SD = 1.45)	-
1	68 (22.5%)	77 (23.8%)	145 (23.2%)	-
2	109 (36.1%)	130 (40.2%)	239 (38.2%)	-
3+	125 (41.4%)	116 (35.9%)	241 (38.6%)	$Chi^2(2) = 2.04; p = 0.36$
Children	-	-	-	-
No	145 (48.0%)	188 (58.2%)	333 (53.3%)	-
Yes	157 (52.0%)	135 (41.8%)	292 (46.7%)	$Chi^2(1) = 6.51; p = 0.01$
Age of Children *	-	-	-	-
Below 1 year	12 (8.3%)	11 (5.9%)	23 (6.9%)	$Chi^2(1) = 0.75; p = 0.39$
1–3 years	27 (18.6%)	14 (7.4%)	41 (12.3%)	$Chi^2(1) = 9.47; p < 0.01$
4–6 years	27 (18.6%)	20 (10.6%)	47 (14.1%)	$Chi^2(1) = 4.30; p = 0.04$
7–10 years	30 (20.7%)	26 (13.8%)	56 (16.8%)	$Chi^2(1) = 2.75; p = 0.10$
11–14 years	34 (23.4%)	32 (17.0%)	66 (19.8%)	$Chi^2(1) = 2.13; p = 0.15$
15–18 years	63 (43.4%)	128 (68.1%)	191 (57.4%)	$Chi^2(1) = 20.32; p < 0.01$

Note. M = Mean; SD = Standard Deviation; * Age of Children can result from different children.

4. Results

Younger employees were more likely to work from home than older employees. Below 40- and above 70-year-old employees made more use of remote work options or were more able to do so because of their job setup. However, there were no significant differences between men and women or regarding the number of people living in the household (see Table 2). Further, it was tested whether having children and the age of the offspring would make a difference.

Analyses revealed that those with children were more likely to work from home than the ones without children. The ones with children in the age group 1 to 6 years were more likely to work from home than to work on-site, which may be related to the need to do so because of closed daycare centers and schools as well as unavailable grandparents and nannies, due to distancing policies to prevent the spread of the COVID-19 virus. Significant differences occurred also with regard to children aged 15–18 years old; those parents were more likely to work on-site, which may also be related to their own age.

To test whether those employees working from home differ from those working on-site with regard to suffering, feeling mentally and/or physically strained from the limitations during the COVID-19 pandemic, and their well-being/health in general, contrasts were computed taking the age effects into account (Figure 2).

A MANCOVA was conducted with age as the covariate and number of individuals living in the household, as well as suffering and mental- and physical-health status, to examine differences due to working from home (factor 1) and gender (factor 2). In this model, only age transpired significant with $F(4, 393) = 7.08$; $p < 0.01$ and $Eta^2 = 0.07$ (see Figure 2), indicating that older individuals were less affected by the COVID-19 pandemic and reported better well-being/mental and physical health. However, after controlling these age differences, working from home or on-site did not significantly interrelate with feeling affected by the COVID-19 pandemic or mental and physical health.

Descriptively, individuals aged 40 (regarding mental aspects) or 50 (regarding physical aspects) seemed to benefit from working from home, whereas the mental health of those aged below 40 appeared better off when maintaining their routine of going to work and not having to work in a home office. However, this was not statistically significant.

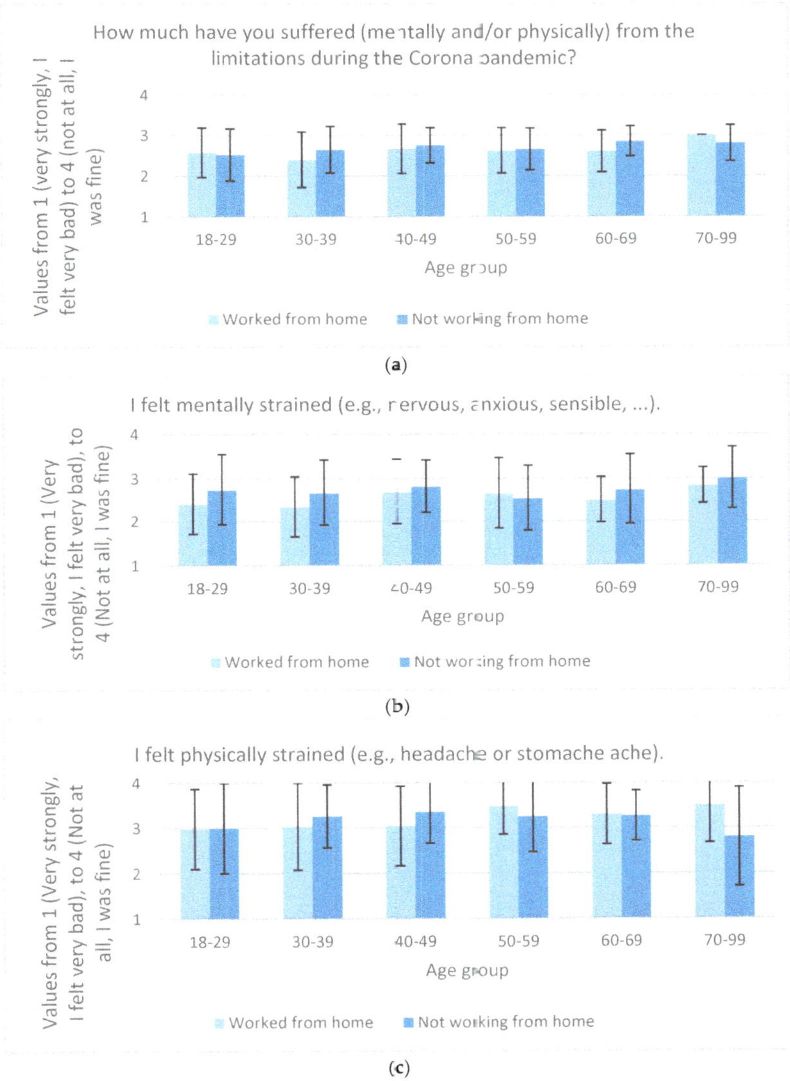

Figure 2. (**a**) General Effect of the Corona Pandemic (Mean and Standard deviations), higher means indicate feeling better (less affected). (**b**) Mean rating of mental health status during the Corona pandemic for different age groups working from home vs. not working from home (Mean and Standard deviations), higher means indicate feeling better. (**c**) Mean rating of physical health status during the Corona pandemic for different age groups working from home vs. not working from home (Mean and Standard deviations), higher means indicate feeling better.

Taken together, individuals' perceptions of mobile work were dependent on their age and the situation at home. Those employees having to manage not only working from home but also running other responsibilities such as family duties, seemed to be burdensome. If such responsibilities do not exist, older individuals especially benefit from the option to work from home and make use of digitization and i4.0 in terms of mobile work.

Surely, more aspects should be explicitly taken into account relating to human factors, such as attitudes to change, perceived anxiety, or stress. With this study, we were only able to investigate the aspects that the study participants were interviewed about. Further aspects could also be more positive, such as beneficial consequences of staying home and feeling safer in the face of a pandemic. General factors could be to save the time from commuting for more recreation, more flexibility, and self-determination, which was found to be positively related to employee job satisfaction. As these aspects were not measured in the study, this should be done in the future.

Concluding, employees seem to be ready to make use of mobile work, i.e., digital and also utilizing i4.0 at home. This intrinsic motivation of employees to digitize should now be actively used as a driver of change in order to successfully and sustainably implement digitization in companies [33]. In parallel, we have investigated the general effects of i4.0 technologies on employees and derived solutions for avoiding the associated negative consequences. In the following, these effects are related to aspects (e.g., selection of digital enablers—see Section 2) that need to be taken into account when making decisions in the context of digital transformation in a company. The decision model leads to a software tool that can help companies to reach a good to an optimal decision.

5. Discussion

As already predicted by Gattiker and Howg in 1990, digitization/i4.0 technology was leading to major job design change and organizational change [34]. In research, terms similar to "i4.0" in relation to technology are not often associated with the challenge in the aspect of the human component on a personal level or the psychological aspects. These challenges occurred when terms such as "digitization", "decentralization", and "Internet of Things" appeared [1]. Also, Bolander "has pointed to numerous potential weaknesses related to the automation of human decision making" [35] (p. 866). There is a lack of empirical studies or analytical studies with regards to workers' negative responses and behaviors in a digitized environment. Moreover, the symptoms occurring in different sectors, stations, or positions vary significantly. Therefore, the question is how to analyze and describe the cause–effect chain of corporate goals using i4.0 tools, the effects on employees, possible symptoms, and corresponding countermeasures. In order to be able to answer this question, a decision-support model was developed as part of this article (see Figure 3).

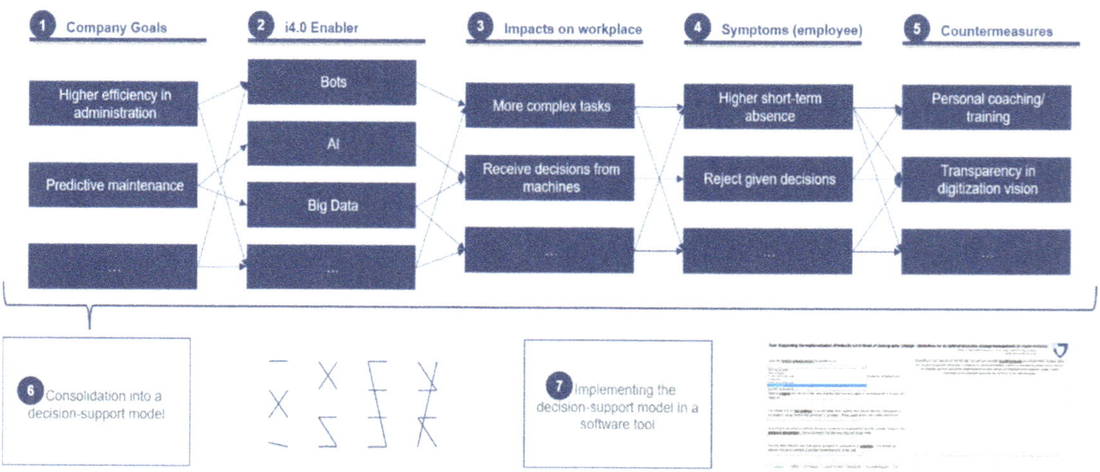

Figure 3. Decision-support model.

The seven steps of the procedure were developed accordingly on the basis of the aforementioned preliminary work as well as on our own experience from various industrial digitization projects [36,37]. The seven steps are explained in more detail below and are intended to pave the way for sustainable change.

5.1. Company Goals

Companies introduce i4.0 applications, tools, and standards very selectively according to their company goals. The selection of technologies is determined both by the requirements of the industry and, in particular, by the corporate strategy or individual objectives derived from it [33,38]. Individual and relevant technologies are explained in the following Section 5.2, with examples. In order to better understand the motivation and link for selected technologies (see also Section 5.6), typical company goals are explained initially. Moore stated that "All organizations benefit from developing a strategy. The most well-developed strategy models come from the private sector and focus on markets, customers, and competition" [39] (p. 183).

Many previous publications explore new business concepts and strategies for adapting to the new industrial revolution. In this context Crnjac et al. [40] discussed the changes that will occur by applying i4.0. The authors provided an overview of several concepts and strategies and an evaluation of business models linked to i4.0 [40]. Furthermore, they defined eight priority areas for action related to selection and implementing i4.0 [40]. Combining the knowledge and standard approaches of strategy consulting [41] with the topics of sustainability, social responsibility [42], and i4.0, we will divide the overarching company objectives into these seven categories:

- Performance targets: profit and return;
- Financial targets: capital structure and enterprise value;
- Product and market goals: market share, sales revenues, and product range;
- Social goals: jobs, personnel development, job satisfaction, income, and pensions;
- Power and prestige goals: independence and corporate image;
- Ecological goals: conservation of resources and environmental protection;
- Innovation and automation: business bots and robotic process automation.

Accordingly, these aspects should be taken into account when the company goals are translated into the further steps, which will be described in the following.

5.2. Industry 4.0/Digital Enablers (with Impacts on Employees)

Certainly, there is a multitude of methods and tools that are counted among the technologies of i4.0. In a huge scientometric survey, scientists found 35 main technologies, which are most frequently discussed in relevant publications. Ranking the technologies based on the number of appearances in these publications in a Pareto chart, 13 key technologies were identified [43]: Internet of Things, Big Data, Additive Manufacturing, Cloud Computing, Autonomous Robots, Virtual Reality and Augmented Reality, Cyber-Physical Systems, Artificial Intelligence and Machine Learning, Smart Sensors, Advanced Simulation, Nanotechnology, Drones, and Biotechnology.

On one hand, the autonomous robots can be industrial robots or robot arms. On the other hand, there are also robots that support employees in carrying out daily work processes. So-called software robots perform tasks mostly similar to humans or by imitating them [44–46]. This process is called Robot Process Automation (RPA). RPA is an approach for automating processes within a wide range of use cases [47]. This method is therefore very well applicable to reaching some of the company goals shown in Section 5.1. For example, some goals include raising profit or achieving higher job satisfaction by letting frequently recurring or boring tasks be performed by RPA.

At this point, the focus should not be on the enumeration and description of i4.0 technologies. It can rather be stated that the toolbox of i4.0 technologies is huge and that each tool or method will have its own impact on the activities of employees. On the one hand, the focus is, of course, on the automation and computerization of workplaces [48–50].

On the other hand, the impact of advanced digital technologies means that some new positions and specializations are associated with new skills and experience [51]. As a result, there will be major changes in the world of employment. For example, using i4.0 technologies, a self-organizing adaptive logistics system can function without significant intervention by employees. Also, the increasing digitization in industrial production can assist employees so that they can assess increasingly complex situations, paying more attention to the key factors and information regarding their tasks (platform Industry 4.0 [52], especially in a time when fast actions are needed, such as during the COVID-19 pandemic). Therefore, the following section is focused on the impact on employees in the course of the introduction of i4.0 technologies.

5.3. Industry 4.0 Impact on Workplaces

It is likely that i4.0 technologies have or will soon have huge implications for both work content and work organization and will be able to change the nature of and manner in which employees work [53]. Despite the impact of i4.0 on older, less-educated, and low-paid workers, the continuous progress will result in computers performing tasks that require a high level of education and training. As a result, those employees in highly skilled jobs may feel threatened by machines and software algorithms capable of sophisticated analysis and decision-making. Continued progress in production automation and the deployment of advanced commercial robots would further limit the possibilities for low-skilled workers. Technological progress is relentless, and machines and computers will someday approach the point where they would match or exceed the ability of the average worker to perform most routine tasks [54]. Frey and Osborne carried out a study on the impact of digitization on jobs in the US. The main finding of their study is that 47% of jobs in the US are at risk of being made redundant by computerization. In the future, robots will not only be able to execute standardized programs but will also be able to perform demanding tasks beyond the routine. As a logical outcome, most of the less skilled human jobs would be eliminated and substituted by technology, which would make the surviving human jobs more complex and wide-ranging [49].

Brynjolfsson and McAfee forecast substantial economic changes resulting from the rapidly growing use of i4.0 technologies but are skeptical about the potential positive effects on jobs and expect competition for jobs to increase. In their estimation, technological developments would eradicate not only routine jobs, but also high-skill jobs that are defined by pattern recognition and cognitive non-routine tasks. They also recommend a number of actions to reduce the negative impact of i4.0 technologies and to counterbalance the job losses resulting from the ever-advancing digital technologies [55,56]. On the opposite side, the Boston Consulting Group anticipated a fairly positive future scenario for the impact of i4.0 in an evaluation carried out exclusively for a German management magazine. It put the potential impact of i4.0 technologies at more than 100,000 new jobs in mechanical engineering and construction over a 10-year period. They based their logic on the recognition that the introduction of i4.0 technologies would necessitate a substantial number of additional staff with dedicated technical expertise [53].

Koh and colleagues [53] reviewed the potential impact of i4.0 on corporate, administrative, and management tasks, as well as activities. On the operating work level, they concluded that "Industry 4.0 is transforming jobs and required skills, which have impacts on the working environment and skills development. With more robots and smart machines involved in the daily operation, the physical and virtual world are fusing together, thus launching transformation in the working environment." [53] (p. 823). Like many others, they see a future with greater job enrichment, as new technology would lead to better decision-making and planning decentralization, with the necessity for higher process integrity and cross-functional approaches. Processes such as quality assurance and maintenance would presumably be subject to increased automation, and the growing complexity would most likely call for a greater need for cross-functional management control and real-time capability.

Accordingly, it can be argued that the impact of i4.0 technologies will have different effects. These are in particular the automation of repetitive standard activities, the assumption of more complex tasks by employees and the associated need for employees to acquire new skills and training, and the strong trend towards IT topics and digitization [57]. At this point, it is important to state that the impacts of i4.0 technologies on employees are largely positive. These technologies provide possibilities to make working life more comfortable. One example is already mentioned above in regard to the Robot Process Automation, which allows employees to be relieved from boring standard tasks. Further samples are also listed in the 13 most-significant technologies in Section 5.2. Robots are capable of significantly improving ergonomics at workplaces in assembly lines [58]. In terms of employee development and qualification, simulations, as well as virtual and augmented reality, can increase learning retention, improve problem-solving capabilities, and enable situations with instant feedback in a virtual training environment [59–61].

5.4. Potential Symptoms of Employees Linked to Industry 4.0

Job insecurity, negative change attitudes/learning anxiety, and technology anxiety may increase with age. One possibility might also be a feeling of insecurity in terms of losing a job due to i4.0 technologies. Job insecurity can be described as the anxiety of any vertical occupational change. Employees might fear that the technological transformation in the company would disqualify them. A case study concluded that changes in job descriptions could occur in different ways, including transferring tasks from one job description to another, by fusing two or more job descriptions, or by adding new tasks to the old job descriptions [62] (p. 65). Compare also Figure 1: Behavior during change processes in seven discrete phases.

Some possible examples of symptoms in the job insecurity category include disengagement in social events of the company and increased activity on employment-oriented platforms, such as LinkedIn, and higher short-term absence [63–65]. The interviewees in the study of Pfaffinger et al. in 2020 mentioned that information technology experts without experience in leadership could attain a high level of position as the importance of information technology in organizations increases [66]. Increased autonomy occurs when workers are required to learn more leadership skills and self-organizing skills. The perception of the notion "being promoted" generally leads to positive feedback; however, one study also pointed out that increased autonomy or responsibility could also be interpreted as increased accountability [67].

To sum up, employees need to learn new skills while taking into consideration the above-mentioned disadvantages that occur while coping with increased autonomy. This can cause a negative change in attitudes or, in its extreme form, lead to developing a "learning anxiety". More independence could bring about advantages of digitization to make use of a crisis situation such as the COVID-19 pandemic. However, our results, like many other similar findings from times prior to the COVID-19 pandemic, might be interpreted in terms of older people avoiding adapting to digitization because of lifelong learning reservations. This might especially be the case if less learning experience was performed in older age during the past years, because no lifelong learning options or no formal training was provided.

Avoidance learning can be described as a self-protecting response that serves to minimize the perceived threats. In the context of hard-skill acquisition, the employees could develop negative feelings towards the technology. Avoidance, then, could be interpreted as the fear of not being able to master it. Examples for symptoms within the learning anxiety category may include declines in motivation, work quality, and morale in the team, as well as increases in intentional or non-intentional mistakes, absenteeism, and interpersonal conflicts. Furthermore, being uncooperative to formal training and slacking to keep up with the development belong to the potential symptoms [31,66].

Learning anxiety is the main source of resistance to giving up previous beliefs. Learning anxiety can be applied as the anxiety that occurs when one is learning skills to perform a new task and role [67]. The triggers of learning anxiety are avoidance learning and lack of time for the trainings necessary to keep up while still performing one's current task [66,68].

Technology anxiety refers to anxiety towards digitized technology or technology in general. One study pointed out that there is a high correlation between technology anxiety and technology acceptance [69]. The causes and triggers of the technology anxiety class are an overwhelming amount of information input from the digital devices, lack of knowledge and information, lack of user experience, complexity of the technology user interface, lack of expertise, guidance, and assistance, or even learning anxiety (class of symptoms defined earlier [66]).

A classical review concluded the correlation of "computer anxiety", another closely defined term in the context of technology anxiety, with test and math anxiety [31]. "Computer anxiety inhibits computer learning in a similar way that test anxiety reduces test performance and math anxiety decreases math achievement." [31] (p. 102). This implied that the symptoms in the technology anxiety class could be identified in relevance to work performance in respect to the expected potential. Many studies also mentioned the lack of knowledge of the technology in reference to technology anxiety [31,66,70]. Examples of symptoms from technology anxiety include avoidance of computers or electronics at work or at home, excessive caution with computers or devices given, as well as attempts to cut short the necessary use of these technologies [30,70].

5.5. Possible Countermeasures

Based on the defined symptom classes, job insecurity, technology anxiety, and learning anxiety, the corresponding countermeasures are described in the following. One possible countermeasure towards job insecurity in a digitized work environment could be the introduction of prevention and social welfare programs. The latter may step in in the case of job loss in provision of emotional security, to a certain extent, and thereby also work preventatively in terms of diminishing the fear of job loss [66].

For possible countermeasures, a hard distinction between job insecurity and technology anxiety is not appropriate since possible countermeasures mostly help with both types of symptoms. Training or coaching soft skills in the workplace can be considered as a countermeasure for the identified possible symptoms. Exemplary skills are managing financial resources and material resources, individuals' management, time management, decision-making, communication, and leadership skills [71].

Trainings and coaching in the workplace could be very effective as a result of positive organizational outcomes [72,73]. Examples of these outcomes are the performances of the individual, team, and organization. Furthermore, outcomes include skill-based performance, such as leadership skills and top-down/bottom-up management [26], learning competencies, as well as affective performance such as attitude and motivation [73]. Especially in coaching, emphasizing one-on-one stresses assisting the relationship between the coach and the coachee, as well as formally defined agreements in setting goals for personal development, fulfilling the agreement through development, focusing on interpersonal and intrapersonal issues, and using the provision of tools and reinforcements for opportunities the coachee needs for development and growth. The findings of Jones et al. in 2015 showed that coaching would be more effective if training was conducted by internal coaches and when multisource feedback was excluded [73]. Therefore, coaching and training are indispensable in the development of employees, and would directly and indirectly mitigate the symptoms. This is also in line with the finding that employees' participation is needed for continuous improvement of companies' affairs [26].

In an effort to minimize learning anxiety, in 2018, Wirth formulated some solutions that included building visions of the future in presence of the change; formal training on learning competencies; and personal mastery. There were some notable aspects in formulating the training and communication. The management had better focus on the entire group that may be involved. The cultural and peer support was deemed necessary in assistance of the change maintenance. In recommendation of personal mastery, the learner in the learning process needed to determine the learning objectives, the approach of learning, and learning pace. All of the above targets create sufficient psychological safety for the learner to accept the new information and balance the threats induced from the disconfirmation of the old information [74,75].

5.6. Consolidation into a Decision-Support Model

In this part, the systematic structuring is described, which should make the previous steps usable for a model application. Recommendations on what to look for and how to deal with specific symptoms of employees depend on which i4.0 enablers can achieve the corresponding company goals. In order to obtain recommendations for the correct handling of affected employees in digital transformation processes, taking into account the associated enablers and the related objectives, a holistic model is intended to provide support. The model we have developed includes the five procedural steps shown in Figure 3 from the corporate goals to the i4.0 enablers, the impact on employees and their possible symptoms, and suitable countermeasures—and thus provides the relevant recommendations for the executive team. There is no universal solution to guide employees perfectly through the digital transformation but there are certain interactions between these five steps. Depending on which of the goals companies pursue, as described in Section 5.1, different enablers must be selected from the large portfolio of i4.0 technologies. Some enablers make a greater contribution to the achievement of objectives and others correspondingly less. Furthermore, there are connections between the i4.0 enablers and the resulting impacts on the employees concerned. As a function of the effects on their activities and the employees' daily work routines, employees will also show different symptoms. Finally, based on which symptoms the employees show, the appropriate countermeasures must be taken to maintain the employees' efficiency.

If, for example, a company aims to increase the satisfaction of its employees by relieving them of boring and repetitive everyday tasks, one possible i4.0 enabler is Robot Process Automation. The effect of this enabler is that manual, time-consuming, or fault-sensitive activities are automated by software robots. On one hand, the effect that can be expected is that fewer mistakes will be made, which will make the employees more satisfied and give them more time for more interesting tasks. On the other hand, since many standard tasks are no longer necessary, employees suddenly have more time for new tasks. These new tasks may be different from the tasks they have been working on so far or they may even be more demanding or more complex. The effect of this change can lead to feeling overwhelmed and a higher perceived stress level. Possible symptoms that may occur due to this situation include declining motivation, declining work quality, interpersonal conflicts, or higher short-term absenteeism.

Unfortunately, the original intention to increase the satisfaction of the employees through Robot Process Automation, in this case, has quickly transformed into the contrary. If this situation arises, it would be counterproductive to reverse the i4.0 enabler once imposed in order to restore the original situation. Instead, it would be necessary to correctly interpret the symptoms of the employees and to take the appropriate countermeasures. In this case, possible countermeasures can be focused training measures, so that employees do not feel overwhelmed by the new, more complex tasks, or an open discussion with the relevant manager.

In this scenario, the manager can grant the employees the necessary time to familiarize themselves with the new tasks and provide experienced employees with support for a certain period of time. This is just one example of the many possible connections between the five steps shown in Figure 3. By meaningfully combining all identified elements in these five steps, the authors create a holistic model for guiding employees through digital transformation processes. The next section will describe how to convert this model into a software tool.

5.7. Implementing the Decision-Support Model in a Software Tool

To support the implementation of i4.0 in times of demographic change, a tool (see Figure 4) is being developed to promote optimal proactive change management. The tool is based on the previously described relationships in the "decision-support model".

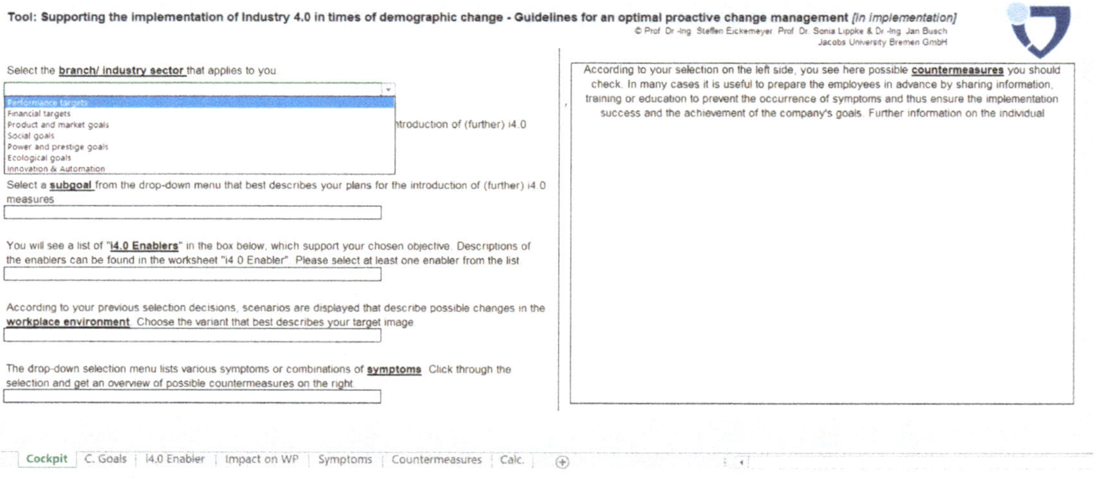

Figure 4. Screenshot of the tool's interface.

When using the tool, the user first selects the branch/industry sector that applies to them, and in the second drop-down menu, the target that best describes the individual plans for the introduction of (further) i4.0 measures. This selection concretizes the options for selection in the form of target areas described in more detail. Accordingly, in the next selection field, a sub-target from the drop-down menu must be selected that best describes the plans for the introduction of (further) i4.0 measures.

6. Conclusions

The positive impact of digitization on business objectives is well known. However, this manuscript specifically reviewed the challenges involved in the introduction of i4.0 technologies. Currently, most companies are still at the beginning of the implementation of measures for universal i4.0 standards and many companies are not yet as far advanced in this process as they would like to be. Ślusarczyk stated in her study, "it occurs that the level of preparation of enterprises to individual dimensions of i4.0 and the ability to use the benefits of i4.0 are lower than expected, taking into account the positive attitude towards new technologies" [76] (p. 238).

However, there are also many advantages of digitization, especially during a crisis such as the COVID-19 pandemic. Additionally, the ongoing and increasing environmental changes due to pollution and required restrictions in mobility can act as a crisis. Human factors such as negative health effects, well-being, anxiety, or stress need to be taken into account when aiming to help companies and their human capital to adapt accordingly within operations and production management.

The model presented in this manuscript, which leads to a software application, aims to identify the potential negative effects of digitization on employees. By knowing the corresponding causes of negative effects, the executive team can act promptly and correctly and thus efficiently manage the desired digital change. It is not only important to systematically identify the appropriate countermeasures for the affected employees, but it is also essential to eliminate the causes. Thus, the executive management must create a clear understanding of the overall change process, starting with the goal (see Section 5.1). They must then define the path to this goal in order to be able to determine which changes are required for which departments and employees, and which i4.0 enabler should be implemented. This planned roadmap must be handed over to the managers in such a way that they understand and support the entire process fully.

Only if the managers support the digital transformation will their employees go along with it and help the company to move forward in i4.0. It is always important to take employees' concerns seriously and to respond to them. To support this, the executive management should communicate directly with teams and the individual members that will face the changes.

A participatory approach could help to find sustainable solutions which translate into long-term success, not only for the company but also for the humans working in it, i.e., preventing losing them due to feeling overwhelmed or not wanted/needed anymore. Especially in times of physical distancing, when working from home is not just in the interest of the company to save costs but also needed in terms of public health, to prevent the spreading of a communicable disease such as COVID-19, appropriate solutions are needed. It can be assumed that even after the complete containment of the COVID-19 pandemic, including the mutations, the wheel of time will not be turned back. In all areas of operations and production management, it is clear that mobile work, as a more general form of working from home, will remain an essential component in the future. In this paper, we aimed to outline a possible approach for this situation in order to improve operations and production management. In parallel, other researchers, such as Ślusarczyk, plan to begin research in the same subject area this year: "In the further perspective, the authors suggest expanding the analysis and conducting additional research that will allow for the identification of the links between specific i4.0 technologies as well as the required competences and skills of employees" [77] (p. 13). Due to the activity of several researchers in this field, further research steps can be expected in the coming years by intensifying the discussion.

In summary, the following research limitations and implications can be noted. Although the survey was limited, it does show the importance of employee age. Therefore, industrial application shows that the tool can be used for consulting. Implications for practice are that the understanding of typical developments, such the age of affected employees, can help companies to take appropriate measures. The software tool can thus serve operational and production management to enable efficient digitization. As an implication in the field of social research, it can be stated that employee well-being can be increased by making companies aware of necessary measures, such as training for individuals and groups early in the digitization process. The benefits of the research underlying the paper for readers and for further research are the review, synthesis, and software tool, which provide suggestions for operations and production management practice and stimulate discussion on this for future research. Future research on the optimization of the decision-support model will be conducted according to Ulrich's [78] "strategy of applied research". Since the developed methodology is focused on the challenges of strategic, tactical, as well as operational decisions, for the solution of which practical knowledge is lacking, it belongs to the applied action sciences [78,79]. The better manageability of reality is in the foreground here. Accordingly, the structure of the decision-support model is expanded by us with each digitization project, and opens questions or difficulties encountered by employees on the way to successful implementation of digitization measures, providing the impetus for further subsequent research work.

Author Contributions: Conceptualization, S.C.E., C.-L.L. and S.L.; methodology, S.C.E. and S.L.; software, S.C.E.; validation, J.B., S.C.E. and S.L.; formal analysis, S.L.; investigation, S.L.; resources, S.C.E.; data curation, S.L.; writing—original draft preparation, S.C.E.; writing—review and editing, J.B., S.C.E. and S.L.; visualization, S.C.E. and S.L.; supervision, J.B.; project administration, S.L.; funding acquisition, S.L. All authors have read and agreed to the published version of the manuscript.

Funding: No funding was received to assist with the preparation of this manuscript. The data collection was funded by the company WELEDA (Project Weleda Trendforschung 2020).

Institutional Review Board Statement: Not applicable as all procedures conducted in this study were in accordance with the ethical standards of the 1964 Helsinki declaration and its later amendments or comparable ethical standards.

Informed Consent Statement: The individuals provided their written informed consent to participate in this study.

Data Availability Statement: The anonymized data is available from the corresponding author.

Acknowledgments: We wish to thank Katharina Crefeld, Janine Graumüller and Frau Sarah Melerski for their support with data collection and management. We also thank Tim Thiwissen for selective support from literature research to proofreading. We acknowledge Ronja Bellinghausen for supporting us with manuscript format editing and proofreading.

Conflicts of Interest: The authors declare that they have no competing interests.

References

1. Majeed, A.; Lee, S. Applications of machine learning and high-performance computing in the era of COVID-19. *Appl. Syst. Innov.* **2021**, *4*, 40. [CrossRef]
2. Newton, C. How COVID-19 is Changing Public Perception of Big Tech Companies. 2020. Available online: https://www.theverge.com/interface/2020/3/26/21193902/tech-backlash-COVID-19-coronavirus-google-facebook-amazon (accessed on 27 April 2020).
3. Kovács, G.; Kuula, M.; Seuring, S.; Blome, C. Operations adding value to society. *Int. J. Oper. Prod. Manag.* **2020**, *40*, 1293–1300. [CrossRef]
4. Gorecky, D.; Schmitt, M.; Loskyll, M.; Zühlke, D. Human-machine-interaction in the industry 4.0 era. In Proceedings of the 2014 12th IEEE International Conference on Industrial Informatics (INDIN), Porto Alegre, Brazil, 27–30 July 2014; pp. 189–294.
5. Dombrowski, U.; Wagner, T. Mental strain as field of action in the 4th industrial revolution. *Proc. CIRP* **2014**, *17*, 100–105. [CrossRef]
6. Yamaguchi, K.; Kosaka, T. Transition of machining system and the latest robot cell. *J. Jpn. Soc. Precis. Eng.* **2012**, *78*, 670–673. (In Japanese) [CrossRef]
7. D'Addona, D.M.; Bracco, F.; Bettoni, A.; Nishino, N.; Carpanzano, E.; Bruzzone, A.A. Adaptive automation and human factors in manufacturing: An experimental assessment for a cognitive approach. *CIRP Annals* **2018**, *67*, 455–458. [CrossRef]
8. Yuan, F.; Woodman, R.W. Innovative behavior in the workplace: The role of performance and image outcome expectations. *Acad. Manag. J.* **2010**, *53*, 323–342. [CrossRef]
9. Janssen, O. Innovative behaviour and job involvement at the price of conflict and less satisfactory relations with co-workers. *J. Occup. Org. Psychol.* **2003**, *76*, 347–364. [CrossRef]
10. Longo, F.; Nicoletti, L.; Padovano, A. Smart operators in industry 4.0: A human-centered approach to enhance operators' capabilities and competencies within the new smart factory context. *Comput. Ind. Eng.* **2017**, *113*, 144–159. [CrossRef]
11. Pacaux-Lemoine, M.P.; Trentesaux, D.; Rey, G.Z.; Millot, P. Designing intelligent manufacturing systems through Human-Machine Cooperation principles: A human-centered approach. *Comput. Ind. Eng.* **2017**, *111*, 581–595. [CrossRef]
12. Kadir, B.A.; Broberg, O.; Conceição, C.S.D. Current research and future perspectives on human factors and ergonomics in Industry 4.0. *Comput. Ind. Eng.* **2019**, *137*, 106004. [CrossRef]
13. Hamzeh, R.; Zhong, R.; Xu, X.W. A survey study on Industry 4.0 for New Zealand manufacturing. In *Procedia Manufacturing, Proceedings of the 46th SME North American Manufacturing Research Conference (NAMRC 46), College Station, TX, USA, 18–22 June 2018*; Wang, L., Ed.; Elsevier: Amsterdam, The Netherlands, 2018; Volume 26, pp. 49–57. [CrossRef]
14. Hilb, M. Toward artificial governance? The role of artificial intelligence in shaping the future of corporate governance. *J. Manag. Gov.* **2020**, *24*, 851–870. [CrossRef]
15. Kagermann, H.; Wahlster, W.; Helbig, J. *Recommendations for Implementing the Strategic Initiative Industrie 4.0: Final report of the Industrie 4.0 Working Group*; Blasco, J., Galloway Trans, H., Hellinger, A., Stumpf, V., Eds.; Forschungsunion & Acatech: Frankfurt, Germany, 2013.
16. Hermann, M.; Pentek, T.; Otto, B. Design principles for industrie 4.0 scenarios. In Proceedings of the 2016 49th Hawaii International Conference on System Sciences (HICSS), Koloa, HI, USA, 5–8 January 2016; pp. 3928–3937. [CrossRef]
17. Klingenberg, C.; Do Vale Antunes, J.A., Jr. Industry 4.0: What Makes it Revolution. Paper Presented at the EurOMA. 2017. Available online: https://www.researchgate.net/publication/319127784_Industry_40_what_makes_it_a_revolution/citations (accessed on 15 October 2021).
18. Oztemel, E.; Gursev, S. Literature review of Industry 4.0 and related technologies. *J. Intell. Manuf.* **2018**, *31*, 127–182. [CrossRef]
19. Venema, S.; Bergström, A.A. Industry 4.0: An Opportunity or a Threat? A Qualitative Study Among Manufacturing Companies. 2018. Available online: http://urn.kb.se/resolve?urn=urn:nbn:se:umu:diva-150008 (accessed on 1 April 2020).

20. Vaidya, S.; Ambad, P.; Bhosle, S. Industry 4.0—A glimpse. In *Procedia Manufacturing, Proceedings of the 2nd International Conference on Materials, Manufacturing and Design Engineering (iCMMD2017), MIT Aurangabad, Maharashtra, India, 11–12 December 2017*; Elsevier: Amsterdam, The Netherlands, 2018; Volume 20, pp. 233–238.
21. Erol, S.; Jäger, A.; Hold, P.; Ott, K.; Sihn, W. Tangible Industry 4.0: A scenario-based approach to learning for the future of production. *Proc. CiRp* **2016**, *54*, 13–18. [CrossRef]
22. Sipsas, K.; Alexopoulos, K.; Xanthakis, V.; Chryssolouris, G. Collaborative maintenance in flow-line manufacturing environments: An Industry 4.0 approach. *Proc. CiRp* **2016**, *55*, 236–241. [CrossRef]
23. Thoben, K.; Wiesner, S.; Wuest, T. "Industrie 4.0" and smart manufacturing-a review of research issues and application examples. *Int. J. Autom. Technol.* **2017**, *11*, 4–16. [CrossRef]
24. Cinquini, P.; Krcmar, H.; Oswald, G.; Welz, B. Skills for Digital Transformation. 2015. Available online: http://www.i17.in.tum.de/fileadmin/w00btn/www/IDT_Skill_Report_2015.pdf (accessed on 15 October 2021).
25. Cinquini, L. Editorial: JMG Symposium on evaluation, performance and governance in the digital age. *J. Manag. Gov.* **2019**, *23*, 847–848. [CrossRef]
26. Galeazzo, A.; Furlan, A.; Vinelli, A. The role of employees' participation and managers' authority on continuous improvement and performance. *Int. J. Op. Prod. Manag.* **2021**, *41*, 34–64. [CrossRef]
27. Kübler-Ross, E. Interviews mit Sterbenden. In *The Change Curve*; Herder Verlag: Freiburg, Germany, 1977.
28. Ganzarain, J.; Errasti, N. Three stage maturity model in SME's toward Industry 4.0. *J. Ind. Eng. Manag.* **2016**, *9*, 1119–1128. [CrossRef]
29. Pessl, E.; Sorko, S.R.; Mayer, B. Roadmap Industry 4.0—Implementation guideline for enterprises. *Int. J. Sci. Technol. Soc.* **2017**, *5*, 193–202. [CrossRef]
30. Brosnan, M.J. *Technophobia: The Psychological Impact of Information Technology*; Routledge: London, UK, 2002.
31. Torkzadeh, G.; Angulo, I.E. The concept and correlates of computer anxiety. *Behav. Inf. Technol.* **1992**, *11*, 99–108. [CrossRef]
32. Lippke, S.; Fischer, M.; Ratz, T. Physical activity, loneliness, and meaning of friendship in young individuals—A mixed-methods investigation prior to and during the COVID-19 pandemic with three cross-sectional studies. *Front. Psychol.* **2021**, *12*, 146. [CrossRef]
33. Hofstra, N.; Dullaert, W.; De Leeuw, S.; Spiliotopoulou, E. Individual goals and social preferences in operational decisions: Behavioural insights from transport planning. *Int. J. Oper. Prod. Manag.* **2019**, *39*, 116–137. [CrossRef]
34. Gattiker, U.E.; Howg, L. Information technology and quality of work life: Comparing users with non-users. *J. Bus. Psychol.* **1990**, *5*, 237–260. [CrossRef]
35. Bolander, T. What do we loose when machines take the decisions? *J. Manag. Gov.* **2019**, *23*, 849–867. [CrossRef]
36. CYLAD. Digital Transformation: 10 Key Essentials for Your Board. 2020. Available online: https://cylad-experts.com/wp-content/uploads/2020/12/Digital-transformation_10-key-essentials-for-your-board_VF.pdf (accessed on 15 October 2021).
37. Trautmann, T. 4 Dimensions for Executives to Watch during a Digital Transformation. 2020. Available online: https://www.cylad-consulting.com/wp-content/uploads/2020/06/CYLAD-Consulting-on-Digital-Transformation-%E2%80%93-4-Dimensions-for-Executives-to-watch.pdf (accessed on 15 October 2021).
38. Arnold, C.; Voigt, K.I. Determinants of industrial internet of things adoption in German manufacturing companies. *Int. J. Innov. Technol. Manag. IJITM* **2019**, *16*, 1950038. [CrossRef]
39. Moore, M.H. Managing for value: Organizational strategy in for-profit, nonprofit, and governmental organizations. *Nonprofit Volunt. Sect. Q.* **2000**, *29*, 183–204. [CrossRef]
40. Crnjac, M.; Veža, I.; Banduka, N. From concept to the introduction of Industry 4.0. *Int. J. Ind. Eng. Manag.* **2017**, *8*, 21.
41. Bhalla, V.; Caye, J.; Dyer, A.; Dymond, L.; Morieux, Y.; Orlander, P. *High-Performance Organizations. The Secrets of Their Success*; Boston Consulting Group BCG: Boston, MA, USA, 2011.
42. Székely, F.; Knirsch, M. Responsible leadership and corporate social responsibility: Metrics for sustainable performance. *Eur. Manag. J.* **2005**, *23*, 628–647. [CrossRef]
43. Bongomin, O.; Ocen, G.G.; Nganyi, E.O.; Musinguzi, A.; Omara, T. Exponential disruptive technologies and the required skills of Industry 4.0: A review. *J. Eng.* **2019**, *2019*. [CrossRef]
44. Lacity, M.; Willcocks, L.; Craig, A. Robotic process automation at Telefónica O_2. *MIS Q. Exec.* **2016**, *15*, 21–35.
45. Moffitt, K.C.; Rozario, A.M.; Vasarhelyi, M.A. Robotic process automation for auditing. *J. Emerg. Technol. Account.* **2018**, *15*, 1–10. [CrossRef]
46. Van der Aalst, W.M.P.; Bichler, M.; Heinzl, A. Robotic process automation. *Bus. Inf. Syst. Eng.* **2018**, *60*, 269–272. [CrossRef]
47. Willcocks, L.; Lacity, M.; Craig, A. The IT Function and Robotic Process Automation, The Outsourcing Unit Working Research Paper Series. Available online: https://eprints.lse.ac.uk/64519/1/OUWRPS_15_05_published.pdf (accessed on 22 November 2021).
48. Wilkesmann, M.; Wilkesmann, U. Industry 4.0—Organizing routines or innovations. *VINE J. Inf. Knowl. Manag. Syst.* **2018**, *48*, 238–254. [CrossRef]
49. Frey, C.B.; Osborne, M.A. The Future of Employment: How Susceptible Are Jobs to Computerisation? Available online: http://www.oxfordmartin.ox.ac.uk/downloads/%20academic/The_Future_of_Employment.Pdf (accessed on 15 October 2021).
50. Sivathanu, B.; Pillai, R. Smart HR 4.0—How industry 4.0 is disrupting HR. *Hum. Resour. Manag. Int. Dig.* **2018**, *26*, 7–11. [CrossRef]

51. Stachová, K.; Papula, J.; Stacho, Z.; Kohnová, L. External partnerships in employee education and development as the key to facing Industry 4.0 challenges. *Sustainability* **2019**, *11*, 345. [CrossRef]
52. Plattform Industrie 4.0: Aspects of the Research Roadmap in Application Scenarios. Available online: https://www.plattform-i40.de/I40/Redaktion/EN/Downloads/Publikation/aspects-of-the-research-roadmap.pdf?__blob=publicationFile&v=10 (accessed on 15 October 2021).
53. Koh, L.; Orzes, G.; Jia, F. The fourth industrial revolution (Industry 4.0): Technologies disruption on operations and supply chain management. *Int. J. Oper. Prod. Manag.* **2019**, *39*, 817–828. [CrossRef]
54. Ford, M. *The Lights in the Tunnel*; CreateSpace Independent Publishing Platform: North Charleston, SC, USA, 2009.
55. Bowles, J. The Computerisation of European Jobs. Who Will Win and Who Will Lose from the Impact of New Technology onto Old Areas of Employment? 2014. Available online: http://www.bruegel.org/nc/blog/detail/article/1394-the-computerisation-of-european-jobs/ (accessed on 20 January 2015).
56. Brynjolfsson, E.; McAfee, A. The second machine age. In *Work, Progress, and Prosperity in a Time of Brilliant Technologies*; W.W. Norton & Company: New York, NY, USA, 2014.
57. Srai, J.S.; Graham, G.; Hennelly, P.; Phillips, W.; Kapletia, D.; Lorentz, H. Distributed manufacturing: A new form of localised production? *Int. J. Oper. Prod. Manag.* **2020**, *40*, 697–727. [CrossRef]
58. Mateus, J.C.; Claeys, D.; Limère, V.; Cottyn, J.; Aghezzaf, E.H. A structured methodology for the design of a human-robot collaborative assembly workplace. *Int. J. Adv. Manuf. Technol.* **2019**, *102*, 2663–2681. [CrossRef]
59. Huang, H.M.; Rauch, U.; Liaw, S.S. Investigating learners' attitudes toward virtual reality learning environments: Based on a constructivist approach. *Comput. Educ.* **2010**, *55*, 1171–1182. [CrossRef]
60. Blümel, E. Global challenges and innovative technologies geared toward new markets: Prospects for virtual and augmented reality. *Proc. Comput. Sci.* **2013**, *25*, 4–13. [CrossRef]
61. Quint, F.; Sebastian, K.; Gorecky, D. A mixed-reality learning environment. *Proc. Comput. Sci.* **2015**, *75*, 43–48. [CrossRef]
62. Rintala, N.; Suolanen, S. The implications of digitalization for job descriptions, competencies and the quality of working life. *Nord. Rev.* **2005**, *26*, 53–67. [CrossRef]
63. Sumitani, A. 8 Signs Employees Are About to Quit: What to Look for If You're Worried. 2020. Available online: https://www.tinypulse.com/blog/employee-behavior-8-red-flags-to-look-out-for (accessed on 22 April 2020).
64. Rampton, J. 25 Signs Your Employee Is About to Quit, According to Research. 2017. Available online: https://www.inc.com/john-rampton/25-signs-your-employee-is-about-to-quit-according-to-research.html (accessed on 22 April 2020).
65. Böckerman, P.; Bryson, A.; Ilmakunnas, P. Does high involvement management improve worker wellbeing? *J. Econ. Behav. Org.* **2012**, *84*, 660–680. [CrossRef]
66. Pfaffinger, K.F.; Reif, J.A.M.; Spieß, E.; Berger, R. Anxiety in a digitalised work environment. *Gruppe. Interaktion. Organisation. Z. Angew. Organisation. GIO* **2020**, *51*, 25–35. [CrossRef]
67. Bamberger, S.G.; Larsen, A.; Vinding, A.L.; Nielsen, P.; Fonager, K.; Nielsen, R.N.; Omland, Ø. Assessment of work intensification by managers and psychological distressed and non-distressed employees: A multilevel comparison. *Ind. Health* **2015**, *53*, 322–331. [CrossRef] [PubMed]
68. Klein, Z.; Shner, G.; Ginat-Frolich, R.; Vervliet, B.; Shechner, T. The effects of age and trait anxiety on avoidance learning and its generalization. *Behav. Res. Ther.* **2020**, *129*, 103611. [CrossRef] [PubMed]
69. Khasawneh, O.Y. Technophobia without boarders: The influence of technophobia and emotional intelligence on technology acceptance and the moderating influence of organizational climate. *Comput. Hum. Behav.* **2018**, *88*, 210–218. [CrossRef]
70. Rosen, L.D.; Sears, D.C.; Weil, M.M. Treating technophobia: A longitudinal evaluation of the computerphobia reduction program. *Comput. Hum. Behav.* **1993**, *9*, 27–50. [CrossRef]
71. Aulbur, W.; Arvind, C.J.; Bigghe, R. Skill Development for Industry 4.0. 2016. Available online: http://www.globalskillsummit.com/Whitepaper-Summary.pdf (accessed on 17 April 2020).
72. Santos, A.; Stuart, M. Employee perceptions and their influence on training effectiveness. *Hum. Resour. Manag. J.* **2003**, *13*, 27–45. [CrossRef]
73. Jones, R.J.; Woods, S.A.; Guillaume, Y.R.F. The effectiveness of workplace coaching: A meta-analysis of learning and performance outcomes from coaching. *J. Occup. Org. Psychol.* **2015**, *89*, 249–277. [CrossRef]
74. Wirth, R. Lewin/Schein's Change Theory. 2018. Available online: https://www.researchgate.net/publication/237112705_LewinSchein\T1\textquoterights_Change_Theory (accessed on 11 April 2020).
75. Senge, P.M. *The Fifth Discipline: The Art and Practice of the Learning Organization*; Doubleday/Currency: New York, NY, USA, 1990.
76. Ślusarczyk, B. Industry 4.0: Are we ready? *Pol. J. Manag. Stud.* **2018**, *17*, 232–248. [CrossRef]
77. Ślusarczyk, B.; Nathan, R.J.; Pypłacz, P. Employee preparedness for industry 4.0 in logistic sector: A cross-national study between Poland and Malaysia. *Soc. Sci.* **2021**, *10*, 258. [CrossRef]
78. Ulrich, H. *Management-Gesammelte Beiträge, Schriftenreihe Unternehmung und Unternehmungsführung*; Paul Haupt: Bern, Switzerland, 1984; ISBN 9783258034461.
79. Ulrich, P.; Hill, W. *Wissenschaftstheoretische Grundlagen der Betriebswirtschaftslehre*; Raffée, H., Abel, B., Eds.; Wissenschaftstheoretische Grundfragen der Wirtschaftswissenschaften: München, Germany, 1979; pp. 161–190.

applied system innovation

MDPI

Review

Digital Twin: Origin to Future

Maulshree Singh [1], Evert Fuenmayor [1], Eoin P. Hinchy [2], Yuansong Qiao [3], Niall Murray [3] and Declan Devine [1,*]

1 Material Research Institute, Athlone Institute of Technology, Athlone, Co., N37 HD68 Westmeath, Ireland; m.singh@research.ait.ie (M.S.); efuenmayor@ait.ie (E.F.)
2 Confirm Centre for Smart Manufacturing, School of Engineering University of Limerick, Limerick, Co., V94 T9PX Limerick, Ireland; Eoin.Hinchy@ul.ie
3 Software Research Institute, Athlone Institute of Technology, Athlone, Co., N37 HD68 Westmeath, Ireland; yuangsongqiao@ait.ie (Y.Q.); nmurray@ait.ie (N.M.)
* Correspondence: ddevine@ait.ie; Tel.: +353-879-137-012

Abstract: Digital Twin (DT) refers to the virtual copy or model of any physical entity (physical twin) both of which are interconnected via exchange of data in real time. Conceptually, a DT mimics the state of its physical twin in real time and vice versa. Application of DT includes real-time monitoring, designing/planning, optimization, maintenance, remote access, etc. Its implementation is expected to grow exponentially in the coming decades. The advent of Industry 4.0 has brought complex industrial systems that are more autonomous, smart, and highly interconnected. These systems generate considerable amounts of data useful for several applications such as improving performance, predictive maintenance, training, etc. A sudden influx in the number of publications related to 'Digital Twin' has led to confusion between different terminologies related to the digitalization of industries. Another problem that has arisen due to the growing popularity of DT is a lack of consensus on the description of DT as well as so many different types of DT, which adds to the confusion. This paper intends to consolidate the different types of DT and different definitions of DT throughout the literature for easy identification of DT from the rest of the complimentary terms such as 'product avatar', 'digital thread', 'digital model', and 'digital shadow'. The paper looks at the concept of DT since its inception to its predicted future to realize the value it can bring to certain sectors. Understanding the characteristics and types of DT while weighing its pros and cons is essential for any researcher, business, or sector before investing in the technology.

Keywords: digital twin; Industry 4.0; digital model; system optimization; predictive maintenance

Citation: Singh, M.; Fuenmayor, E.; Hinchy, E.P.; Qiao, Y.; Murray, N.; Devine, D. Digital Twin: Origin to Future. *Appl. Syst. Innov.* **2021**, *4*, 36. https://doi.org/10.3390/asi4020036

Academic Editor: Félix Jesús García Clemente

Received: 30 April 2021
Accepted: 21 May 2021
Published: 24 May 2021

Publisher's Note: MDPI stays neutral with regard to jurisdictional claims in published maps and institutional affiliations.

Copyright: © 2021 by the authors. Licensee MDPI, Basel, Switzerland. This article is an open access article distributed under the terms and conditions of the Creative Commons Attribution (CC BY) license (https://creativecommons.org/licenses/by/4.0/).

1. Introduction

In the era of Industry 4.0, when several industries are experiencing a digital transformation, Digital Twin (DT) is considered no less than a linchpin for gaining a competitive and economic advantage over competitors. DT saw its origins in the aerospace industry, and it is expected to revolutionize other industries [1]. The main applications of DT in different sectors are designing/planning, optimization, maintenance, safety, decision making, remote access, and training, among others. It can be a great tool for companies to increase their competitiveness, productivity, and efficiency [2]. DT has the ability to link physical and virtual worlds in real time, which provides more a realistic and holistic measurement of unforeseen and unpredictable scenarios [3].

The value DT brings to any sector, by reducing time to market, optimizing operations, reducing maintenance cost, increasing user engagement, and fusing information technologies, is indisputable [4]. The global market of DT was estimated at USD 3.1 billion in 2020 [5] and is expected to grow exponentially in succeeding years. The outbreak of the COVID-19 pandemic has changed the way production and maintenance are looked at, which has accelerated the adoption of Digital Twins [6]. Thus, it becomes essential to understand what the implications of DT implementations could or should be, depending on the nature of the industry where it is adopted.

This paper provides a detailed account of different types of DT and their characteristics. In literature, DT is often described and reviewed within a manufacturing context [2,7–14], whereas this review looks at DT as a technology beyond just manufacturing as well as consolidates different types of DTs at one place, which has not been reported before. This work aims to provide an overview of the current state-of-the-art of DT as well as a classification system of what is to be considered DT based on key required functionalities. The paper intends to analyze the role and advantages of DT and how can it bring value to the existing systems. Academia defines DT in several different ways depending on their field of research; therefore, this paper will also look at the origin and different definitions of DT throughout the literature to create a clearer picture of what can be referred to as DT and what should not be. The paper proposes a consolidated definition of DT that can be used irrespective of any industry and its application.

The paper is divided into four sections. Section 2 explains the origins of DT and its early applications while also exploring its definitions throughout the literature. Section 3 investigates the advantages and characteristics of DT and how it has been classified into different categories based on different criteria by different authors. The future applications of DT and the challenges associated with it are discussed in Section 4 of the paper.

2. Digital Twin in Literature

2.1. History of Digital Twin

Even though DT technology has gained massive popularity in the past couple of years, the concept is not entirely new. Its concept came into being in relation to Product Lifecycle Management (PLM) in 2002 at the University of Michigan by Michael Grieves [15]. The proposed model has three components: real space, virtual space, and linking mechanism for the flow of data/information between the two; the model was then referred to as 'Mirrored Spaces Model' [16]. A similar concept in which software models mimic reality from information input from the physical world was imagined by David Gelernter in 1991 and was called 'Mirror Worlds' [17]. In 2003, Kary Främling et al. also proposed "an agent-based architecture where each product item has a corresponding 'virtual counterpart' or agent associated with it" as a solution to the inefficiency of transfer of production information via paper for PLM [18]. By 2006, the name of the conceptual model proposed by Grieves was changed from 'Mirrored Spaces Model' to 'Information Mirroring Model' [15,19]. The model put emphasis on the linking mechanism between two spaces being bidirectional and having multiple virtual spaces for a single real space where alternate ideas or designs can be explored (Figure 1). Due to the limitations of the technologies, such as low computing power, low or no connectivity of devices with the internet, data storage and management, underdeveloped machine algorithms, etc., DT had no practical applications at the time.

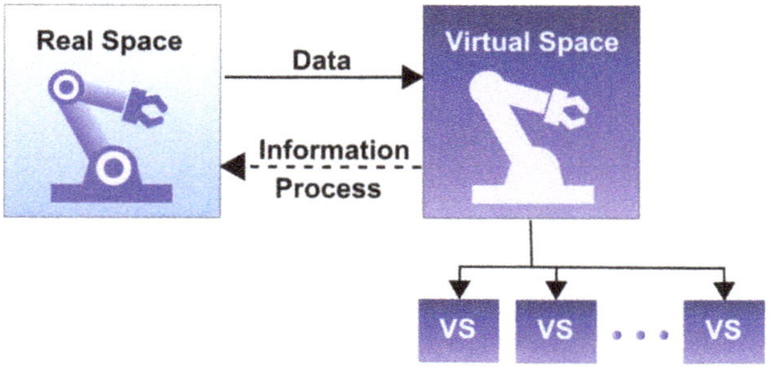

Figure 1. Mirrored Spaces Model/Information Mirroring Model as proposed by Michael W. Grieves. (Adapted from [16,19].)

The name 'Digital Twin' (DT) first appears in NASA's draft version of the technological roadmap in 2010 [20]. In the NASA roadmaps, DT was also referred to as 'Virtual Digital Fleet Leader'. NASA was the first association to forge the definition of DT; it was described as "an integrated multi-physics, multi-scale, probabilistic simulation of a vehicle or system that uses the best available physical models, sensor updates, fleet history, etc., to mirror the life of its flying twin". Even though the first mention of DT is in the 2010 roadmap, NASA had used a similar concept before for the Apollo program, where two identical space vehicles were built to mirror each other [21,22]. Soon, the US Air Force followed the footsteps of NASA and used DT technology for the design, maintenance, and prediction of their aircraft [23–25]. The idea was to use DT to simulate physical and mechanical properties of the aircraft to forecast any fatigue or cracks in the structure, thus prolonging the remaining useful life of the aircraft. E. Tuegel [23] and B. T. Gockel et al. [25] in their papers defined DT only for the aircraft and called it 'Airframe Digital Twin' or ADT, which was a computational model to manage the aircraft over its entire lifecycle. Besides monitoring, DT was also proposed for sustainable space exploration as well as for future generations of aerospace vehicles [26]. The timeline of the evolution of DT can be seen in Figure 2.

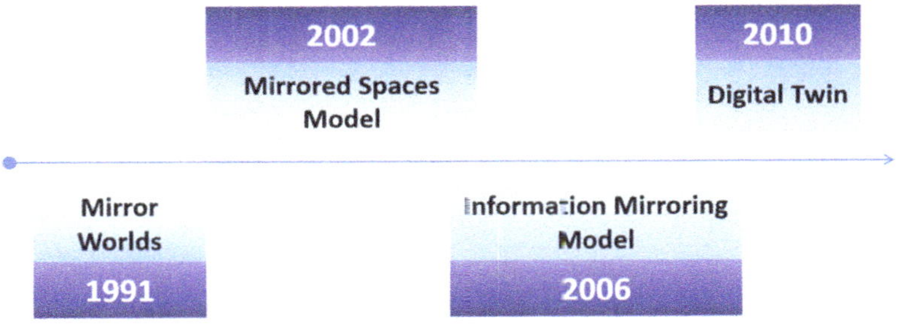

Figure 2. Timeline of evolution of Digital Twin.

2.2. Digital Twin Description in Literature

Since the first-ever definition published by NASA, different authors have described DT in their own terms and based on its application. The definitions in the literature have referred to DT as a virtual or ***digital model*** [21,23,27–35], ***layout*** [36], ***counterpart*** [7,9,35,37], ***doppelganger*** [38], ***clone*** [39], ***footprint*** [40], ***software analogue*** [41], ***representation*** [42–46], ***information constructs*** [2,47], or ***simulation*** [20,26,48–51] of its physical counterpart. The first few definitions of DT that came out described DT only in respect to the aerospace/aeronautics industry which then ventured into manufacturing, which is why the words such as ***'aircraft'*** [23,25,30], ***'vehicle'*** [20,26,48], or ***'airframe'*** [25,29] were replaced by ***'system'*** [2,7,21,22,34–36,40,50–52], ***'machine'*** [28,53], ***'product'*** [22,32,36,37,44–47,51], ***'object'*** [9,36,41–43], ***'entity'*** [2,9], ***'asset'*** [31,33,39,40], ***'device'*** [41], or ***'process'*** [9,21,32,33,36,45,46,52,54]. Ever since DT found its application in industries beyond manufacturing, the definition has also moved from just being restricted to non-living entities or industrial products, and can now be used to describe even complex biological systems as humans and trees, among others [43,55].

One thing that binds most definitions of DT other than being a virtual representation of a physical object is the bidirectional transfer or sharing of data between the physical counterpart and the digital one, including quantitative and qualitative data (related to material, manufacturing, process, etc.), historical data, environmental data, and most importantly, real-time data. Using these data, DT can perform such tasks as:

- In-depth analysis of physical twin;
- Design and validation of new or existing product/process;
- Simulate the health conditions of physical twin;

- Increase safety and reliability of physical twin;
- Optimization of part, product, process, or production line;
- Track the status of physical twin throughout its lifetime;
- Predict the performance of physical twin;
- Real-time control over physical twin.

Definitions of DT tend to overlook its longevity; however, some authors consider DT as a cradle-to-gravel model, meaning that DT can be used over the entire life cycle, from the time of inception of the product until its disposal [23,33,44,50]. However, Grieves and Vickers [47], who conceptualize the idea of DT, have defined a type of DT that is created even before its physical twin exists (see Section 3.3). Martin and Nadja [56] in their review found eleven papers in which DT prequels the physical twin. Grieves and Vickers also suggest that DT technology can have information related to the safe decommissioning of the product during its disposal phase. In addition, after the product is disposed, DT of one generation can help in the design and production of the next generation [57].

It is clear from the literature that DT is different from computer models (CAD/CAE) and simulation. Even though many organizations use the term 'Digital Twin' synonymously to 3D model, a 3D model is only a part of DT [33]. DT uses data to reflect the real world at any given point of time and thus can be used for observing and understanding the performance of the system and for its predictive maintenance [58]. Computer models, just like DT, are also used for the generic understanding of a system or for making generalized predictions, but they are rarely used for accurately representing the status of a system in real time. A lack of real-time data makes these models or simulations static, which means that they do not change or cannot make new predictions unless new information is fed to them [59]. However, having real-time data is not enough for DT to operate—the data also need to be loaded automatically to DT and the flow from physical to digital should be bidirectional. Liu et al. [11] reported that more than half of the papers they reviewed were describing and/or studying 'Digital Model' or 'Digital Shadow' despite authors claiming it to be Digital Twin. W. Kritzinger et al. [2] also found the similar result that there is more literature on 'Digital Model' or 'Digital Shadow' than on 'Digital Twin' (see Section 3.3 for Digital Model' and 'Digital Shadow').

There are other concepts such as 'Product Avatar' or 'Digital Thread' which are also sometimes interchangeably used or get confused with DT. Just like 'Digital Twin', the term 'Digital Thread' was also emanated from the aerospace industry, where it was used to describe an integrated system engineering process used for managing the entire process digitally. It included 3D CAD models, model-based engineering, BoM, manufacturing processes, assembly logistics, delivery systems, etc. [50]. Digital Thread has also been described as the communication framework that consolidates the asset's data and allows seamless data flow; in other words, Digital Thread provides the right information at the right time to the right place [31]. Therefore, as compared to DT, which is a real-time virtual representation of its physical twin, Digital Thread is just the record of the information on the physical twin throughout its lifetime [60]. The major difference between Product Avatar and DT comes from the fact that they have been derived from two different research lines and thus have different capabilities and purpose; a Product Avatar is a digital counterpart of a smart product that lets its user use the attributes and services of that smart product for its entire lifecycle [9].

Due to the presence of a plethora of definitions in the literature, there is no consensus of what can actually be described as DT or not. Thus, giving academics and businesses permission to use the term DT loosely and conveniently to their needs creates confusion between different terminologies related to the digitalization of industries. A sudden influx in the number of publications related to 'Digital Twin' also indicates that interest is increasing exponentially, as it happens with any new promising technology.

In order to simplify the confusion around different terminologies used to describe DT, a definition of DT is proposed here that can be applied irrespective of the industry or its application:

"A Digital Twin is a dynamic and self-evolving digital/virtual model or simulation of a real-life subject or object (part, machine, process, human, etc.) representing the exact state of its physical twin at any given point of time via exchanging the real-time data as well as keeping the historical data. It is not just the Digital Twin which mimics its physical twin but any changes in the Digital Twin are mimicked by the physical twin too."

3. Advantages, Characteristics and Types of Digital Twin

Since the first introduction of DT, its popularity has increased as more and more researchers started focusing their research on DT. Figure 3 shows the exponential growth of the number of publications found on Scopus as well as ScienceDirect (limited to the English language) that contain the term 'Digital Twin' in the article title, abstract, or as keywords from 2011 to 2020. The increase in the publications is quite recent, i.e., 2016 onwards, as it can be clearly seen in Figure 3.

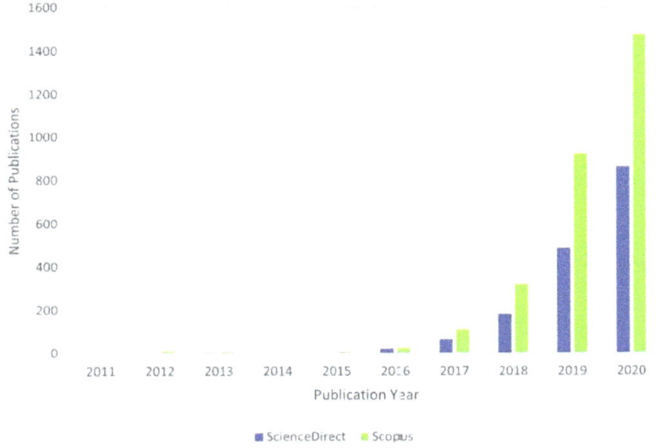

Figure 3. Number of Digital Twin-related publications by year from 2011 to 2020 on Scopus and ScienceDirect.

The growing popularity of DT can also be observed as it has been considered as a strategic technology trend for three consecutive years (2017–2019) by Gartner [61–63], future technology in the field of aerospace and defense by Lockheed Martin [64], and one of the defining technology of next decade by Forbes [65]. According to Juniper Research, by 2021, the global DT market was expected to be USD 12.7 billion [66]. The global DT market share, previously dominated by early adopters of the technology i.e., North America and Europe (>50%), is expected to expand in the Asia Pacific at a compound annual growth rate of 40.0% over the next 5 years [67].

As the DT market is growing exponentially, every sector or business wants to invest in the technology. However, to leverage the DT technology for any sector, it is important to understand the value it can bring to the business It is also beneficial to understand the different characteristics and types of DT in order to choose the correct type of DT that can complement the business and can provide maximum profits.

3.1. Advantages of Digital Twin

The main reason DT technology is seen as the cornerstone in Industry 4.0 is its plethora of advantages, including the reduction of errors, uncertainties, inefficiency, and expenses in any system or process. It also removes all the silos in processes or organizations that otherwise work in isolation within compartments and divisions in more traditional industrial structures. Some of the advantages reported for DT include:

- ***Speed prototyping as well as product re-designing***: Since simulations allow the investigation of a number of scenarios, the design and analysis cycles shorten, making the whole process of prototyping or re-designing easier and faster. Once implemented, DT can be used in different stages of the product design process, from conceptualizing the idea of the product to its testing [68]. Besides that, it also creates an opportunity where the customization of each product based on users' needs and usage data is possible [69]. Since the DT is connected to its physical twin throughout its lifetime, the comparison between the actual and predicted performance can be made, allowing engineer/product designers to reconsider their assumptions on which the product was designed [70].
- ***Cost-effective:*** Due to DT involving mostly virtual resources for its creation, the overall cost of prototyping decreases with time. In traditional prototyping, redesigning a product is time-consuming as well as expensive because of the use of physical materials and labour, and on top of that, a destructive test means the end of that costly prototype, whereas using DT, products can be recreated and put through destructive tests without any additional material cost. Thus, assuming even if the cost is equal at the start, the physical costs keep increasing as inflation rises but the virtual cost decreases significantly as time progresses (Figure 4) [47]. DT allows the testing of products under different operating scenarios, including destructive scenarios, without any additional costs. Moreover, DT can reduce operating costs and extend the life of equipment and assets once implemented.
- ***Predicting Problems/System Planning:*** Using DT, we can predict the problems and errors for future states of its physical twin, providing us an opportunity to plan the systems accordingly. Due to the real-time data flowing between the physical asset and its DT, it can predict problems at different stages of the product lifecycle. This is beneficial especially for products that have multiple parts, complex structures, and are made up of multiple materials such as aircraft, vehicles, factory equipment, etc., because as the complexity of any product increases, it gets harder to predict component failures using conventional methods [68].
- ***Optimizing Solutions and Improved Maintenance:*** The traditional methods of maintenance are based on heuristic experience and worst-case scenarios rather than on the specific material, structural configuration, and usage of an individual product, making them reactive rather than proactive [26]. However, DT can foresee defects and damage in the manufacturing machine or system and thus can schedule the maintenance of the product in advance. By simulating different scenarios, DT provides the best possible solution or maintenance strategy that makes the maintenance of the product/system much easier. In addition, the constant feedback loop between DT and its physical counterpart can be used to validate and optimize the system's process all the time.
- ***Accessibility:*** The physical device can be controlled and monitored remotely using its DT. Unlike physical systems, which are restricted by their geographical location, virtual systems such as DT can be widely shared and can be remotely accessed [47]. Remote monitoring and controlling of equipment and systems becomes a necessity in a situation where local access is limited, like during the COVID-19 pandemic when lockdowns have been enforced by governments and working remotely or non-contact is the only viable option [71].
- ***Safer than the Physical Counterpart:*** In industries such as oil and gas or mining where the working conditions are extreme and hazardous, the capability of DT to remotely access its physical twin, as well as its predictive nature, can reduce the risk of accidents and hazardous failures. However, DT's advantage of accessing remotely is not limited to preventing accidents. During the global COVID-19 pandemic, not having human contact and in-person monitoring is also a way to guarantee safety. According to a recent Gartner survey, almost one-third of companies are using DT amidst the global COVID-19 pandemic to increase the safety of employees and customers through remote monitoring [72].

- ***Waste Reduction:*** Using DT to simulate and test product or system prototypes in a virtual environment significantly reduces wastage. Prototype designs can be probed and scrutinized virtually, under a variety of different test scenarios, to finalize the final product design prior to manufacture. This not only saves on material wastage but also reduces development costs and time to market.
- ***Documentation and Communication:*** To create a DT, it is important to synchronize data scattered across different software applications, databases, hard copies, etc., which simplifies the process of accessing and maintaining the data in one place [33]. DT enable a better understanding of system reactions and thus it can be used to document and communicate the behaviour and mechanisms of the physical twin [32].
- ***Training:*** DT can be used to develop more efficient and illustrative safety training programmes than the traditional one [73]. Before working on a high-risk site or hazardous machinery, operators can be trained using a DT to reduce the dangers, as exposing and educating them about different processes or scenarios will make them confident in dealing with the same situations in person. For example, mining is a high-risk environment where new employees can be trained using DT on machinery operations, as well as how to deal with emergency scenarios [74]. DT can also be a great tool in closing the knowledge gap from experienced workers to newcomers.

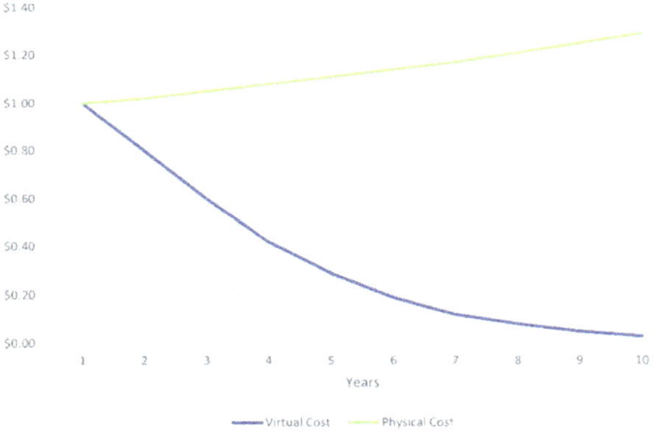

Figure 4. Real vs. virtual costs of prototyping. (Adapted from [47].)

3.2. Characteristics of Digital Twin

Depending on the type of DT, it can possess distinctive properties from others, but regardless, all DTs have a few characteristics in common:

- ***High-fidelity:*** A DT needs to be a near-identical copy of its physical counterpart in terms of appearance, contents, functionality, etc., with a very high degree of accuracy. A super-realistic digital model helps DT in mimicking every aspect of its physical twin. Ultra-high fidelity computer models are considered the backbone of the DT [48]. This level of detail allows DT simulation and prediction tools to be more reliable when presented with a set of alternative actions or scenarios [21].
- ***Dynamic:*** The physical is dynamic, meaning it changes with respect to time. Thus, a DT also needs to change as the physical system changes. This is achieved through the seamless connection and continuous exchange between the physical and virtual worlds. The data exchanging can be dynamic data, historical static data, as well as descriptive static data [9]. DT has been described as a 'living model in 3D' [38]. The objective of DT being dynamic is to mirror the physical twin and its behaviour realistically in the digital world [33].
- ***Self-evolving:*** A DT evolves along with its physical counterpart throughout its life cycle. Any changes in either the physical or Digital Twin are reflected in its counterpart,

creating a closed feedback loop [33]. A DT is self-adapting and self-optimizing with the help of the data collected by physical twin in real time, thus maturing along with its physical counterpart throughout its lifetime [9].

- ***Identifiable:*** Every physical asset needs to have its own DT. During different stages of the product lifecycle, the data and information related to it evolve and so does the model, including 3D geometric models, manufacturing models, usage models, functional models, etc. Due to the existence of such models created for DT, a DT can be uniquely identified from its physical twin or vice versa anywhere in the globe and for the entirety of its life cycle [75].
- ***Multiscale and Multiphysical:*** DT, being the virtual copy of its physical twin, needs to incorporate the properties of the physical twin at multiple scales or levels. Thus, the virtual model in DT is based on macroscopic geometric properties of the physical twin such as shape, size, tolerance, etc., as well as on microscopic properties such as surface roughness, etc. In other words, DT contains the set of information about the physical twin from micro atomic level to macro geometric level [47]. DT is also multiphysical because, besides the aforementioned geometric properties, the model is also based on physical properties of the physical twin such as structural dynamics models, thermodynamic models, stress analysis models, fatigue damage models, and material properties of physical twin such as stiffness, strength, hardness, fatigue strength, etc. [76].
- ***Multidisciplinary:*** Industry 4.0 revolves around many disciplines, and DT being the backbone of Industry 4.0 sees the fusion of disciplines such as computer science, information technology, and communications; mechanical, electrical, electronic, and mechatronic engineering; automation and industrial engineering; and system integration physics, just to name a few [34].
- ***Hierarchical:*** The hierarchical nature of DT comes from the fact that the different components and parts that make up the final product all have their corresponding DT model, e.g., DT of an aircraft is comprised of rack DT, DT of the flight control system, DT of the propulsion system, etc. [23]. Therefore, a DT can be seen as a series of integrated submodels [29].

3.3. Classification of Digital Twin

Digital Twins (DTs) can be classified into different types based on different criteria such as when the DT is created, level of integration, its applications, hierarchy, and maturity level. Different authors have come up with their own nomenclature of DT types based on these criteria.

3.3.1. DT Creation Time

According to Grieves and Vickers [47], there are two types of DT based on when is it developed during the life cycle of the product—before the prototype is created, i.e., at the designing phase, or after the product is ready, i.e., at the production phase. Both types of DTs are integrated and operated for multiple usages on a platform they called Digital Twin Environment (DTE).

- ***Digital Twin Prototype (DTP):*** DTP can be described as a DT that contains the set of data/information that is essential to create or manufacture a physical copy from the virtual version. This includes BOM (bill of materials), design files, CAD models, etc. The product cycle will start from the creation of DTP, which can be put through several tests, even the destructive ones, before creating its physical twin. In addition, DTP helps us in identifying and avoiding unpredictable and undesirable scenarios that are difficult to identify with traditional prototyping. Once DTP is complete and validated, its physical twin can be manufactured in the real world. The accuracy of the simulation/model will determine the quality of the physical twin.
- ***Digital Twin Instance (DTI):*** This type of DT is connected to its physical counterpart throughout its life cycle. DTI came into being during the production phase. Once

a physical system has been built, the data from the real space are sent to the virtual space and vice versa to monitor and predict the system behaviour. With these data, it can be found out if the system is depicting the predicted desirable behaviour or not, as well as if the predicted undesirable scenarios has been successfully eliminated. Since the linkage between both the systems is bidirectional, any changes in one will be duplicated on the other. A collection of DTIs is called Digital Twin Aggregate (DTA) by the authors.

3.3.2. Level of Integration

Based on the integration level of DTs, Kritzinger et al. [2] divided them into three subcategories (Figure 5):

- ***Digital Model:*** In this type of DT, the data between the physical and digital object are exchanged manually, due to which any changes in the state of the physical object are not reflected in the digital one directly, and vice versa.
- ***Digital Shadow:*** The data from the physical object flow to the digital automatically, but this is still manual the other way around. As a result, any change in the physical object can be seen in its digital copy, but not vice versa.
- ***Digital Twin:*** In this type of DT, there is an automatic bidirectional flow of data between the physical and digital object. Therefore, the changes in either object, physical or digital, directly lead to changes in the other.

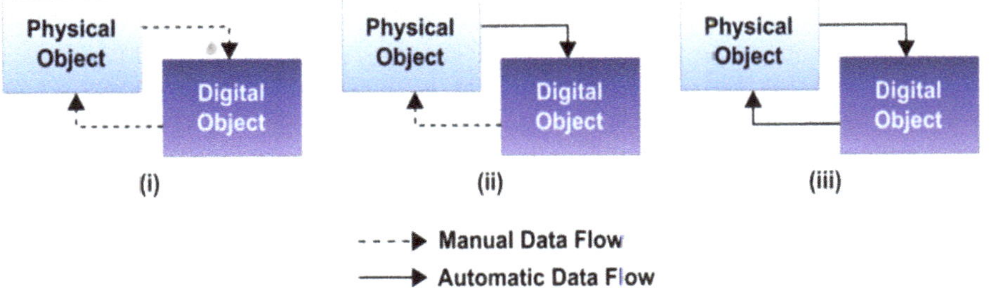

Figure 5. Types of DT based on level of integration (adapted from [2]). **(i)** Digital Model; **(ii)** Digital shadow/Static Digital model; **(iii)** Digital Twin/Dynamic Digital Model.

Centre for Digital Built Britain [77] describes Digital Shadow as 'Static Digital Model' and Digital Twin as 'Dynamic Digital Model' as the data in the dynamic one are fed live, which can be used for a real-time control mechanism, whereas in the static model, data are added with time, making it more suitable for strategic planning.

3.3.3. Application

DT can also be categorized depending on their applications. The two broad applications of a DT are prediction and interrogation [47]. A ***Predictive DT***, as the name suggests, predicts future behaviour and performance of its physical counterpart whereas an ***Interrogative DT*** is used to interrogate the current or past state of its physical counterpart irrespective of its location. DTs can also be divided depending on if the focus of application is on product, process, or performance [46,78]:

- ***Product DT:*** It is used for prototyping as it analyses the product under different conditions and makes sure that the next physical product is behaving as planned. By virtually validating the product, the prototyping can be rapid as the total development time is reduced and there is no longer a need to develop multiple of them.
- ***Production DT:*** It is used for validating the processes by simulating and then analyzing them even before the actual production. This helps in developing an efficient

production methodology under different conditions. The data from Product and Production DT can be used together for monitoring and maintenance of the machinery.

- ***Performance DT:*** It is used for decision-making processes by capturing, Aggregating, and analyzing data from smart products and plants. Since Performance DT includes performances of both product and production, it optimizes the operations depending on the availability of plant resources, which creates an opportunity to improve on the Production and Product DTs via a feedback loop.

3.3.4. Hierarchy

From a hierarchal perspective, DT can be divided into three different levels as well (Figure 6), according to the magnitude involved in manufacturing [75]:

- ***Unit level:*** It is the smallest participating unit in manufacturing and can be a piece of equipment, material, or environmental factors. Unit-level DT is based on the geometric, functional, behavioural, and operational model of unit-level physical twin.
- ***System level:*** It is an amalgamation of several unit-level DTs in a production system such as production line, shop floor, factory, etc. Interconnectivity and collaboration among multiple unit-level DT lead to a wider flow of data and better resource allocation. A complex product, e.g., aircraft, can also be considered as system-level DT.
- ***System of Systems (SoS) level***: A number of system-level DT are connected together to form SoS-level DT, which helps in collaborating different enterprises or different departments with an enterprise such as supply chain, design, service, maintenance, etc. In other words, SoS-level DT integrates different phases of the product throughout its life cycle.

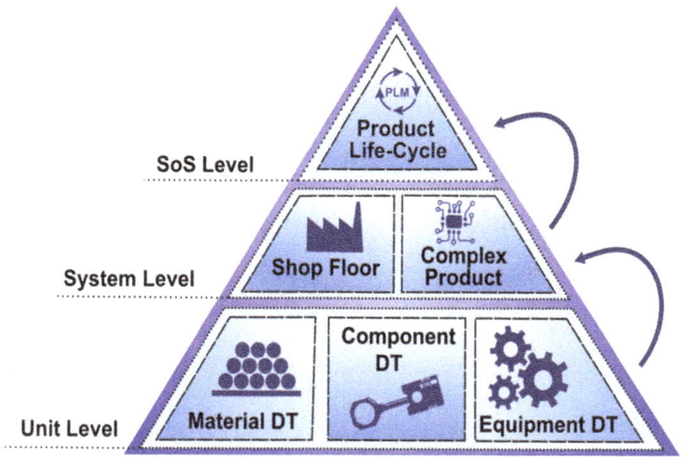

Figure 6. Hierarchical levels of DT in manufacturing. (Adapted from [75].)

The hierarchy of DT can also be classified as ***(i) Part/Component twin, (ii) Product/Asset twin, (iii) System twin, and (iv) Process twin*** [79,80], with part twin being the simplest. More sophisticated and comprehensive systems/processes can be achieved by putting together the lower-level twins.

3.3.5. Level of Maturity/Sophistication

Based on the sophistication level of DT, i.e., the quantity and quality of data obtained from the physical twin and its environment, DTs can be grouped into [81]:

- ***Partial DT:*** It contains a small number of data points, e.g., pressure, temperature, humidity, etc., which is useful in determining the connectivity and functionality of DT.

- *Clone DT:* It contains all significant and relevant data from the product/system that can be used for making prototypes and categorizing development phases.
- *Augmented DT:* It utilizes data from the asset along with its historical data and at the same time derives and correlates the useful data using algorithms and analytics.

The sophistication level of DT can be improved with the accumulation of bigger sets of data over times of operation. For Azad M. Madni et al. [58], the level of maturity of DT is not just limited to data but it also encompasses the sophistication level of the virtual representation/model. On this basis, DT is divided into four levels:

- *Pre-Digital Twin:* This is level 1, where DT is created prior to the physical asset for the purpose of making decisions on prototype designs to reduce any technical risk and resolve issues upfront by using a generic system model.
- *Digital Twin:* Level 2 incorporates the data from the physical asset related to its performance, health, and maintenance. The virtual system model uses these data to assist high-level decision-making in the design and development of the asset, along with scheduling maintenance. The data transfer at this level is bidirectional.
- *Adaptive Digital Twin:* This level 3 provides an adaptive user interface between physical and Digital Twin, and has the capability to learn from the preferences and priorities of human operators using supervised machine learning. Using this DT, real-time planning and decision-making during operations is possible.
- *Intelligent Digital Twin:* In addition to the features from level 3, level 4 has unsupervised machine-learning capability, making it more autonomous than level 3. It can recognize patterns in the operational environment and using that along with reinforced learning allows for a more precise and efficient analysis of the system.

4. Future and Challenges of Digital Twin

The technologies composing DT such as Internet of Things (IoT), Industrial Internet of Things (IIoT), Artificial Intelligence (AI), big data, simulation, and cloud computing, among others, have been on a path of constant evolution, and thus it can be assumed that DT will continue to evolve in parallel to these technologies. Therefore, the true potential of DT is still hidden behind the fog of its future form. This is evident by the estimates about the DT market worldwide. The global market of DT is expected to grow at the rate of 58% and in just six years, i.e., by 2026, it is expected to reach USD 48.2 billion [5]. The COVID-19 pandemic is one of the key factors driving the growth of the DT market as it is expected that post-pandemic, industry will be pushing for the further digitization of processes.

Humans are an integral part of any industry and need to be taken into account when developing any DT. IEEE (The Institute of Electrical and Electronics Engineers) believes that in the future, DT will become an indispensable part of machine development as well as of the human–machine symbiotic relationship [82]. Some researchers see 'Digital Triplet' as the next evolution of DT [83–85]. In addition to physical and cyber worlds, Digital Triplet also includes the 'intelligent activity world' where humans will solve problems by using DT. Unlike DT, where systems and process are automated, in Digital Triplet human interaction with the system and process is also accounted for, thus creating value from data using human intelligence and knowledge. The aim of 'Digital Triplet' is to support engineering activities throughout a product's life cycle including design, manufacturing, use, maintenance, remanufacturing, and circulation of resources by integrating all three worlds, just like DT [83,84].

4.1. Future Applications of Digital Twin

DT technology has been leveraged in both the aerospace and manufacturing sectors, but there are several other sectors where DT is still in its infancy, such as agriculture, construction, automobile, and healthcare. One of the driving factors pushing the demand for DT in several sectors is the COVID-19 pandemic in 2020. The demand for DT has increased in healthcare and pharmaceutical industries as well as in manufacturing industries due to this viral outbreak [5].

In manufacturing, DT is expected to become a central capability in MBSE (model-based systems engineering) in the future, the reason being that using DT MBSE can be applied to the entire life cycle of the system. DT can help MBSE in penetrating new markets such as manufacturing, construction, and real estate [58]. DT is expected to improve future non-destructive testing by integrating it into the manufacturing process through online monitoring and make it capable of decision-making [86]. In the future, it can be possible that the manufactured products will have their own DTs and we will see more cases of connected DTs.

In the automotive industry, General Electric [87] believes that with advances in predictive maintenance and analytics, DT technology will be widely implemented throughout the industry, from individual car owners to manufacturers. For car owners, the maintenance of the car will become a lot easier as unexpected breakdowns will be a thing of the past and the car will be able to book maintenance appointments itself. The mechanics/service industry will be able to provide faster and more efficient solutions as they will have all the right information about the car at their fingertips even before the appointment. Besides managing their clients, DT will help garages in managing their inventory and supply chain more efficiently. As for car dealers, DT can take care of monitoring the performance and health of the entire car fleet by itself along with helping in making crucial business and financial decisions based on depreciation and asset performance management. All the data created by the future cars and garages can be analyzed, improving the entire automotive value chain. IBM (International Business Machines Corporation) [38] too claims that in coming years, DT technology will save considerable amounts of time and money in the automotive industry, as DT will be built long before the cars hit the assembly line. This will provide a data pool on motor type, suspension, chassis, aerodynamics, and even different types of drivers who will interact with the vehicle, enabling designers and engineers to design and model an ideal product and also to observe and analyze the performance of the vehicle even before it hits the road. DT technology can also be extended to motorbikes, where if a bike's part breaks then instead of ordering a generic replacement part, the bike's DT can order a tailor-made, generatively designed replacement part that can be made using 3D printing technology [88].

The global COVID-19 pandemic has given DT a platform to grow in the healthcare industry. The authors who developed 'Cardio Twin' believe that their platform can prevent Ischemic heart disease (IHD) and stroke in the future [89]. As DT technology advances in the medical field, we can expect that every person on the planet will be able to receive extremely personalized medical treatment and care in cost-effective way [90].

DT is in its early stages of development in the agriculture industry. DT can be exploited in farming for (1) storing and collecting data, (2) categorizing actions in complex workflow, (3) automated data analysis, (4) learning and measuring the content and capacity of the soil, (5) simulating the crop outcome, (6) predicting the weather, and (7) recognizing the stress on resources by factors such as invasive plants and animals, soil quality, pollution, etc. [91].

As space technology advances, it will become important for engineers and scientists to make these explorations more sustainable so that we can explore further into the universe. The Commonwealth Scientific and Industrial Research Organization (CSIRO) [92], Australia's national science agency, which has been working on DT solutions for mining operations, is envisioning to build DTs to aid mining processes in future space exploration by collaborating with the Australian and international space agencies to make space operations more sustainable.

More and more national and regional governments are gaining interest in creating DTs of their own cities. These cities with their own DT will be better planned and managed. The administration will be able to serve its stakeholders better and resolve the problems faster. These cities' DTs have added advantage in cases of natural disasters such as earthquakes, cyclones, and tsunamis, among others. DTs of cities can help in cases of emergency such as global pandemics, as multiple cities will be able to share the local strategy and status via DT in a timely manner [93].

In future, DT technology can be combined with other technologies for better results, e.g., mixed reality with DT technology can be used for better visualization in DT as well as for remotely supervised inspections [86]. Besides that, DT is finding its application in the education sector. Sepasgoza [94] showed how combining DT with augmented and virtual reality (AR/VR) can be used for developing digital pedagogy for architecture, engineering, and construction students, which can be valuable for educators delivering remote classes, either in emergencies such as pandemics or as a part of regular online-based learning.

4.2. Challenges to Implement Digital Twin

Realizing a mammoth technology like DT comes with its own challenges. The challenges that arise with developing a DT depends on its scale and complexity but there are some barriers with the technology that are common to all. DT technology being in its infancy stages means that even though it has great potential, its implementation carries complications that can be either engineering and technology-related or can be commercial [95]. Such complications include ambiguity surrounding the definition or concept of DT, lack of appropriate tools, expensive investments, data-related issues, lack of rules and regulations, etc. The reader, in the following paragraphs, can find a compilation of the most common themes currently preventing a streamlined adaptation by industries of DT technologies.

4.2.1. Novelty of Technology

As DT is still an emerging technology, there is a lack of clear understanding about the value it can bring to individuals, businesses, or industries. Incompetency on the part of technical and practical knowledge is also hindering the progress of the technology. There is also a lack of case studies of successful practices or business models implementing DT into company activities or realistic estimations on the costs involved in this implementation [96].

Several technologies come together to make DT a reality such as 3D simulations, IoT/IIoT, AI, big data, machine learning, and cloud computing. Since these technologies themselves are in developing phases, it impedes the evolution of DT. The infrastructure to implement DT needs to be improved to enhance the efficacy of the technology. There is a need for further research in technologies such as high-performance computing technology, machine learning technology, real-time virtual-real interactive technology, intelligent perception and connection technology, among others, in order to implement DT [68].

Along with the infrastructure, there is a need for supporting software. There are a plethora of software packages providing DT solutions such as Predix by General Electronics (GE), Azure Digital Twin by Microsoft, PTC, 3D Experience by Dassault Systèmes, ABB LTD, Watson by IBM, Digital Enterprise Suite by Siemens, etc. [97], which makes it harder to identify and chose the platform that can deliver the most appropriate service based on the specific needs of the interested industries/business.

4.2.2. Time and Cost

One of the biggest challenges DT needs to overcome to reach its full potential is the high cost associated with its implementation. The whole process of developing ultra-high-fidelity computer models and its simulation of processes to create a DT is a time-consuming and labour-intensive exercise that also requires a huge amount of computational power to run, thus making DT an expensive investment [49]. On top of that, embedding the existing system with sensors for data collection along with the requirement of high-performance IT infrastructure, which includes hardware as well as software for storing and processing that data, contribute to the additional cost. Gartner analyst and expert Marc Halpern has also shown concern over cost and time-related aspects of DT at the PDT Europe conference in Gothenburg, Sweden, saying that bringing DT concepts together can take more time and resources than one can imagine [98]. A paper published by West and Blackburn gives a glimpse of the scale of cost and time invested in bringing DT into reality. The authors claim that it could cost trillions of dollars and hundreds of years to completely implement Digital Threads/Digital Twins of weapons systems for the U.S. Air Force, making it impractical to fully realize the technology [99]. This makes it very crucial for industries to perform cost-benefit analysis before implementing DT.

4.2.3. Lack of Standards and Regulations

As there are a plethora of DT models and architectures available in the literature, there is a need for defining a consistent framework for DT throughout the industry that includes shared and mutual understanding of interfaces and standardization for uniformity together with efficient design of data flow to make accessibility of data easier without compromising its security [100]. Standardizing models, interfaces, protocols, and data is essential for efficient third-party communication, product and human safety, data security, and integrity, especially in industries such as aerospace, automobile, healthcare, etc. [95]. Besides that, standards and standards-based interoperability need to be developed to address the social and organizational challenges unfolded by digital transformation within industries [101]. A lack of device communication and data collection standards can compromise the quality of data being processed for DT, which will be reflected on its performance [102]. Since the technologies including big data and AI are also still in their infancy, the laws and regulations around them are yet to be formalized. There is a standardized framework ISO 23247 (Digital Twin Manufacturing Framework) under development which is aimed at providing guidelines and methods for developing and implementing Digital Twins in the manufacturing sector. This framework will have four parts: (i) Overview and general principles, (ii) Reference architecture, (iii) Digital representation of physical manufacturing elements, and (iv) Information exchange [103].

4.2.4. Data Related Issues

As DT technology deals with the data, one of the biggest concerns that arises is about privacy, confidentiality, transparency, and ownership of these data.

Owning and sharing data is influenced by company policies as well as by the mindset of people and society about data ownership, thus putting a limitation on DT that is beyond the complexities of technology and engineering [95]. Not having proper policies in place regarding sharing the data internally (within the organization) or externally (stakeholders across the supply chain) can lead to data silos within different departments of an organization, which can be detrimental to the value chain [95] as data silos lead to inconsistency and synchronization issues [104]. Another possible issue that needs to be considered is how to share the data among different DTs, i.e., data interoperability. In a setting where there are multiple DTs at different hierarchical levels, each generating a different type of data and one feeding on the other DT can create a complicated relationship between data set that causes data interoperability issues [105]. Cybersecurity cannot be neglected when it comes to handling the data. On one hand, where having data silos can affect the overall performance of DT, not having them makes DT more vulnerable to cybercrime [105].

Another major challenge regarding data involved in DT is its convergence and governance. Projects involving big data are likely to fail due to lack of data governance and management to tackle the challenges related to big data, which include identifying and accessing data, transforming data from different sources, poor quality of data, translation loss, etc. [104].

4.2.5. Life-Cycle Mismatching

An additional concern over DT technology is related to the products that have long life cycles such as buildings, aircraft, ships, machinery, or even cities. The life cycles of such products are far longer than the validity of the software used for designing or simulating the DT as well as for storing and analyzing the data for DT [106]. This means that there is a high risk, in the future at some point of time, of either the formats used by software becoming obsolete or becoming locked with the same vendor for new versions of software or other authoring tools [105,107].

5. Conclusions

Though the concept of DT is decades old, its real impact has only come into being in recent years. The digital market is and will continue to grow as more and more end-use industries adapt the technology owing to its potential in reducing operational costs and time, increasing the productivity of the existing system, improving maintenance, easing accessibility, creating a safer work environment, and other purposes yet to be realized.

DT technology, when combined with the other technologies such as AR/VR, mixed reality, additive manufacturing, 3D printing, etc., as well as with other DTs, will open doors for new applications and potential. Although this technology comes with its own challenges, the benefits are far greater. This paper identified different types of DTs to better understand and distinguish what can and cannot be described as DT to edge closer to consensus on its definition and limitations. Identifying and understanding the potential of DT in any sector and complementing it with the pertinent type of DT provides an opportunity for developing Industry 4.0 tools that offer numerous advantages, from simulation and prediction capabilities to record-keeping and forensic troubleshooting. This paper looked at the advantages offered by implementing this technology and contrasted them with the challenges. The benefits involve reductions in cost, increased outputs, remote access, and streamlining of services and operations, all of which can be obtained by operating the system digitally instead of physically without any additional material or investment cost. It also gave a glimpse of the future of different industries with DT as one of their core technologies.

DT as a technology is in its infancy, meaning it is far from reaching its full potential. Identifying and addressing the challenges of DT is crucial in leveraging the technology in different sectors. A lot of challenges associated with DT can be attributed to the novelty of the technology: lack of consensus on its definition and its value, lack of standards and regulations, lack of competent engineers and technicians, and lack of supporting software. There are issues of data security and ownership associated with DT that demand more attention since data serve as the foundation for DT. In addition to addressing the present challenges related to DT, there is also a need to broaden our horizons and identify the problems that DT as a technology can face or create in the future. Understanding the holistic view on DT—which involves what qualifies as DT, its characteristics, its advantages, how can it be implemented, and its challenges—is essential if we want to unleash the true potential of the technology.

Author Contributions: Conceptualization, M.S. and E.F.; writing—original draft preparation, M.S.; writing—review and editing, E.F. and E.P.H.; writing—review, Y.Q. and D.D.; supervision, E.F., D.D., and N.M. All authors have read and agreed to the published version of the manuscript.

Funding: This research was funded by grant number 16/RC/3918 and the APC was funded by Science Foundation Ireland.

Institutional Review Board Statement: Not Applicable.

Informed Consent Statement: Not Applicable.

Conflicts of Interest: The authors declare no conflict of interest.

References

1. Tao, F.; Zhang, M.; Nee, A.Y.C. Chapter 2—Applications of Digital Twin. In *Digital Twin Driven Smart Manufacturing*; Tao, F., Zhang, M., Nee, A.Y.C., Eds.; Academic Press: Cambridge, MA, USA, 2019; pp. 29–62.
2. Kritzinger, W.; Karner, M.; Traar, G.; Henjes, J.; Sihn, W. Digital Twin in manufacturing: A categorical literature review and classification. *IFAC-PapersOnLine* **2018**, *51*, 1016–1022. [CrossRef]
3. Parrott, A.; Warshaw, L. *Industry 4.0 and the Digital Twin: Manufacturing Meets its Match*; Deloitte University Press: New York, NY, USA, 2017; pp. 1–17.

4. Tao, F.; Zhang, M.; Nee, A.Y.C. Chapter 1—Background and Concept of Digital Twin. In *Digital Twin Driven Smart Manufacturing*; Tao, F., Zhang, M., Nee, A.Y.C., Eds.; Academic Press: Cambridge, MA, USA, 2019; pp. 3–28.
5. MarketsandMarkets. *Digital Twin Market by Technology, Type (Product, Process, and System), Application (Predictive Maintenance, and Others), Industry (Aerospace & Defense, Automotive & Transportation, Healthcare, and others), and Geography—Global Forecast to 2026*; MarketsandMarkets: Pune, India, 2020; p. 177.
6. Erol, T.; Mendi, A.F.; Doğan, D. Digital Transformation Revolution with Digital Twin Technology. In Proceedings of the 2020 4th International Symposium on Multidisciplinary Studies and Innovative Technologies (ISMSIT), Istanbul, Turkey, 22–24 October 2020; pp. 1–7.
7. Negri, E.; Fumagalli, L.; Macchi, M. A review of the roles of digital twin in cps-based production systems. *Procedia Manuf.* **2017**, *11*, 939–948. [CrossRef]
8. Tao, F.; Zhang, H.; Liu, A.; Nee, A.Y. Digital twin in industry: State-of-the-art. *IEEE Trans. Ind. Inform.* **2018**, *15*, 2405–2415. [CrossRef]
9. Barricelli, B.R.; Casiraghi, E.; Fogli, D. A Survey on Digital Twin: Definitions, Characteristics, Applications, and Design Implications. *IEEE Access* **2019**, *7*, 167653–167671. [CrossRef]
10. Cimino, C.; Negri, E.; Fumagalli, L. Review of digital twin applications in manufacturing. *Comput. Ind.* **2019**, *113*, 103130. [CrossRef]
11. Liu, M.; Fang, S.; Dong, H.; Xu, C. Review of digital twin about concepts, technologies, and industrial applications. *J. Manuf. Syst.* **2020**, *58*, 346–361. [CrossRef]
12. Lu, Y.; Liu, C.; Kevin, I.; Wang, K.; Huang, H.; Xu, X. Digital Twin-driven smart manufacturing: Connotation, reference model, applications and research issues. *Robot. Comput. Integr. Manuf.* **2020**, *61*, 101837. [CrossRef]
13. He, B.; Bai, K.-J. Digital twin-based sustainable intelligent manufacturing: A review. *Adv. Manuf.* **2020**, *9*, 1–21. [CrossRef]
14. Magomadov, V. The digital twin technology and its role in manufacturing. In Proceedings of the IOP Conference Series: Materials Science and Engineering, Krasnoyarsk, Russia, 16–18 April 2020; p. 032080.
15. Grieves, M. Origins of the Digital Twin Concept. 2016. Available online: https://www.researchgate.net/publication/307509727_Origins_of_the_Digital_Twin_Concept. (accessed on 25 September 2020).
16. Grieves, M.W. Product lifecycle management: The new paradigm for enterprises. *Int. J. Prod. Dev.* **2005**, *2*, 71–84. [CrossRef]
17. Gelernter, D. *Mirror Worlds: Or: The Day Software Puts the Universe in a Shoebox. How it Will Happen and What it Will Mean*; Oxford University Press: Oxford, UK, 1993.
18. Främling, K.; Holmström, J.; Ala-Risku, T.; Kärkkäinen, M. Product agents for handling information about physical objects. *Rep. Lab. Inf. Process. Sci. Ser. B TKO-B* **2003**, *153*.
19. Grieves, M. Back to the future: Product lifecycle management and the virtualization of product information. In *Product Realization*; Springer: Boston, MA, USA, 2009; pp. 1–13.
20. Shafto, M.; Conroy, M.; Doyle, R.; Glaessgen, E.; Kemp, C.; LeMoigne, J.; Wang, L. Draft modeling, simulation, information technology & processing roadmap. *Technol. Area* **2010**, *11*.
21. Rosen, R.; Von Wichert, G.; Lo, G.; Bettenhausen, K.D. About the importance of autonomy and digital twins for the future of manufacturing. *IFAC-PapersOnLine* **2015**, *48*, 567–572. [CrossRef]
22. Boschert, S.; Rosen, R. Digital twin—The simulation aspect. In *Mechatronic Futures*; Springer: Cham, Switzerland, 2016; pp. 59–74.
23. Tuegel, E. The airframe digital twin: Some challenges to realization. In Proceedings of the 53rd AIAA/ASME/ASCE/AHS/ASC Structures, Structural Dynamics and Materials Conference 20th AIAA/ASME/AHS Adaptive Structures Conference 14th AIAA, Honolulu, HI, USA, 23–26 July 2012; p. 1812.
24. Tuegel, E.J.; Ingraffea, A.R.; Eason, T.G.; Spottswood, S.M. Reengineering aircraft structural life prediction using a digital twin. *Int. J. Aerosp. Eng.* **2011**, *2011*. [CrossRef]
25. Gockel, B.T.; Tudor, A.H.; Brandyberry, M.D.; Penmetsa, R.C.; Tuegel, E.J. Challenges with Structural Life Forecasting Using Realistic Mission Profiles. In Proceedings of the 53rd AIAA/ASME/ASCE/AHS/ASC Structures, Structural Dynamics and Materials Conference, Honolulu, HI, USA, 23–26 July 2012; p. 1813.
26. Glaessgen, E.; Stargel, D. The digital twin paradigm for future NASA and US Air Force vehicles. In Proceedings of the 53rd AIAA/ASME/ASCE/AHS/ASC Structures, Structural Dynamics and Materials Conference, Honolulu, HI, USA, 23–26 July 2012; p. 1818.
27. Majumdar, P.K.; FaisalHaider, M.; Reifsnider, K. Multi-physics response of structural composites and framework for modeling using material geometry. In Proceedings of the 54th AIAA/ASME/ASCE/AHS/ASC Structures, Structural Dynamics, and Materials Conference, Boston, MA, USA, 8–11 April 2013; p. 1577.
28. Lee, J.; Lapira, E.; Bagheri, B.; Kao, H.-A. Recent advances and trends in predictive manufacturing systems in big data environment. *Manuf. Lett.* **2013**, *1*, 38–41. [CrossRef]
29. Hollkamp, J.J. Modeling vibratory damage with reduced-order models and the generalized finite element method. *J. Sound Vib.* **2014**, *333*, 6637–6650.
30. Bielefeldt, B.; Hochhalter, J.; Hartl, D. Computationally efficient analysis of SMA sensory particles embedded in complex aerostructures using a substructure approach. In Proceedings of the ASME 2015 Conference on Smart Materials, Adaptive Structures and Intelligent Systems, Colorado Springs, CO, USA, 21–23 September 2015.
31. Leiva, C. Demystifying the digital thread and digital twin concepts. *Ind. Week. August* **2016**, *1*, 2016.

32. Redelinghuys, A.; Basson, A.; Kruger, K. A six-layer digital twin architecture for a manufacturing cell. In Proceedings of the International Workshop on Service Orientation in Holonic and Multi-Agent Manufacturing, Bergamo, Italy, 11–12 June 2018; pp. 412–423.
33. LaGrange, E. Developing a Digital Twin: The Roadmap for Oil and Gas Optimization. In Proceedings of the SPE Offshore Europe Conference and Exhibition, Aberdeen, UK, 3–6 September 2019.
34. Bajaj, M.; Cole, B.; Zwemer, D. Architecture to geometry-integrating system models with mechanical design. In Proceedings of the AIAA SPACE, Long Beach, CA, USA, 13–16 September 2016; p. 5470.
35. Bazilevs, Y.; Deng, X.; Korobenko, A.; Lanza di Scalea, F.; Todd, M.; Taylor, S. Isogeometric fatigue damage prediction in large-scale composite structures driven by dynamic sensor data. *J. Appl. Mech.* **2015**, *82*. [CrossRef]
36. Demkovich, N.; Yablochnikov, E.; Abaev, G. Multiscale modeling and simulation for industrial cyber-physical systems. In Proceedings of the 2018 IEEE Industrial Cyber-Physical Systems (ICPS), St. Petersburg, Russia, 15–18 May 2018; pp. 291–296.
37. Ríos, J.; Hernández, J.C.; Oliva, M.; Mas, F. Product Avatar as Digital Counterpart of a Physical Individual Product: Literature Review and Implications in an Aircraft. In Proceedings of the ISPE CE, Delft, The Netherlands, 20–23 July 2015; pp. 657–666.
38. Wired Brand Lab; Digital Twin: Bridging the Physical-Digital Divide. 2017. Available online: https://www.ibm.com/blogs/internet-of-things/iot-digital-twin-enablers/ (accessed on 11 October 2020).
39. Banerjee, A.; Dalal, R.; Mittal, S.; Joshi, K.P. Generating digital twin models using knowledge graphs for industrial production lines. *UMBC Inf. Syst. Dep.* **2017**. [CrossRef]
40. Mayani, M.G.; Svendsen, M.; Oedegaard, S. Drilling Digital Twin Success Stories the Last 10 Years. In Proceedings of the SPE Norway One Day Seminar, Bergen, Norway, 18 April 2018.
41. Volodin, V.; Tolokonskii, A. Concept of instrumentation of digital twins of nuclear power plants units as observers for digital NPP I&C system. *J. Phys. Conf. Ser.* **2019**, *1391*, 012083.
42. Canedo, A. Industrial IoT lifecycle via digital twins. In Proceedings of the Eleventh IEEE/ACM/IFIP International Conference on Hardware/Software Codesign and System Synthesis, Pittsburgh, PA, USA, 2–7 October 2016; p. 1.
43. Schluse, M.; Rossmann, J. From simulation to experimentable digital twins: Simulation-based development and operation of complex technical systems. In Proceedings of the 2016 IEEE International Symposium on Systems Engineering (ISSE), Edinburgh, UK, 3–5 October 2016; pp. 1–6.
44. Schroeder, G.N.; Steinmetz, C.; Pereira, C.E.; Espindola, D.B. Digital twin data modeling with automationml and a communication methodology for data exchange. *IFAC-PapersOnLine* **2016**, *49*, 12–17. [CrossRef]
45. Rauch, L.; Pietrzyk, M. Digital twins as a modern approach to design of industrial processes. *J. Mach. Eng.* **2019**, *19*. [CrossRef]
46. Fergani, O. The digital twin of production, the ultimate tool to achieve first-time-right in metal additive manufacturing. In *Progress in Digital and Physical Manufacturing: Proceedings of ProDPM'19*; Almeida, H.A., Vasco, J.C., Eds.; Springer: Cham, Switzerland, 2020; pp. xxvii–xxx.
47. Grieves, M.; Vickers, J. Digital twin: Mitigating unpredictable, undesirable emergent behavior in complex systems. In *Transdisciplinary Perspectives on Complex Systems*; Springer: Cham Switzerland, 2017; pp. 85–113.
48. Reifsnider, K.; Majumdar, P. Multiphysics stimulated simulation digital twin methods for fleet management. In Proceedings of the 54th AIAA/ASME/ASCE/AHS/ASC Structures, Structural Dynamics, and Materials Conference, Boston, MA, USA, 8–11 April 2013; p. 1578.
49. Gabor, T.; Belzner, L.; Kiermeier, M.; Beck, M.T.; Neitz, A. A simulation-based architecture for smart cyber-physical systems. In Proceedings of the 2016 IEEE International Conference on Autonomic Computing (ICAC), Wuerzburg, Germany, 17–22 July 2016; pp. 374–379.
50. Kraft, E.M. The air force digital thread/digital twin-life cycle integration and use of computational and experimental knowledge. In Proceedings of the 54th AIAA Aerospace Sciences Meeting, San Diego, CA, USA, 4–8 January 2016; p. 0897.
51. Zhang, H.; Liu, Q.; Chen, X.; Zhang, D.; Leng, J. A digital twin-based approach for designing and multi-objective optimization of hollow glass production line. *IEEE Access* **2017**, *5*, 26901–26911. [CrossRef]
52. Zheng, Y.; Yang, S.; Cheng, H. An application framework of digital twin and its case study. *J. Ambient Intell. Humaniz. Comput.* **2019**, *10*, 1141–1153. [CrossRef]
53. Lee, J.; Lapira, E.; Yang, S.; Kao, A. Predictive manufacturing system-Trends of next-generation production systems. *IFAC Proc. Vol.* **2013**, *46*, 150–156. [CrossRef]
54. Vachálek, J.; Bartalský. L.; Rovný, O.; Šišmišová, D.; Morháč, M.; Lokšík, M. The digital twin of an industrial production line within the industry 4.0 concept. In Proceedings of the 21st international conference on process control (PC), Strbske Pleso, Slovakia, 6–9 June 2017; pp. 258–262.
55. El Saddik, A. Digital twins: The convergence of multimedia technologies. *IEEE Multimed.* **2018**, *25*, 87–92. [CrossRef]
56. Enders, M.R.; Hoßbach, N. Dimensions of Digital Twin Applications-A Literature Review. In Proceedings of the 2019 Americas Conference on Information Systems (AMCIS), Cancún, México, 15–17 August 2019.
57. Application of Digital Twin in Industrial Manufacturing. Available online: https://www.futurebridge.com/industry/perspectives-mobility/application-of-digital-twin-in-industrial-manufacturing/ (accessed on 13 October 2020).
58. Madni, A.M.; Madni, C.C.; Lucero, S.D. Leveraging digital twin technology in model-based systems engineering. *Systems* **2019**, *7*, 7. [CrossRef]

59. Maloney, A. The Difference between a Simulation and a Digital Twin. 2019. Available online: https://blogs.sw.siemens.com/mindsphere/the-difference-between-a-simulation-and-a-digital-twin/#:~{}:text=While%20a%20digital%20twin%20begins,from%20its%20real%2Dworld%20counterpart.&text=Because%20it%20dynamically%20changes%20based,matures%20through%20the%20product%20lifecycle. (accessed on 13 October 2020).
60. Miskinis, C. What Does a Digital Thread Mean and How it Differs from Digital Twin. 2018. Available online: https://www.challenge.org/insights/digital-twin-and-digital-thread/#:~{}:text=Simply%20stated%20the%20digital%20twin,its%20creation%20to%20its%20removal. (accessed on 13 October 2020).
61. Gartner Identifies the Top 10 Strategic Technology Trends for 2017. Available online: https://www.gartner.com/en/newsroom/press-releases/2016-10-18-gartner-identifies-the-top-10-strategic-technology-trends-for-2017 (accessed on 12 October 2020).
62. Gartner Identifies the Top 10 Strategic Technology Trends for 2018. Available online: https://www.gartner.com/en/newsroom/press-releases/2017-10-04-gartner-identifies-the-top-10-strategic-technology-trends-for-2018 (accessed on 12 October 2020).
63. Gartner Identifies the Top 10 Strategic Technology Trends for 2019. Available online: https://www.gartner.com/en/newsroom/press-releases/2018-10-15-gartner-identifies-the-top-10-strategic-technology-trends-for-2019 (accessed on 12 October 2020).
64. Qi, Q.; Tao, F.; Zuo, Y.; Zhao, D. Digital twin service towards smart manufacturing. *Procedia Cirp.* **2018**, *72*, 237–242. [CrossRef]
65. Marr, B. These 25 Technology Trends Will Define the Next Decade. Available online: https://www.forbes.com/sites/bernardmarr/2020/04/20/these-25-technology-trends-will-define-the-next-decade/#3153a5329e3b (accessed on 12 October 2020).
66. D'mello, A. Spend on Digital Twins to Reach $12.7bn by 2021, as Solutions Offer IoT Investment RoI. 2020. Available online: https://www.iot-now.com/2020/06/02/103204-spend-on-digital-twins-to-reach-12-7bn-by-2021-as-solutions-offer-iot-investment-roi/ (accessed on 12 October 2020).
67. *Digital Twin: Market Estimates & Trend Analysis To 2025*; Grand View Research: San Francisco, CA, USA, 2018; p. 80.
68. Tao, F.; Cheng, J.; Qi, Q.; Zhang, M.; Zhang, H.; Sui, F. Digital twin-driven product design, manufacturing and service with big data. *Int. J. Adv. Manuf. Technol.* **2018**, *94*, 3563–3576. [CrossRef]
69. Liu, Q.; Zhang, H.; Leng, J.; Chen, X. Digital twin-driven rapid individualised designing of automated flow-shop manufacturing system. *Int. J. Prod. Res.* **2019**, *57*, 3903–3919. [CrossRef]
70. Qi, Q.; Tao, F. Digital twin and big data towards smart manufacturing and industry 4.0:360 degree comparison. *IEEE Access* **2018**, *6*, 3585–3593. [CrossRef]
71. Shen, W.; Yang, C.; Gao, L. Address business crisis caused by COVID-19 with collaborative intelligent manufacturing technologies. *IET Collab. Intell. Manuf.* **2020**, *2*, 96–99. [CrossRef]
72. Gartner Survey Reveals 47% of Organizations Will Increase Investments in IoT Despite the Impact of COVID-19. 2020. Available online: https://www.gartner.com/en/newsroom/press-releases/2020-10-29-gartner-survey-reveals-47-percent-of-organizations-will-increase-investments-in-iot-despite-the-impact-of-covid-19- (accessed on 13 October 2020).
73. Kaarlela, T.; Pieskä, S.; Pitkäaho, T. Digital Twin and Virtual Reality for Safety Training. In Proceedings of the 2020 11th IEEE International Conference on Cognitive Infocommunications (CogInfoCom), Mariehamn, Finland, 23–25 September 2020; pp. 115–120.
74. Miskinis, C. How are Mining Will be Improved Using Digital Twin Simulations. 2018. Available online: https://www.challenge.org/insights/digital-twin-in-mining/ (accessed on 13 October 2020).
75. Tao, F.; Qi, Q.; Wang, L.; Nee, A. Digital twins and cyber–physical systems toward smart manufacturing and industry 4.0: Correlation and comparison. *Engineering* **2019**, *5*, 653–661. [CrossRef]
76. Wang, Z. Digital Twin Technology. In *Industry 4.0-Impact on Intelligent Logistics and Manufacturing*; IntechOpen: London, UK, 2020.
77. Lamb, K. *Principle-Based Digital Twins: A Scoping Review*; Centre for Digital Built Britain: Cambridge, UK, 2019.
78. Digital Twin. Available online: https://www.plm.automation.siemens.com/global/en/our-story/glossary/digital-twin/24465 (accessed on 12 October 2020).
79. Parks, M. Digita Twinning: Types of Digital Twins. Method: Mouser's Technology & Solutions Journal. 2018. Available online: https://eu.mouser.com/applications/digital-twinning-types/ (accessed on 12 October 2020).
80. Woods, D. Why Digital Twins Should Be the CEO's Best Friend. 2018. Available online: https://www.forbes.com/sites/danwoods/2018/07/18/why-digital-twins-should-be-the-ceos-best-friend/?sh=11dfe0a63c75 (accessed on 12 October 2020).
81. Exosite. The 3 Types of Digital Twin Maturity Models. 2017. Available online: https://blog.exosite.com/3-digital-twin-maturity-models. (accessed on 12 October 2020).
82. Saracco, R. Digital Twins: Advantages & Issues of a Powerful Emerging Technology. 2018. Available online: https://cmte.ieee.org/futuredirections/2018/06/14/digital-twins-advantages-issues-of-a-powerful-emerging-technology/ (accessed on 12 October 2020).
83. Umeda, Y.; Ota, J.; Kojima, F.; Saito, M.; Matsuzawa, H.; Sukekawa, T.; Takeuchi, A.; Makida, K.; Shirafuji, S. Development of an education program for digital manufacturing system engineers based on 'Digital Triplet'concept. *Procedia Manuf.* **2019**, *31*, 363–369. [CrossRef]
84. Umeda, Y.; Ota, J.; Shirafuji, S.; Kojima, F.; Saito, M.; Matsuzawa, H.; Sukekawa, T. Exercise of digital kaizen activities based on 'digital triplet'concept. *Procedia Manuf.* **2020**, *45*, 325–330. [CrossRef]
85. Gichane, M.M.; Byiringiro, J.B.; Chesang, A.K.; Nyaga, P.M.; Langat, R.K.; Smajic, H.; Kiiru, C.W. Digital Triplet Approach for Real-Time Monitoring and Control of an Elevator Security System. *Designs* **2020**, *4*, 9. [CrossRef]

86. Liu, Z.; Meyendorf, N.; Mrad, N. The role of data fusion in predictive maintenance using digital twin. In Proceedings of the AIP Conference Proceedings, Provo, UT, USA, 16–21 July 2017; p. 020023.
87. Sherry, D. The Car Battery Just Calle. 2017. Available online: https://www.linkedin.com/pulse/car-battery-just-called-deborah-sherry (accessed on 14 October 2020).
88. Coors-Blankenship, J. Taking Digital Twins for a Test Drive with Tesla, Apple. 2020. Available online: https://www.industryweek.com/technology-and-iiot/article/21130033/how-digital-twins-are-raising-the-stakes-on-product-development (accessed on 14 October 2020).
89. Martinez-Velazquez, R.; Gamez, R.; El Saddik, A. Cardio Twin: A Digital Twin of the human heart running on the edge. In Proceedings of the 2019 IEEE International Symposium on Medical Measurements and Applications (MeMeA), Istanbul, Turkey, 26–28 June 2019; pp. 1–6.
90. The Future of Digital Twins. 2020. Available online: https://www.nec.com/en/global/insights/article/2020022510/index.html#:~{}:text=Digital%20twins%20are%20now%20seeing,robust%20and%20more%20cost%20effective (accessed on 14 October 2020).
91. MentorMate. The Future of Farming: 7 Ways A Digital Twin Can Be Applied to Agriculture. 2020. Available online: https://medium.com/@MentorMate/the-future-of-farming-7-ways-a-digital-twin-can-be-applied-to-agriculture-595a1750c453 (accessed on 12 October 2020).
92. Nicholls, J. Unleashing the Potential of Digital Twins. 2020. Available online: https://www.csiro.au/en/work-with-us/industries/mining-resources/resourceful-magazine/issue-20/digital-twins (accessed on 13 October 2020).
93. Pang, J.; Li, J.; Xie, Z.; Huang, Y.; Cai, Z. Collaborative City Digital Twin for Covid-19 Pandemic: A Federated Learning Solution. *arXiv* **2020**, arXiv:2011.02883.
94. Sepasgozar, S.M. Digital Twin and Web-Based Virtual Gaming Technologies for Online Education: A Case of Construction Management and Engineering. *Appl. Sci.* **2020**, *10*, 4678. [CrossRef]
95. Singh, S.; Shehab, E.; Higgins, N.; Fowler, K.; Tomiyama, T.; Fowler, C. *Challenges of Digital Twin in High. Value Manufacturing*; 0148-7191; SAE Technical Paper; SAE International: Warrendale, PA, USA, 2018.
96. Simchenko, N.; Tsohla, S.; Chyvatkin, P. IoT & Digital Twins Concept Integration Effects on Supply Chain Strategy: Challenges and Effects. *Int. J. Sup. Chain. Mgt. Vol.* **2019**, *8*, 803.
97. Bulygina, I. Product Owner Talks: 20 Digital Twins Solution Providers. 2017. Available online: https://www.dashdevs.com/blog/product-owner-talks-20-digital-twins-service-companies/ (accessed on 15 October 2020).
98. Ogewell, V. Digital Twins: Beware of Naive Faith in Simplicity. 2018. Available online: https://www.engineering.com/PLMERP/ArticleID/16272/Digital-Twins-Beware-of-Naive-Faith-in-Simplicity.aspx (accessed on 15 October 2020).
99. West, T.D.; Blackburn, M. Is Digital Thread/Digital Twin Affordable? A Systemic Assessment of the Cost of DoD's Latest Manhattan Project. *Procedia Comput. Sci.* **2017**, *114*, 47–56. [CrossRef]
100. Wagner, R.; Schleich, B.; Haefner, B.; Kuhnle, A.; Wartzack, S.; Lanza, G. Challenges and potentials of digital twins and Industry 4.0 in product design and production for high performance products. *Procedia CIRP* **2019**, *84*, 88–93. [CrossRef]
101. Wyckoff, A.; Pilat, D. *Key Issues for Digital Transformation in the G20*; OECD: Berlin, Germany, 2017.
102. Common Pitfalls to Digital Twin Implementation. Available online: https://www.compunneldigital.com/infographic/common-pitfalls-to-digital-twin-implementation/#gref (accessed on 12 October 2020).
103. Shao, G.; Helu, M. Framework for a digital twin in manufacturing: Scope and requirements. *Manuf. Lett.* **2020**, *24*, 105–107. [CrossRef] [PubMed]
104. Tyagi, P.; Demirkan, H. Data Lakes: The Biggest Big Data Challenges. *Anal Mag.* **2016**. [CrossRef]
105. Gagné, M.R. Digital Twins, Another Reason to Worry about the iot and Data Security. 2020. Available online: https://irishtechnews.ie/digital-twins-iot-and-data-security/ (accessed on 12 October 2020).
106. Digital Twin: 5 Challenges for 7 Benefits. 2019. Available online: https://www.ingenium-magazine.it/en/digital-twin-6-sfide-per-7-benefici/ (accessed on 12 October 2020).
107. Goasduff, L. Confront Key Challenges to Boost Digital Twin Success. 2018. Available online: https://www.gartner.com/smarterwithgartner/confront-key-challenges-to-boost-digital-twin-success/ (accessed on 12 October 2020).

applied system innovation

MDPI

Article

Implementing Circular Economy Strategies in Buildings—From Theory to Practice

Kamel Mohamed Rahla [1,*], Ricardo Mateus [1,2] and Luís Bragança [1,2]

1 Institute for Sustainability and Innovation in Structural Engineering (ISISE), University of Minho, 4800-058 Guimarães, Portugal; ricardomateus@civil.uminho.pt (R.M.); braganca@civil.uminho.pt (L.B.)
2 Civil Engineering Department, School of Engineering, University of Minho, 4800-058 Guimarães, Portugal
* Correspondence: k.m.rahla@civil.uminho.pt

Abstract: Population growth, along with a rapid urban expansion, is imposing a heavy pressure on the planet's finite resources. It is widely acknowledged that the building industry consumes large amounts of raw materials while generating waste and emissions. To set apart economic growth from environmental repercussions, the Circular Economy (CE) arose as an innovative paradigm that can offer a fast-track towards a sustainable built environment. This paper will tackle a research gap that academia and policymakers often highlighted, which is how can we apply CE to assets that are predominantly meant to be demolished and their resources wasted when they reach their end-of-life. Globally, the paradigm aims at erasing the waste concept, relying on renewable and regenerative sources, and keeping the materials, components, and systems in use at their highest value as long as possible. The concept's implementation would attempt to consider the built environment as a closed-loop system wherein resources are viewed as a scarce commodity. Although the CE seems straightforward, translating the circular thinking to the building level might be a hardship. The following paper will attempt to shed light on how to promote CE in buildings that will ultimately lead to healthier, more efficient, and more sustainable cities on a broader scale. The proposed framework considers CE implementation strategies throughout the building's lifecycle and mainly deals with three innovative aspects: wise resource management, building design approaches, and digitalization of the building industry. In this sense, this study will explore these game-changing factors that are considered paramount to concretize the concept in practice and provide a smooth pathway for CE uptake in buildings.

Keywords: circular economy; circular building; implementation strategies; design strategies; circular resource flows

Citation: Rahla, K.M.; Mateus, R.; Bragança, L. Implementing Circular Economy Strategies in Buildings—From Theory to Practice. *Appl. Syst. Innov.* **2021**, *4*, 26. https://doi.org/10.3390/asi4020026

Academic Editor: Evangelos Bellos

Received: 4 February 2021
Accepted: 31 March 2021
Published: 7 April 2021

Publisher's Note: MDPI stays neutral with regard to jurisdictional claims in published maps and institutional affiliations.

Copyright: © 2021 by the authors. Licensee MDPI, Basel, Switzerland. This article is an open access article distributed under the terms and conditions of the Creative Commons Attribution (CC BY) license (https://creativecommons.org/licenses/by/4.0/).

1. Introduction

The building industry plays a vital role in the economic prosperity and social welfare [1]. However, economic progress has led to significant environmental damage, and the construction industry accounts for 33% of greenhouse gas emissions, 40% of resource consumption, and 40% of waste generation [2]. Former sustainability-related practices have come short in terms of managing the end-of-life scenario of resources by mainly focusing on recycling rather than adopting a holistic approach that would enable more significant environmental benefits [3,4].

With the emergence of the Circular Economy (CE) concept, professionals, academia, and policymakers have regarded the paradigm as a suitable response to this resource-intensive sector. The building sector is one of the five priority sectors in the European CE package, which may prompt this particular branch into developing new eco-technologies [5,6]. The CE is increasingly gaining recognition within the construction industry [7]. It is considered an innovative practice to foster sustainability in a systematic way and move away from the basic linear economy model, which consists of "extract, use, and landfill". It encourages the reduction of raw materials inputs, relies on renewable sources, and eliminate waste from the system [8]. Recent studies have focused on recycling Construction and Demolition Waste (C&DW) with little consideration on the reuse of products, resulting in the reduction of reclaimed materials [9,10]. The avoidance of C&DW is becoming more adopted, whereas further studies must be conducted mainly at the design stages to ensure better end-of-life scenarios [11–13]. While, typically, contemporary buildings are not designed for deconstruction [14–17], the concept of reusing their assets after their end-of-life should be considered for achieving higher levels of environmental performance with less material input [18]. Others have also stressed the significance of the building stock, which represents the most critical part of the produced (human-made) wealth of industrialized societies [19,20], and have underlined the urgent need to decrease material use and increase the quality and durability of the building stock as a resource [2].

Nevertheless, implementing CE strategies in buildings promises to be hindered by the ambiguity surrounding the concept. Despite the fact that the research output contributing to the development of a circular built environment is considerably increasing [21,22], there is still a lack of a thorough study that entails various facets of CE in buildings [7]. The present research will describe CE background, strategies, and approaches that can be applied in the construction industry according to recent studies, in an attempt to link theory to practice and provide better insights on the concept's uptake at the building level.

2. Materials and Methods

This research sought to pinpoint key-strategies that are in line with the idea of CE and that can be implemented in buildings. Numerous studies have already tackled this aspect; however, the focus was mainly put on a single feature (e.g., materials and components, energy, or design strategies) or merely related to C&DW practices [7]. This exploratory research will provide a better understanding to a common research question, which is how CE principles can be applied to buildings. This is done through analyzing CE-related articles and reports related to resource management, design strategies, Building Information Modelling (BIM) and Materials Passports applications, and case-studies to provide a framework that holistically takes into account CE principles throughout the building's lifecycles.

In this context, the article will outline CE principles and the origins of the paradigm (Section 3), describe how resources should be used and managed from a CE perspective in a closed-loop system (Section 4), identify CE design strategies (Section 5), highlight the need for digitizing the building sector (Section 6), showcase practical case-studies wherein CE strategies have been implemented (Section 7), and, finally, conclude with a framework, summary, limitations, and further recommendations (Section 8).

3. CE Background and Principles in the Built Environment

The need for better management of resources that are becoming scarcer, and dire environmental impacts has urged academia and policymakers to provide a new pattern to move away from the economic model adopted since the industrial revolution. The idea of CE was refined throughout the last century to be widely acknowledged during the previous decade. Nevertheless, the roots of CE are still ambiguous and cannot precisely be pinned to a single research. One common agreement is that CE is deeply rooted in numerous schools of thought.

Walter Stahel, an architect and industrial analyst, together with Genevieve Reday, released the research report named *"The potential for substituting Manpower for Energy,"* which focuses on an economy in loops and its outcomes on job creation, economic competitiveness, resource savings, and waste prevention [23]. Stahel also worked on a new approach to production and processes, the "closed-loop". It aims at promoting product-life extension, long-life goods, reconditioning activities, and waste prevention. His latest work focuses on "Performance economy" [24] where he claims the importance of selling services instead of products, which is considered a key feature of CE.

The Cradle-to-Cradle concept was developed by the German chemist and visionary Michael Braungart and the American architect Bill McDonough [25]. The Cradle-to-Cradle concept considers all materials involved in industrial processes as nutrients from two categories: technical and biological. This framework deals with the design for effectiveness and positive impacts while reducing negative ones resulting from design for efficiency [26].

The industrial ecology focuses on material and energy flows through the industrial systems [27]. The concept arose as a solution to resource scarcity and high material cost. This approach intends to create closed-loop strategies that promote waste as an input to eliminate the notion of undesirable by-products while promoting regenerative use of resources [28]. Consequently, wastes and resources stream in a circular way between the different ecosystem components with renewable energy supplying those cycles [29].

Back in the 1970s, the American professor, John Tillman Lyle, challenged his students to come up with an idea for society by adopting the following approach: "daily actives are based on the value of living within the limits of available renewable resources without environment degradation" [30]. The idea of regenerative design relies on developing buildings and cities to regenerate ecosystems [31]. This concept has been formulated earlier for agriculture and has its roots in bioregionalism and permaculture, but Lyle expanded it to the entire social-ecological system to broaden its ability to enhance its potential [32].

Among the front-runners in terms of CE implementation, the Chinese government started to show an increasing interest in the concept by implementing explicit policies back in 2002. Later on, research on CE increased exponentially through the CE promotion law in 2009, to stimulate cleaner production, develop eco-industrial parks, and cope with the rapid urbanization [33,34]. In Europe, the Ellen MacArthur Foundation's creation and the release of the action plan by the European Commission accelerated the transition towards the CE by promoting renewable energy, designing out waste and closed-loop systems to retain material's value circulating in the economy.

Initially, the CE-related Chinese literature framed the concept around the 3R's principles, Reduce, Reuse, and, Recycle [35–37]. Wherein, "Reduce" refers to the action of minimizing inputs and outputs such as raw materials and waste, "Reuse" is the operation of using a product again for the same purpose when it reaches its end-of-life, and "Recycle" is the process of recovering waste to manufacture a new product.

The 3R's principles were later extended to a 9R's framework to encompass more actions and achieve a transition towards CE more effectively. The R-list includes three key-strategies to increase circularity and innovation in product design [38]. The first strategy stresses the need for wiser product manufacturing and includes three actions:

Refuse: Depreciate a product with dire impacts and proposing a different one with identical or better functions and fewer impacts;
Rethink: Intensify the product use and adopt smarter strategies as sharing economy or products with multiple functions; and,
Reduce: Decrease virgin materials and energy consumption while enhancing efficiency.
The second strategy encourages product lifespan extension and consists of:
Reuse: Reuse a discarded product that keeps the same functions by another user;
Repair: Fix a damaged product to give back its initial performance;
Refurbish: Renovate an outdated product to make it as a new one;
Remanufacture: Make a product using parts from a damaged product that had the same functions; and,
Repurpose: Make a product using parts from a damaged product that had different functions.

The last and least favored strategy comprises:

Recycle: Include, into the manufacturing process of a product, materials that reached their end-of-life use to make materials with same, higher (upcycle), or lower (downcycle) qualities; and,
Recover: A process of retrieving heat, electricity, or fuel from non-recyclable materials by incineration.

The Ellen MacArthur Foundation outlined three principles to embrace CE [39]:

Principle 1: "to preserve and enhance natural capital by controlling finite stocks and balancing renewable resource flows."

Principle 2: "to optimize resource yields by circulating products, components, and materials at the highest utility at all times in both technical and biological cycles."

Principle 3: "to foster system effectiveness by revealing and designing out negative externalities."

Scaling up the CE principles to buildings holds the promise to attenuate negative environmental impacts by keeping materials and components in use and guarantying their reuse when the building reaches its end-of-life. Designing for a CE implies designing for adaptability, flexibility, and disassembly to enable reversibility and salvage the value of building's products. This practice will decrease the use of raw materials and reduce waste generation up from the design stage.

The European commission has recently released a report entitled "Circular Economy principles for buildings design" wherein the macro-objectives "Resource efficient and circular material life cycles" of the assessment methodology Level(s) has been linked to three CE strategies, which are durability and extended lifespans of building materials, adaptability, and efficient waste management [40]. In the same context, the Ellen MacArthur foundation outlined in their report two different cycles for reintroducing materials to the loop and optimizing resource consumption, technical and biological [41]. Biological materials are those elements that can be put back to the biosphere safely at their end-of-life. Technical materials are human-made elements that be reused, repaired refurbished, recycle, and incinerated at their end-of-life. From these principles, the Circular Building emerged as a new practice that embed every aspect of the concept. it can be defined as "a building that is designed, planned, built, operated, maintained, and deconstructed in a manner consistent with CE principles" [42].

Nevertheless, merging CE principles to the built environment is hampered by the inherent complexity of the concept, given the fact that buildings are predominantly constructed to be demolished at their end-of-life instead of deconstructing them or adapting them to users' needs. Moreover, design strategies and materials and components selection are crucial strategies to ensure an optimal CE implementation. Preference is given to low embodied carbon and energy materials with higher quality that can embrace reversibility without compromising building's performances and user's comfort.

4. CE Strategies for Energy, Materials, and Water

Natural resources are substantially used and consumed throughout the buildings' life stages as this industry is responsible for around 30% of water use and 40% of raw materials extraction and energy consumption [2,43,44]. By adopting particular CE strategies, savings may occur by creating proper systems to retain value and keep the resources flowing in a circular manner [45]. This approach would close materials and components, energy, and water loops and minimize the associated potential environmental impacts [46].

The life cycle of circular buildings should be a closed-loop system wherein components and materials are optimally used and retained at their highest value (Figure 1). The technical cycle consists of selecting materials and components that can be maintained to extend their service-life by reusing, refurbishing, repairing, and remanufacturing [21]. At the same time, recycling and incineration are final strategies. While, on the other hand, the biological cycle involves natural materials that can be biodegradable or compostable at their end of life (e.g., bio-based materials) [47].

According to Pomponi and Moncaster [42] circular buildings encompass green and sustainability strategies. The environmental footprint of materials and components should be as minimal as possible. It is preferable to select locally sourced materials to reduce emissions due to transportation and stimulate the local economy. Low embodied energy and carbon materials that are abundant, renewable, and pure are fundamental strategies to achieve eco-friendly structures with a minimal input of raw materials within a circular flow system.

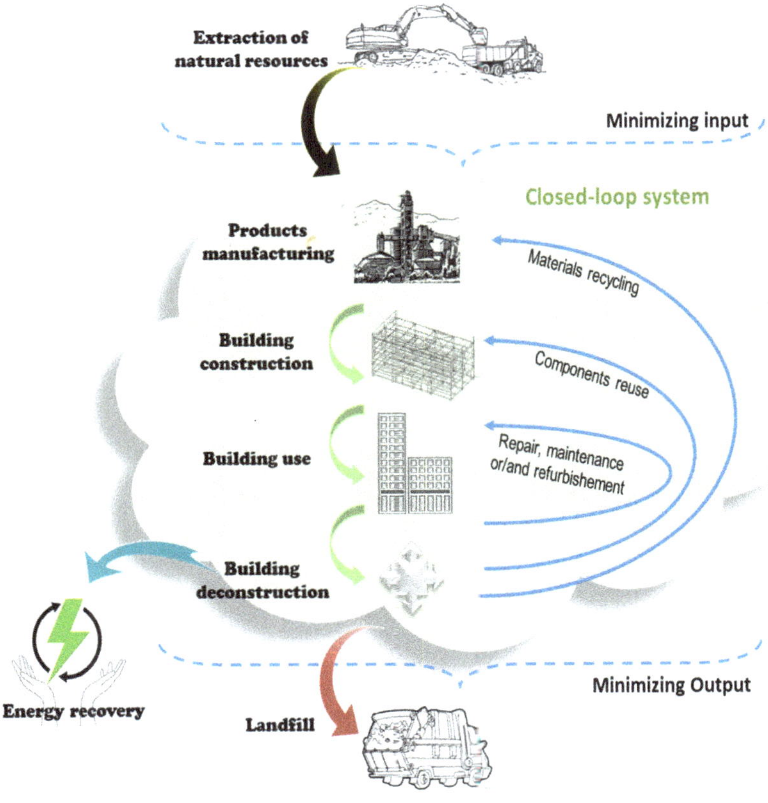

Figure 1. Closing materials and components loops in buildings.

Circular building design entails adaptability, flexibility, and deconstruction at its core. Materials and components need to endure numerous life cycles, depending on the use purpose. Therefore, durability and resilience are two features that need to be taken into account when selecting materials and components to ensure no loss in quality or value over time and use-cycles.

Throughout the building's lifecycle, energy is substantially used and consumed, from materials extraction and processing to the building's construction, operation, and demolition/deconstruction phases [48]. Recent policies regarding Nearly Zero-Energy Buildings (NZEB) challenge practitioners to produce buildings with high energy efficiency [49]. In this context, circular buildings can achieve energy neutrality and even produce an excess of energy through design and on-site renewable energy production [50,51]. The overall energy consumption of buildings is mainly allocated to building operations, followed by embodied energy [52]. The use of materials and components that are energy-intensive increases the buildings embodied energy and, therefore, the life cycle energy consumption. In Europe, energy consumption related to building products ranges from 5% to 10% [53]. Another critical parameter to reduce energy consumption during the use phase is enhancing the building's thermal insulation. High-quality materials are needed to provide the necessary services and thermal comfort for the users [54]. Hence, materials selection and contextual building design and layout should be considered to reduce energy consumption throughout the building's life-cycle.

With the global urbanization, water is getting scarcer since this resource is encountering an increasing demand, and many researchers have predicted a freshwater crisis by the end of this century [55]. In all buildings' life cycle, water is considerably used and consumed by all building stakeholders. Similarly to energy consumption, water is most consumed during the operation phase by building users, followed by buildings' embodied water [56]. Water is used to manufacture several building materials, such as concrete and cement-based materials, ceramics, coatings, and steel. Selecting low embodied water materials will eventually decrease the building's water footprint. Fidar et al. [57] investigated water-efficient microcomponents' use to reduce the water consumption in residential buildings and concluded that there is a linear relationship between water consumption and energy use [57,58]. The building design should also be adapted to the context of its location to embrace more water efficiency. Reaching water efficiency in circular buildings can be achieved during the use-phase by harvesting rainwater and reusing greywater.

5. Design for CE

A critical aspect of the CE thinking is to provide the building's materials, components, and systems a second life of use when the building itself reaches its end-of-life. A specific set of design strategies should be followed by adopting a holistic approach to enable reversibility, adaptability, and flexibility. This section will cover different design strategies that could be put under the umbrella of designing for CE (DfCE).

Over the past decades, several studies have highlighted the multiple benefits of building's disassembly [59–64]. These studies were driven by the environmental and economic benefits resulting from designing for further reuse and recycling to cut down demolition costs and slow down resource consumption pressure. However, with CE's emergence, these practices gained more attention as they matched the circular thinking for designing out waste and creating a closed-loop resource flow.

In this sense, Crowther [59] outlined a total of 27 principles for Design for Disassembly (DfD) that can be categorized as reducing the amount of input and reliance on safe and healthy secondary materials, standardizing the connection between materials and components while considering disassembly at the end-of-life, and retaining information regarding all materials and components involved in a building [59]. Similarly, Ciarimboli & Guy [65] framed DfD principles around ease of deconstruction of building elements, eradicating chemical links and relying on mechanical connections that are accessible, dry construction, providing guidelines to support safe deconstruction, and selecting suitable materials for the process.

One of the most crucial parameters in DfD is the choice of the construction technology and how building elements will be assembled and disassembled. Materials should be eligible for embracing CE principles such as reuse, refurbish, repair, and higher purity to limit quality loss during the assembly/disassembly process. Unlike steel and timber, concrete has been overlooked when it comes to DfD, given the numerous challenges to disassemble concrete elements [60]. Offsite constructions and modularity are important features to consider while assembling components. In his book *"Building in layers"* Brandt's work has been paramount to current strategies to DfCE [66]. The author described buildings as separate but somewhat interlinked layers with their own technical and functional lifespan. Brand's widely-known model includes six layers with different life-spans [66,67]:

1—Site: the location of the building;

2—Structure: the skeleton of the building including the foundation and load-bearing elements;

3—Skin: Building elements in contact with the external environment such as façade and roof;

4—Services: the pipe, wires, energy, and heating systems;

5—Space: the internal fit-out like walls and floors;

6—Stuff: The rest of the internal fit-out, including the furniture and lighting.

Schmidt III & Austin [68] stretched this building system decomposition to cover the Surroundings and Social as a new 8S model.

Adaptability is another characteristic that has been linked to DfCE [69,70]. Addis and Schouten [71] stressed the difference between adaptability and flexibility and defined a flexible building as "a building that has been designed to allow easy rearrangement of its internal fit-out and arrangement to suit the changing needs of occupants," whereas an adaptable building is "a building that has been designed with thought of how it might be easily altered to prolong its life." On the other hand, Moffatt and Russel [72] included flexibility as a sub-strategy to design for adaptability, convertibility, and expandability. According to the authors, adaptable buildings should be maintainable, versatile, simple in design, and upgradable to accommodate user's desires with minimum quality loss and environmental impacts [72]. In the same context, Durmisevic & Brouwer [73] described three dimensions of transformation: spatial transformation, structural transformation, and element and material transformation.

Disassembling sections and parts from buildings into components and allowing their reassembling in a new combination is a better alternative to destroying buildings and systems from economic and environmental standpoints [45,64,74]. Practitioners and academia need to learn from the past to design for the future and adopt a DfCE strategy, since CE is deeply rooted on several schools of thought and consequently englobes numerous design strategies (Figure 2). It is noteworthy that designers should also consider resource flows in buildings throughout its life cycle. As discussed in the previous section, a neat selection of materials with low embodied energy and carbon with minimal water use throughout their life cycle is crucial to keep positive environmental impacts. Water and energy should be considered during the design phase to produce buildings that optimise resource flows without compromising overall quality and comfort for the building users.

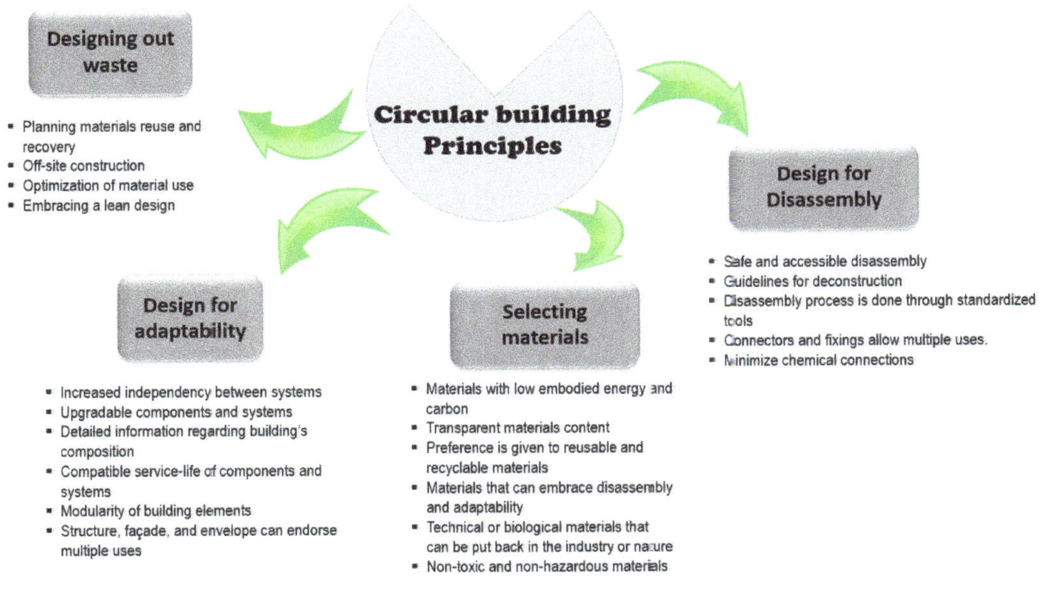

Figure 2. Design for Circular Economy.

6. Data Relevance for CE in Buildings

With the technological advances, data are becoming an essential asset for the construction industry to manage buildings' performances from the design stages until the end-of-life. Specific information is required to ensure the maintenance of a building and the technical life of all in-use materials. Tracking resource flows will allow building stakeholders to get deeper insights into the resource input-output. For the time being, the most significant challenge on the urban level is the generation of knowledge and data on the material composition of buildings [75]. Digitizing the built environment is regarded as a fast-track approach to implementing CE in the building sector. Keeping track of materials through specific datasets will support materials reuse once they reach their end-of-life, which will smoothen value retention. For materials and components, related information can be stored in a "Material passport" that will be shared with all building's stakeholders [76]. The Material Passport (MP) was developed during the EU-funded project "BAMB" to support circular building design, smoothen circular building materials selection, and extend their lives while keeping their value [77].

Comprehensive and accurate data can enhance decision-making, manage financial risk, and monitor environmental impacts. The MP provides qualitative and quantitative information regarding physical, chemical, and biological properties of materials composition, allowing practitioners to select building products that are safe to humans and ecosystems whilst taking into account CE principles. MP comprises different hierarchy levels, from materials, components, products, systems, and buildings, to describe specific features for value recovery and reversibility [78].

The Building Information Management (BIM) is a computational methodology that digitally models buildings to provide the construction stakeholders with valuable information to aid decision-making throughout the building's life cycle [79]. Understanding the purpose of BIM will support applying the planned strategy throughout the asset's life cycle [80,81]. In the context of CE, BIM can be considered as an approach to minimize waste throughout the building's life cycle [82,83]. However, BIM is seldomly used for managing buildings when they reach their end-of-life [82]. Recent studies have attempted to link MP to BIM to compile data on building materials characteristics and assess their recyclability and environmental impacts [75,76] or reuse waste as a novel construction material [84]. Nonetheless, research regarding MPs' integration into BIM is still in its early stage and requires more profound commitment from the construction industry. Digitizing the building industry will still face several obstacles such as the amount of data that must be generated, their quality, and the reliability of the used data, which will require standardization, collaboration, and transparency from the building stakeholders along the value chain [35].

7. Real-Life CE Implementation and Case-Studies

This section presents some success stories of CE implementation in buildings to provides concrete examples of current worldwide CE practices. One of the first projects to adopt CE strategies is the project Circular Building in London developed by Arup, which was driven by the challenge of applying CE principles to the built environment [85]. The building's design was inspired by the Brand's 6S model consisting of site, structure, skin, services, space plan, and stuff. For the structure, reclaimed steel was used as a material that can be reusable after deconstructing the prototype and put back into the loop. The project designers tried to put together a structure with simple and accessible bolts that will not hinder disassembly and embrace flexibility and adaptability. Regarding the building's skin, engineers used a softwood that has the needed durability to sustain different use-cycles. All the products involved in the building's layers were sustainably sourced, and most of them contain recovered materials and are eligible for reuse or recycling at their end-of-life. These materials are healthy both to humans and the environment and have a low embodied carbon and energy. The circular building was equipped with sensors to generate valuable data to control and adjust the indoor environment quality, namely lighting, temperature, and air quality to measure critical parameters.

The first building project to rely on MP was the Brummen Town Hall [86]. The building was designed for future disassembly over a historic structure from the 19th century building restored beforehand. The materials used are in line with the CE principles and can be further dismantled to be relocated and used into a new structure. As the building was designed for a service life of 20 years, the materials passport plays a crucial role in this project, namely, to keep track of involved materials and components to facilitate their reuse.

Likewise, the renovation of the headquarters of the energy grid company Alliander, located in Duiven, The Netherlands, relied on MP to store valuable information regarding their origin and their further reuse [51,86]. Additionally, the building is energy positive, which means that it generates more energy than it requires. It has solar panels and underground water for thermal storage. The project used a minimum quantity of raw materials and used recycled wood and steel. Another example of materials reuse is the Quay Quarter Tower redesign located in Sydney, Australia, by reusing an existing structure. Half of the building resources are reclaimed from an existing structure, which substantially decreased the associated time, costs, and environmental impacts.

Several case-studies across the globe attempted to concretize CE principles and provide real-life examples of how the construction industry can apply the paradigm shift. However, numerous challenges were noticed along the way. For instance, selecting the right building materials and components proved to be crucial for designers to embed circularity along the value chain [69,86].

In spite of the recent CE-related studies at the building level, intensive research is needed to accurately quantify the environmental, economic, and social benefits of applying CE in these entities. Considering an office building as a case-study, Eberhardt et al. [64] used a simplified allocation method to measure environmental impacts of disassembly and concluded that several parameters greatly influenced the findings such as the type of materials used, the reuse cycles, and building's service life. Similarly, Brambilla el al. [18] assessed and compared the environmental impacts of different structural composite floor systems and found out that one that was designed for disassembly was identified as the most environmentally friendly compared to the conventional scenario. These positive results have been further supported by Minunno et al. [4] who found out that a building designed for disassembly allows for reusing 62% of its mass, which will result in a reducing 88% of the emissions.

Although numerous studies have highlighted the environmental benefits of CE in buildings, several obstacles and barriers may slow down the momentum. A lack of governmental support through financial support and CE-related instruments is often highlighted as the main barrier for a greater CE uptake [22,87–89]. Additionally, construction professionals have encountered a lack of collaboration and obstacles to gather relevant data from materials suppliers [86]. Furthermore, these types of innovative projects require close management to keep the expected objective aligned. Workers, clients, team members and other building stakeholders will need to have a different mindset to embed circularity throughout the project life cycle.

8. Conclusions

With the population growth and the rapid urbanization, the building industry is imposing heavy environmental damage to the biosphere by consuming massive amounts of resources and generating considerable waste. With the emergence of the CE, construction stakeholders are considering adopting the paradigm and implementing its principles in building practices. Embracing these principles will require a paradigm shift towards sustainable production and use, a predictable end-of-life scenario of buildings, and a holistic approach to create a closed-loop system and adopt a wise resource management.

The objective of this paper is to explore the current theory regarding CE and develop a framework that introduces tangible strategies throughout the building's lifecycle (Figure 3).

Adopting CE principles in the built environment can reduce the consumption of resources and waste production, retain the value of resources as long as possible within the system, and reintroduce resources into the use phase through particular strategies. One critical approach to embed circularity is to select suitable construction materials with low embodied energy and carbon that can enable deconstruction and other design strategies in line with the CE principles. Circular building design should encompass low energy and water footprints and optimize the use of these resources in the operation phase. Energy efficiency practices and local and renewable energy sources are valuable approaches to consider along with CE principles to attain energy neutrality. Similarly, water use and consumption can be minimized through an adequate building design and reuse and harvesting systems for rainwater and wastewater. A valuable approach would be monitoring resource use and consumption in the operational phase, which can provide relevant insights to buildings actors and allow to adjust resource flows and reach higher levels of resource efficiency.

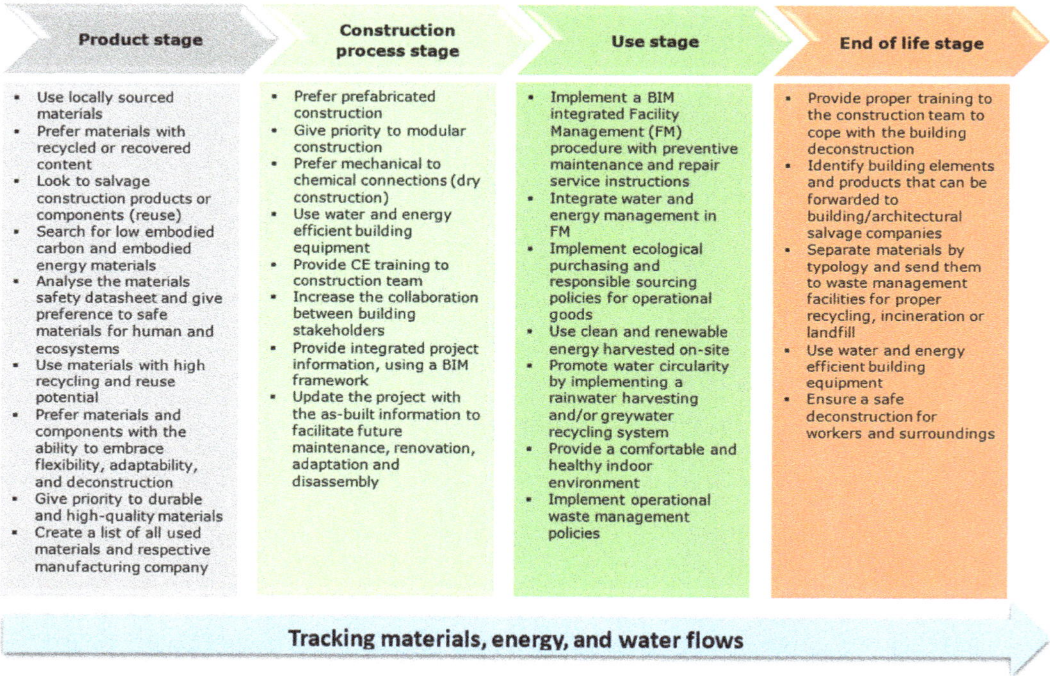

Figure 3. A framework for CE implementation throughout the building's life stages.

Several studies highlighted the need for design strategies that can enable material reusability and recyclability during the past two decades. With the increasing interest that CE is gaining, these particular design strategies have been encapsulated under the paradigm's principles. They matched the same objectives of value retention throughout several uses. To ensure the quality of materials and components and describe their specific characteristics, several studies have proposed integrating Materials Passports to the Building Information Management, which will allow building stakeholders to track materials, understand their origins, and assess their quality.

Finally, the CE aims to redefine the construction industry and challenge designers to reconsider how they design buildings and take into account their end-of-life scenario. Although numerous case-studies went beyond recycling to apply CE core principles, there are still several obstacles along the way (e.g., lack of economic incentives, lack of governmental support, the misconception regarding reused and recycled materials). These obstacles will eventually delay the transition towards a circular built environment and require a particular understanding of inevitable trade-offs.

Although, this study attempted to outline a framework for CE implementation in buildings, some limitations need to be addressed. The study can further be enhanced through a case-study to consider the local context or with a more sophisticated methodology, which will allow a broader scope and more concrete conclusions. Further research is also needed to quantify the economic and social benefits of implementing CE strategies in buildings.

Author Contributions: Conceptualization, K.M.R.; validation, K.M.R., R.M. and L.B.; investigation, K.M.R.; writing—original draft preparation, K.M.R.; writing—review and editing, R.M. and L.B.; supervision, R.M. and L.B.; funding acquisition, L.B. All authors have read and agreed to the published version of the manuscript.

Funding: This research was funded by University of Minho, grant number UMINHO/BID/2020/71.

Institutional Review Board Statement: Not applicable.

Informed Consent Statement: Not applicable.

Data Availability Statement: All the retrieved journal articles and reports can be found in Scopus, Web Of Science, and Google Scholar.

Conflicts of Interest: The authors declare no conflict of interest.

References

1. Hossain, M.U.; Thomas, N.S. Influence of waste materials on buildings' life cycle environmental impacts: Adopting resource recovery principle. *Resour. Conserv. Recycl.* **2019**, *142*, 10–23. [CrossRef]
2. Ness, D.A.; Xing, K. Toward a Resource-Efficient Built Environment: A Literature Review and Conceptual Model. *J. Ind. Ecol.* **2017**, *21*, 572–592. [CrossRef]
3. Rahla, K.M.; Mateus, R.; Bragança, L. Comparative sustainability assessment of binary blended concretes using Supplementary Cementitious Materials (SCMs) and Ordinary Portland Cement (OPC). *J. Clean. Prod.* **2019**, *220*, 445–459. [CrossRef]
4. Minunno, R.; O'Grady, T.; Morrison, G.M.; Gruner, R.L. Exploring environmental benefits of reuse and recycle practices: A circular economy case study of a modular building. *Resour. Conserv. Recycl.* **2020**, *160*, 104855. [CrossRef]
5. European Commission. Closing the Loop—An EU Action Plan for the Circular Economy. Available online: https://eur-lex.europa.eu/legal-content/EN/TXT/?uri=CELEX:52015DC0614 (accessed on 21 March 2021).
6. Kanters, J. Circular Building Design: An Analysis of Barriers and Drivers for a Circular Building Sector. *Buildings* **2020**, *10*, 77. [CrossRef]
7. Hossain, U.; Ng, S.T.; Antwi-Afari, P.; Amor, B. Circular economy and the construction industry: Existing trends, challenges and prospective framework for sustainable construction. *Renew. Sustain. Energy Rev.* **2020**, *130*, 109948. [CrossRef]
8. Kirchherr, J.; Reike, D.; Hekkert, M. Conceptualizing the circular economy: An analysis of 114 definitions. *Resour. Conserv. Recycl.* **2017**, *127*, 221–232. [CrossRef]
9. Adams, K.T.; Osmani, M.; Thorpe, T.; Thornback, J. Circular economy in construction: Current awareness, challenges and enablers. *Proc. Inst. Civ. Eng. Waste Resour. Manag.* **2017**, *170*, 15–24. [CrossRef]
10. Ghisellini, P.; Ji, X.; Liu, G.; Ulgiati, S. Evaluating the transition towards cleaner production in the construction and demolition sector of China: A review. *J. Clean. Prod.* **2018**, *195*, 418–434. [CrossRef]
11. Joensuu, T.; Edelman, H.; Saari, A. Circular economy practices in the built environment. *J. Clean. Prod.* **2020**, *276*, 124215. [CrossRef]
12. Gorgolewski, M. *Resource Salvation: The Architecture of Reuse*; John Wiley & Sons: Hoboken, NJ, USA, 2017.
13. Eberhardt, L.C.M.; Birkved, M.; Birgisdottir, H. Building design and construction strategies for a circular economy. *Arch. Eng. Des. Manag.* **2020**, 1–21. [CrossRef]
14. Densley Tingley, D.; Cooper, S.; Cullen, J. Understanding and overcoming the barriers to structural steel reuse, a UK perspective. *J. Clean. Prod.* **2017**, *148*, 642–652. [CrossRef]
15. Eckelman, M.J.; Brown, C.; Troup, L.N.; Wang, L.; Webster, M.D.; Hajjar, J.F. Life cycle energy and environmental benefits of novel design-for-deconstruction structural systems in steel buildings. *Build. Environ.* **2018**, *143*, 421–430. [CrossRef]
16. Rios, F.C.; Chong, W.K.; Grau, D. Design for Disassembly and Deconstruction—Challenges and Opportunities. *Procedia Eng.* **2015**, *118*, 1296–1304. [CrossRef]
17. Bertino, G.; Kisser, J.; Zeilinger, J.; Langergraber, G.; Fischer, T.; Österreicher, D. Fundamentals of Building Deconstruction as a Circular Economy Strategy for the Reuse of Construction Materials. *Appl. Sci.* **2021**, *11*, 939. [CrossRef]

18. Brambilla, G.; Lavagna, M.; Vasdravellis, G.; Castiglioni, C.A. Environmental benefits arising from demountable steel-concrete composite floor systems in buildings. *Resour. Conserv. Recycl.* **2019**, *141*, 133–142. [CrossRef]
19. Schiller, G. Urban infrastructure: Challenges for resource efficiency in the building stock. *Build. Res. Inf.* **2007**, *35*, 399–411. [CrossRef]
20. Arora, M.; Raspall, F.; Cheah, L.; Silva, A. Residential building material stocks and component-level circularity: The case of Singapore. *J. Clean. Prod.* **2019**, *216*, 239–248. [CrossRef]
21. Rahla, K.; Mateus, R.; Bragança, L. Selection Criteria for Building Materials and Components in Line with the Circular Economy Principles in the Built Environment—A Review of Current Trends. *Infrastructures* **2021**, *6*, 49. [CrossRef]
22. Munaro, M.R.; Tavares, S.F.; Bragança, L. Towards circular and more sustainable buildings: A systematic literature review on the circular economy in the built environment. *J. Clean. Prod.* **2020**, *260*, 121134. [CrossRef]
23. Stahel, W.R.; Reday-Mulvey, G. *Jobs for Tomorrow: The Potential for Substituting Manpower for Energy*; Vantage Press: Erith, UK, 1981.
24. Stahel, W.R. *The Performance Economy*; Springer: Berlin/Heidelberg, Germany, 2010.
25. McDonough, W.; Braungart, M. *Remaking the Way We Make Things: Cradle to Cradle*; North Point Press: New York, NY, USA, 2002; ISBN 1224942886 104.
26. Ellen MacArthur Foundation. *The Circular Model—Brief History and Schools of Thought*; Ellen MacArthur Foundation: Cowes, UK, 2013; pp. 1–4.
27. Sharfman, M.; Graedel, T.E.; Allenby, B.R. Industrial Ecology. *Acad. Manag. Rev.* **1995**, *20*, 1090. [CrossRef]
28. Despeisse, M.; Baumers, M.; Brown, P.; Charnley, F.; Ford, S.; Garmulewicz, A.; Knowles, S.; Minshall, T.; Mortara, L.; Reed-Tsochas, F.; et al. Unlocking value for a circular economy through 3D printing: A research agenda. *Technol. Forecast. Soc. Chang.* **2017**, *115*, 75–84. [CrossRef]
29. Erkman, S. Industrial ecology: An historical view. *J. Clean. Prod.* **1997**, *5*, 1–10. [CrossRef]
30. Cal Poly Pomona. History of Lyle Center. Available online: https://env.cpp.edu/rs/history-lyle-center (accessed on 2 October 2018).
31. Lyle, J.T. *Regenerative Design for Sustainable Development*; John Wiley & Sons: Hoboken, NJ, USA, 1996.
32. Du Plessis, C. Towards a regenerative paradigm for the built environment. *Build. Res. Inf.* **2012**, *40*, 7–22. [CrossRef]
33. Geng, Y.; Fu, J.; Sarkis, J.; Xue, B. Towards a national circular economy indicator system in China: An evaluation and critical analysis. *J. Clean. Prod.* **2012**, *23*, 216–224. [CrossRef]
34. Xue, B.; Chen, X.-P.; Geng, Y.; Guo, X.-J.; Lu, C.-P.; Zhang, Z.-L.; Lu, C.-Y. Survey of officials' awareness on circular economy development in China: Based on municipal and county level. *Resour. Conserv. Recycl.* **2010**, *54*, 1296–1302. [CrossRef]
35. Rahla, K.M.; Bragança, L.; Mateus, R. Obstacles and barriers for measuring building's circularity. *IOP Conf. Ser. Earth Environ. Sci.* **2019**, *225*, 012058. [CrossRef]
36. Su, B.; Heshmati, A.; Geng, Y.; Yu, X. A review of the circular economy in China: Moving from rhetoric to implementation. *J. Clean. Prod.* **2013**, *42*, 215–227. [CrossRef]
37. Ghisellini, P.; Cialani, C.; Ulgiati, S. A Review on Circular Economy: The Expected Transition to a Balanced Interplay of Environmental and Economic Systems. *J. Clean. Prod.* **2016**, *114*, 11–32. [CrossRef]
38. Potting, J.; Hekkert, M.; Worrell, E. Circular Economy: Measuring Innovation in the Product Chain—Policy Report. Available online: http://www.pbl.nl/sites/default/files/cms/publicaties/pbl-2016-circular-economy-measuring-innovation-in-product-chains-2544.pdf (accessed on 21 March 2021).
39. Ellen MacArthur Foundation. *Growth Within: A Circular Economy Vision for a Competitive Europe*; Ellen MacArthur Foundation: Cowes, UK, 2015; p. 100.
40. European Commission. Circular Economy Principles for Buildings Design. Available online: https://ec.europa.eu/docsroom/documents/39984 (accessed on 21 March 2021).
41. Ellen MacArthur Foundation. *Towards a Circular Economy—Economic and Business Rationale for an Accelerated Transition*; Ellen MacArthur Foundation: Cowes, UK, 2012; p. 97.
42. Pomponi, F.; Moncaster, A. Circular economy for the built environment: A research framework. *J. Clean. Prod.* **2017**, *143*, 710–718. [CrossRef]
43. Benachio, G.L.F.; Freitas, M.D.C.D.; Tavares, S.F. Circular economy in the construction industry: A systematic literature review. *J. Clean. Prod.* **2020**, *260*, 121046. [CrossRef]
44. Mannan, M.; Al-Ghamdi, S.G. Environmental impact of water-use in buildings: Latest developments from a life-cycle assessment perspective. *J. Environ. Manag.* **2020**, *261*, 110198. [CrossRef] [PubMed]
45. Foster, G. Circular economy strategies for adaptive reuse of cultural heritage buildings to reduce environmental impacts. *Resour. Conserv. Recycl.* **2020**, *152*, 104507. [CrossRef]
46. Mercader-Moyano, P.; Esquivias, P. Decarbonization and Circular Economy in the Sustainable Development and Renovation of Buildings and Neighbourhoods. *Sustainability* **2020**, *12*, 7914. [CrossRef]
47. Stegmann, P.; Londo, M.; Junginger, M. The circular bioeconomy: Its elements and role in European bioeconomy clusters. *Resour. Conserv. Recycl.* **2020**, *6*, 100029. [CrossRef]
48. Hossain, M.U.; Ng, S.T. Critical consideration of building's environmental impact assessment towards adoption of circular economy: An analytical review. *J. Clean. Prod.* **2018**, *205*, 763–780. [CrossRef]

49. D'Agostino, D.; Zangheri, P.; Castellazzi, L. Towards Nearly Zero Energy Buildings in Europe: A Focus on Retrofit in Non-Residential Buildings. *Energies* **2017**, *10*, 117. [CrossRef]
50. Leising, E.; Quist, J.; Bocken, N. Circular Economy in the building sector: Three cases and a collaboration tool. *J. Clean. Prod.* **2018**, *176*, 976–989. [CrossRef]
51. Merrild, H.; Jensen, K.G.; Sommer, J. *Building a Circular Future*; GXN: Kobenhavn, Denmark, 2016.
52. Ramesh, T.; Prakash, R.; Shukla, K. Life cycle energy analysis of buildings: An overview. *Energy Build.* **2010**, *42*, 1592–1600. [CrossRef]
53. European Commission. *Resource Efficiency Opportunities in the Building Sector*; European Commission: Brussels, Belgium, 2014.
54. Shoubi, M.V.; Shoubi, M.V.; Bagchi, A.; Barough, A.S. Reducing the operational energy demand in buildings using building information modeling tools and sustainability approaches. *Ain Shams Eng. J.* **2015**, *6*, 41–55. [CrossRef]
55. United Nations. *Revision of World Urbanization Prospects 2018*; Department of Economic and Social Affairs of the United Nations: Washington, DC, USA, 2018.
56. Thornback, J.; Snowden, C.; Anderson, J.; Foster, C. *Water Efficiency: The Contribution of Construction Products*; The Open University: Milton Keynes, UK, 2015.
57. Fidar, A.; Memon, F.; Butler, D. Environmental implications of water efficient microcomponents in residential buildings. *Sci. Total. Environ.* **2010**, *408*, 5828–5835. [CrossRef] [PubMed]
58. Hackett, M.; Gray, N. Carbon Dioxide Emission Savings Potential of Household Water Use Reduction in the UK. *J. Sustain. Dev.* **2009**, *2*, 36. [CrossRef]
59. Crowther, P. Design for Disassembly—Themes and Principles RAIA/BDP Environ. Environment Design Guide. 2005, pp. 1–7. Available online: https://www.jstor.org/stable/26149108?seq=1#metadata_info_tab_contents (accessed on 7 April 2021).
60. Salama, W. Design of concrete buildings for disassembly: An explorative review. *Int. J. Sustain. Built Environ.* **2017**, *6*, 617–635. [CrossRef]
61. Thormark, C. *Recycling Potential and Design for Disassembly in Buildings*; Lund Institute of Technology, Lund University: Lund, Sweden, 2001; p. 104.
62. Crowther, P. *Design for Disassembly: An Architectural Strategy*; Queensland University of Technology: Brisbane, Australia, 1999.
63. Durmisevic, E. *Transformable Building Structures*; University of Sarajevo: Sarajevo, Bosnia and Herzegovina, 2006.
64. Eberhardt, L.C.M.; Birgisdóttir, H.; Birkved, M. Life cycle assessment of a Danish office building designed for disassembly. *Build. Res. Inf.* **2019**, *47*, 666–680. [CrossRef]
65. Ciarimboli, N.; Guy, B. *Design for Disassembly in the Built Environment: A Quide to Cloosed-Loop Design and Building*; The Pennsylvania State University: State College, PA, USA, 2005.
66. Brand, S. *How Buildings Learn: What Happens After They're Built*; Penguin: London, UK, 1995.
67. Schmidt, R.I.; Deamer, J.; Austin, S. Understanding adaptability through layer dependencies. In Proceedings of the 18th International Conference on Engineering Design (ICED 11), Impacting Society through Engineering Design, Copenhagen, Denmark, 15–19 August 2011; pp. 1–12.
68. Schmidt, I.I.I.R.; Austin, S. *Adaptable Architecture: Theory and Practice*; Routledge: Abingdon, UK, 2016.
69. Cheshire, D. *Building Revolutions: Applying the Circular Economy to the Built Environment*; Routledge: Abingdon, UK, 2016.
70. Minunno, R.; O'Grady, T.; Morrison, G.M.; Gruner, R.L.; Colling, M. Strategies for Applying the Circular Economy to Prefabricated Buildings. *Buildings* **2018**, *8*, 125. [CrossRef]
71. Addis, W.; Schouten, J. *Principles of Design for Deconstruction to Facilitate Reuse and Recycling*; Ciria: London, UK, 2004.
72. Moffatt, S.; Russell, P. Assessing the Adaptability of Buildings. IEA Annex 31. 2001. Available online: https://www.researchgate.net/profile/Amir_Causevic/post/Is_there_any_scientific_method_for_assessing_the_flexibility_and_adaptability_of_buildings/attachment/59d61f2dc49f478072e97900/AS%3A271749810196482%401441801611675/download/APS-CA+Assessing+the+Adaptability+of+Buildings.pdf (accessed on 7 April 2021).
73. Durmisevic, E.; Brouwer, P.J. Design Aspects of Decomposable Building Structures. Proc CIB Task Gr 39-Deconstruction Meet. Available online: http://www.4darchitects.nl/download/TG39_2002.pdf (accessed on 7 April 2021).
74. Campbell, A. Mass timber in the circular economy: Paradigm in practice? *Proc. Inst. Civ. Eng. Eng. Sustain.* **2018**, *172*, 141–152. [CrossRef]
75. Honic, M.; Kovacic, I.; Rechberger, H. Improving the recycling potential of buildings through Material Passports (MP): An Austrian case study. *J. Clean. Prod.* **2019**, *217*, 787–797. [CrossRef]
76. Honic, M.; Kovacic, I.; Sibenik, G.; Rechberger, H. Data and stakeholder management framework for the implementation of BIM-based Material Passports. *J. Build. Eng.* **2019**, *23*, 341–350. [CrossRef]
77. Heinrich, M.; Lang, W. *Materials Passports—Best Practice*; Technical University of Munich: Munich, Germany, 2019.
78. Luscuere, L.; Mulhall, D Circularity information management for buildings: The example of materials passports. In *Designing for the Circular Economy*; Charter, M., Ed.; Routledge: Abingdon, UK, 2019; pp. 369–379.
79. Carvalho, J.P.; Bragança, L.; Mateus, R. A Systematic Review of the Role of BIM in Building Sustainability Assessment Methods. *Appl. Sci.* **2020**, *10*, 4444. [CrossRef]
80. Shou, W.; Wang, J.; Wang, X.; Chong, H.Y. A Comparative Review of Building Information Modelling Implementation in Building and Infrastructure Industries. *Arch. Comput. Methods Eng.* **2015**, *22*, 291–308. [CrossRef]

81. Carvalho, J.P.; Bragança, L.; Mateus, R. Optimising building sustainability assessment using BIM. *Autom. Constr.* **2019**, *102*, 170–182. [CrossRef]
82. Charef, R.; Emmitt, S. Uses of building information modelling for overcoming barriers to a circular economy. *J. Clean. Prod.* **2021**, *285*, 124854. [CrossRef]
83. Won, J.; Cheng, J.C. Identifying potential opportunities of building information modeling for construction and demolition waste management and minimization. *Autom. Constr.* **2017**, *79*, 3–18. [CrossRef]
84. Wandiga, C.A. Methodological review: Socio-cultural analysis criteria for BIM modeling and material passport tracking of agriwaste as a building construction raw material. *MRS Energy Sustain.* **2020**, *7*, 1–14. [CrossRef]
85. ARUP. Stuart Smith: The Circular Building the Most Advanced Reusable Building Yet. Available online: https://www.arup.com/perspectives/the-circular-building (accessed on 7 April 2021).
86. Ellen MacArthur Foundation. Circularity in the Built Environment: Case Studies. Available online: https://www.ellenmacarthurfoundation.org/assets/downloads/Built-Env-Co.Project.pdf (accessed on 7 April 2021).
87. Campbell-Johnston, K.; Cate, J.T.; Elfering-Petrovic, M.; Gupta, J. City level circular transitions: Barriers and limits in Amsterdam, Utrecht and The Hague. *J. Clean. Prod.* **2019**, *235*, 1232–1239. [CrossRef]
88. Mahpour, A. Prioritizing barriers to adopt circular economy in construction and demolition waste management. *Resour. Conserv. Recycl.* **2018**, *134*, 216–227. [CrossRef]
89. Górecki, J.; Núñez-Cacho, P.; Corpas-Iglesias, F.A.; Molina, V. How to convince players in construction market? Strategies for effective implementation of circular economy in construction sector. *Cogent Eng.* **2019**, *6*. [CrossRef]

MDPI
St. Alban-Anlage 66
4052 Basel
Switzerland
Tel. +41 61 683 77 34
Fax +41 61 302 89 18
www.mdpi.com

Applied System Innovation Editorial Office
E-mail: asi@mdpi.com
www.mdpi.com/journal/asi

www.ingramcontent.com/pod-product-compliance
Lightning Source LLC
Chambersburg PA
CBHW040749250926
51850CB00007BB/1448
* 9 7 8 3 0 3 6 5 4 4 0 1 4 *